Lewis Acid–Lewis Base Interactions

Mechanisms and Related Phenomena

Other Titles by the Author

Hydrogen Bonding - New Insights

Understanding Hydrogen Bonds: Theoretical and Experimental Views

Lewis Acid–Lewis Base Interactions

Mechanisms and Related Phenomena

Sławomir J Grabowski

Ikerbasque - Basque Foundation for Science, Spain
University of the Basque Country (UPV/EHU), Spain
Donostia International Physics Center (DIPC), Spain

World Scientific

NEW JERSEY • LONDON • SINGAPORE • BEIJING • SHANGHAI • TAIPEI • CHENNAI

Published by

World Scientific Publishing Co. Pte. Ltd.

5 Toh Tuck Link, Singapore 596224

USA office: 27 Warren Street, Suite 401-402, Hackensack, NJ 07601

UK office: 57 Shelton Street, Covent Garden, London WC2H 9HE

Library of Congress Cataloging-in-Publication Data
Names: Grabowski, Sławomir J., author.
Title: Lewis acid-Lewis base interactions : mechanisms and related phenomena /
Sławomir J Grabowski, Ikerbasque-Basque Foundation for Science, Spain, University of the Basque Country (UPV/EHU), Spain, Donostia International Physics Center (DIPC), Spain.
Description: New Jersey : World Scientific, [2025] | Includes bibliographical references and index.
Identifiers: LCCN 2024048532 | ISBN 9789819801435 (hardcover) |
ISBN 9789819801442 (ebook for institutions) | ISBN 9789819801459 (ebook for individuals)
Subjects: LCSH: Acid-base chemistry. | Lewis acids. | Lewis bases.
Classification: LCC QD477 .G73 2025 | DDC 546/.24--dc23/eng/20250116
LC record available at https://lccn.loc.gov/2024048532

British Library Cataloguing-in-Publication Data
A catalogue record for this book is available from the British Library.

For any available supplementary material, please visit
https://www.worldscientific.com/worldscibooks/10.1142/14059#t=suppl

Desk Editor: Shaun Tan Yi Jie

Typeset by Stallion Press
Email: enquiries@stallionpress.com

To my wife and daughters

Introduction

"If, in some cataclysm, all of scientific knowledge were to be destroyed, and only one sentence passed on to the next generations of creatures, what statement would contain the most information in the fewest words? I believe it is the *atomic hypothesis* (or the atomic *fact*, or whatever you wish to call it) that *all things are made of atoms — little particles that move around in perpetual motion, attracting each other when they are a little distance apart, but repelling upon being squeezed into one another*. In that one sentence, you will see, there is an *enormous* amount of information about the world, if just a little imagination and thinking are applied." This is a quote from *The Feynman Lectures on Physics*.[1] One can see that, in this statement made by R.P. Feynman, several physical phenomena are included as the most important ones for the next generation, among them one about interactions between atoms that are in motion. However, "if just a little imagination and thinking are applied", this statement also concerns interactions between molecules, ions or other chemical species. These interactions in various analyses are very often restricted to the nearest contacts between terminal atoms (centres) of chemical species.

This book is devoted to interactions; it deals with specific, but probably the most common, types of interactions, i.e., Lewis acid — Lewis base interactions. Someone may ask, "why another book?", if many other monographs on interactions have appeared so far. The following important books where interactions are analysed from a theoretical point of view may be mentioned. The book of I.G. Kaplan, *Intermolecular Interactions: Physical Picture, Computational Methods and Model Potentials*,[2] is devoted to terms resulting from the decomposition of the energy of interaction; the ways of calculating energies of intermolecular interactions are also discussed. *The Theory of Intermolecular Forces* by A.J. Stone is an important monograph where *ab initio* methods applied in the analysis of interactions are discussed; the perturbation theory of interactions is also a very important part of this book.[3] One can also mention *Non-covalent Interactions in Quantum Chemistry and Physics: Theory and Applications*[4] edited by A. Otero de la Roza and G.A. DiLabio as well as *Intermolecular Forces*[5] edited by S. Scheiner. In the latter, various kinds of interactions related to centres being in contact are discussed; in particular, different kinds of the hydrogen bond are described such as the dihydrogen bond, as well as chalcogen and pnicogen bonds. Similarly, *Non-covalent Interactions: Theory and Experiment*[6] by P. Hobza and K. Müller-Dethlefs describes briefly few, selected theoretical and experimental methods for the analysis of different interactions; mainly hydrogen and halogen bonds are considered.

There are several other monographs where interactions are related to other topics or phenomena. *Intermolecular Interactions in Crystals: Fundamentals of Crystal Engineering*[7] edited by J.J. Novoa is oriented, as the title suggests, on interactions in crystal structures; different kinds of interactions that occur in crystals are analysed. *Noncovalent Interactions in Catalysis*[8] edited by K.T. Mahmudov and others is focused on the role of interactions in catalysis, similar to *Non-Covalent Interactions in the Synthesis and Design of New Compounds*[9] edited by A.M. Maharramov and others that is related to processes of synthesis. There are numerous monographs on selected kinds of interactions; for those particularly devoted to the hydrogen bond, one can mention those of G.A. Jeffrey[10,11] or S. Scheiner,[12] and it is worth mentioning the recent monograph concerning this kind of interaction[13] or of the monographs on the halogen bond.[14,15]

The abovementioned selected examples of monographs are mainly devoted to theoretical methods. There are also numerous important and sound review articles where theoretical approaches that are applied to describe different kinds of intermolecular forces are analysed. Bearing in mind such a large number of studies concerning these phenomena, one may ask what the reason is for publishing another book devoted to this topic.

It is worth pointing out that this book is restricted to Lewis acid — Lewis base interactions. The meaning of the term is explained in the first chapter of this book; however, it may be indicated briefly here that such interactions involve electron charge transfers from the Lewis base unit to the Lewis acid. Such basic and acidic sites (one-centre or multi-centre) have to be well determined for this class of interactions. Thus, very weak interactions, which mostly refer to the dispersion forces,[16] are not discussed in this book.

It should be noted here that a lot of important and sound studies in recent years have emerged that concern new experimental and theoretical methods and techniques applied to analyse interactions; hence these methods are described in this book. The theoretical techniques include the Non-Covalent Interactions (NCI) approach,[17] the Quantum Theory of Atoms in Molecules (QTAIM),[18] the Car-Parrinello molecular dynamics,[19] and the Electron Localization Function (ELF) method,[20] and its various variants. The experimental methods related to the determination of crystal structures are also described in this book since they are useful in numerous studies concerning inter- and intramolecular interactions.[21]

This book contains seven chapters concerning various kinds of interactions and approaches that are useful to analyse them. The first chapter covers the definition of Lewis acid — Lewis base interactions. The nature of interactions is described, and the concepts of their formation are presented, among them the σ-hole and π-hole.[22] Besides, this chapter introduces short descriptions of different kinds of interactions; characteristics of hydrogen, halogen, chalcogen bonds and of other interactions are included.

The second chapter concerns theoretical approaches. It includes a description of different methods; starting from the *ab initio* methods, through their various approximations, to the description of other theoretical approaches that are often applied to characterise inter- and intramolecular interactions, such as those mentioned above, as well as various schemes of the decomposition of the energy of interaction,[23] the Natural Bond Orbital (NBO) approach,[24] and others. Parameters resulting from these methods are described in this chapter such as energetic and geometrical ones, for example binding energy, deformation energy, interaction energy, characteristics of bond critical points, and distances corresponding to interatomic contacts. Other properties are defined and analysed, such as atomic charges, occupation of orbitals, orbital-orbital interactions, etc.

The third chapter concerns experimental methods that are useful to analyse interactions. Mainly X-ray and neutron diffraction methods are described since they often provide complete information on structures of molecules and their arrangements in crystals.[21,25]

The fourth chapter describes different kinds of the hydrogen bond since it is the interaction that has been analysed most often and that is the most important in numerous chemical and biochemical processes. Also, those interactions that often possess numerous properties of the hydrogen bond but are problematic to be classified as hydrogen bonds according to the accepted definitions are presented.[26]

The fifth chapter concerns different classifications of interactions; in particular, the σ-hole and π-hole bonds are discussed. The mechanisms of formation of hydrogen bonds and other interactions, the description of interactions as preliminary stages of reactions and catalysis, the interrelations between the Valence-Shell Electron-Pair Repulsion (VSEPR) model [27] and various interactions are presented in the sixth chapter, while special attention is devoted to intramolecular interactions in the seventh chapter.

The above descriptions clearly show what is unique in this book; the great development of studies on various interactions needs actualisation, hence new insights are presented here. New techniques, mainly theoretical approaches, were introduced in recent studies, thus their use and significance are described here. In this book there is also a unique attempt to present compact and coherent descriptions of the mechanisms of the formation of various interactions, with comparison to the hydrogen bond; it is shown that the interactions are preliminary stages of numerous reactions and that they lead to changes of conformations that is in agreement with the VSEPR model.

Sławomir J. Grabowski
San Sebastian, October 2024

References

1. Feynman, R.P.; Leighton, R.B.; Sands, M. *The Feynman Lectures on Physics. Volume I: Mainly Mechanics, Radiation and Heat.* Basic Books, New York, USA, 1963.
2. Kaplan, I.G. *Intermolecular Interactions: Physical Picture, Computational Methods and Model Potentials.* John Wiley & Sons, Chichester, England, 2006.
3. Stone, A. *The Theory of Intermolecular Forces: Second Edition.* Oxford University Press, Oxford, United Kingdom, 2013.
4. Otero de la Roza, A.; DiLabio, G.A. (Eds.). *Non-Covalent Interactions in Quantum Chemistry and Physics: Theory and Applications.* Elsevier, Amsterdam, Netherlands, 2017.
5. Scheiner, S. (Ed.). *Noncovalent Forces.* Springer, Heidelberg, Germany, 2015.
6. Hobza, P.; Müller-Dethlefs, K. *Non-Covalent Interactions: Theory and Experiment.* Royal Society of Chemistry, Cambridge, United Kingdom, 2010.
7. Novoa, J.J. (Ed.). *Intermolecular Interactions in Crystals: Fundamentals of Crystal Engineering.* Royal Society of Chemistry, London, United Kingdom, 2018.
8. Mahmudov, K.T.; Kopylovich, M.N.; Guedes da Silva, M.F.C.; Pombeiro, A.J.L. (Eds.). *Noncovalent Interactions in Catalysis.* Royal Society of Chemistry, London, United Kingdom, 2019.
9. Maharramov, A.M.; Mahmudov, K.T.; Kopylovich, M.N.; Pombeiro, A.J.L. (Eds.). *Non-Covalent Interactions in the Synthesis and Design of New Compounds.* John Wiley & Sons, New Jersey, USA, 2016.
10. Jeffrey, G.A.; Saenger, W. *Hydrogen Bonding in Biological Structures.* Springer, Berlin, Germany, 1991.
11. Jeffrey, G.J. *An Introduction to Hydrogen Bonding.* Oxford University Press, New York, USA, 1997.
12. Scheiner, S. *Hydrogen Bonding: A Theoretical Perspective.* Oxford University Press, New York, USA, 1997.
13. Grabowski, S.J. *Understanding Hydrogen Bonds: Theoretical and Experimental Views.* Royal Society of Chemistry, Cambridge, United Kingdom, 2021.
14. Metrangolo, P.; Resnati, G. (Eds.). *Halogen Bonding: Fundamentals and Applications.* Springer, Berlin Heidelberg, Germany, 2008.
15. Metrangolo, P.; Resnati, G. (Eds.). *Halogen Bonding: Impact on Material Chemistry and Life Sciences. Volumes I and II.* Springer, Heidelberg New York Dordrecht London, 2015.
16. Leach, A.R. *Molecular Modelling: Principles and Applications Second Edition.* Pearson Education Limited, Harlow, United Kingdom, 2001.

17. Johnson, E.R.; Keinan, S.; Mori-Sánchez, P.; Contreras-García, J.; Cohen, A.J.; Yang, W. Revealing noncovalent interactions. *J. Am. Chem. Soc.* **2010**, *132*, 6498–6506.
18. Bader, R.F.W. *Atoms in Molecules: A Quantum Theory*. Oxford University Press, Oxford, United Kingdom, 1990.
19. Car, R.; Parrinello, M. Unified approach for molecular dynamics and density-functional theory. *Phys. Rev. Lett.* **1985**, *55*, 2471–2474.
20. Silvi, B.; Savin, A. Classification of chemical bonds based on topological analysis of electron localization functions. *Nature* **1994**, *371*, 683–686.
21. Wilson, C.C. *Single Crystal Neutron Diffraction from Molecular Materials*. World Scientific, Singapore, 2000.
22. Politzer, P.; Murray, J.S.; Clark, T. Halogen bonding: An electrostatically-driven highly directional noncovalent interaction. *Phys. Chem. Chem. Phys.* **2010**, *12*, 7748–7758.
23. Pendas, A.M.; Blanco, M.A.; Francisco, E. The nature of the hydrogen bond: A synthesis from the interacting quantum atoms picture. *J. Chem. Phys.* **2006**, *125*, 184112.
24. Weinhold, F.; Landis, C. *Valency and Bonding: A Natural Bond Orbital Donor — Acceptor Perspective*. Cambridge University Press, Cambridge, United Kingdom, 2005.
25. Desiraju, G.R.; Vittal, J.J.; Ramanan, A. *Crystal Engineering: A Textbook*. World Scientific, Singapore, 2011.
26. Grabowski, S.J. Hydrogen bond types which do not fit accepted definitions. *Chem. Comm.* **2024**, 60, 6239–6255.
27. Gillespie, R.J.; Hargittai, I. *The VSEPR Model of Molecular Geometry*. Dover Publications, Mineola, New York, USA, 2012.

About the Author

Dr. Sławomir J. Grabowski is Ikerbasque Research Professor at University of the Basque Country (UPV/EHU) and Donostia International Physics Center (DIPC), Spain. He has published more than 200 peer-reviewed papers, 15 book chapters and 3 books (one as author and two as editor), with a h-index of 58. His main interests are in non-covalent interactions, crystallography and quantum chemistry. He holds a MSc and PhD from the University of Warsaw, Poland and a DSc from the Technical University of Łódź, Poland. He is a member of the editorial boards of *Journal of Physical Organic Chemistry* and *Computational and Theoretical Chemistry.*

Contents

Chapter I

Which Interactions are of the Lewis Acid — Lewis Base Type?

Abstract: The definition of Lewis acid — Lewis base interactions as well as their properties are discussed; several energetic characteristics of interactions are presented. The Molecular Mechanics approach is roughly described to consider physical meanings of interaction energy terms; the decomposition interaction energy scheme is also briefly examined. The σ-hole and π-hole concept is described; particularly the positive and negative areas of the electrostatic potential are discussed as indicators of the Lewis acid and Lewis base sites, respectively. Several examples of σ-hole and π-hole bonds are presented and characteristics of hydrogen, halogen, chalcogen bonds and other numerous interactions are indicated.

Keywords: Lewis acid — Lewis base interactions, energy of interaction, binding energy, deformation energy, Molecular Mechanics, interaction energy terms, σ-hole and π-hole concept, electrostatic potential.

I.1 The Definition of Lewis Acid – Lewis Base Interactions

It is important to indicate here which interactions are defined as Lewis acid-base and why such interactions are a subject of this book. These issues were roughly presented in the Introduction but it is also crucial to explain them extensively to justify the scope of this and next chapters.

It is worth quoting here an excerpt from the article of Vogel and Drago: "The interactions between Lewis acids (A) and bases (B) are fundamental in chemistry and biochemistry. The scope of these donor-acceptor interactions include classical Lewis acid-base reactions (e.g., charge-transfer complexes), all coordination chemistry, Brønsted acid-base interactions, hydrogen bonding, displacement reactions, nucleophilic and electrophilic attacks, and nonradical atom transfer reactions. The ubiquitous nature of donor-acceptor interactions in such diverse areas as synthesis, surface science, catalysis, and the collection and proper interpretation of thermodynamic data as well as their importance in the isolobal concept emphasises that a fundamental understanding of Lewis acid-base interactions must be an essential component of the practicing chemist's knowledge."[1] Thus one can see that these are very important interactions in numerous reactions and processes. The question boils down to the basic and elementary problem that practically concerns definitions, i.e., which chemical species are defined as Lewis acids and which as Lewis bases? One can refer here to early statements of Lewis who has proposed that "a basic substance is one which has a lone pair of electrons which may be used to complete the stable group of another atom, and that an acid substance is one which can employ a lone pair from another molecule in completing the stable group of one of its own atoms."[2] However, it is also worth mentioning that the terms are very similar or even indistinguishable from nucleophilic and electrophilic reagents, the terms that were introduced by Ingold.[3] In some studies it is pointed out that the Lewis acidity-basicity and electrophilicity-nucleophilicity are related to the thermodynamic and kinetic aspects of processes, respectively, but their meanings are practically the same. In this book the terms Lewis acid and Lewis base are used in relation to the electron-poor (electron-accepting) and electron-rich (electron-donating) regions, respectively, and not only strictly to an existence or a lack of lone electron pairs.

Another issue related to Lewis acidity-basicity concerns acid-base reactions. According to the Brønsted definition, an acid can split off a proton while a base is able to attach a proton.[4,5] There are numerous studies on Brønsted acid-base equilibria, particularly in solutions but not only. The Lewis definition seems to be useful for more scenarios. While it concerns reactions in a variety of liquid, solid, and gaseous media, similar to the Brønsted approach, it covers a broader spectrum of chemical species since the acidity and basicity are not restricted only to the proton-detaching and proton-attaching substances, respectively. It was stated that "all substances which have lone pairs of electrons, capable of employment in onium formation, are basic substances. On the other hand substances like hydrogen ion, iodous ion, silicon dioxide, sulfur trioxide and boron trichloride are acid substances, since the hydrogen ion will accept a pair of electrons to form its stable group of two, and the remaining substances will accept pairs of electrons to complete their stable groups of eight."[2] One can see that these Lewis acid-base reactions, or interactions, take place in order to preserve the octet rule, or the doublet in the case of the hydrogen atom.[6] The latter topic is a subject of special attention later in this book (see Chapter VI).

Let us consider a few examples of the Lewis acid — Lewis base interactions. Figure I.1a presents an example of the BF_3-NH_3 Lewis acid-base adduct. Similar adducts were analysed early on by Lewis,[7] who considered boron trimethyl and boron trihalides where boron may complete its octet structure by interacting with species that donate the lone electron pair. In other words, strong Lewis acid properties of the boron centre result from its electron structure; the boron is an electron-deficient centre in the BH_3 unit, in the boron trihalides, and in numerous other compounds, as it has six electrons in the outer shell. This characteristic, that the centre does not obey the octet rule because of the deficiency of electrons, is known as hypovalency,[8] in contrast to hypervalency where the number of electrons exceeds eight and thus the octet rule is also not obeyed.[9] In the BF_3-NH_3 complex, the ammonia species is an electron donor that leads to the completion of eight electrons in the valence shell of boron. The B···N

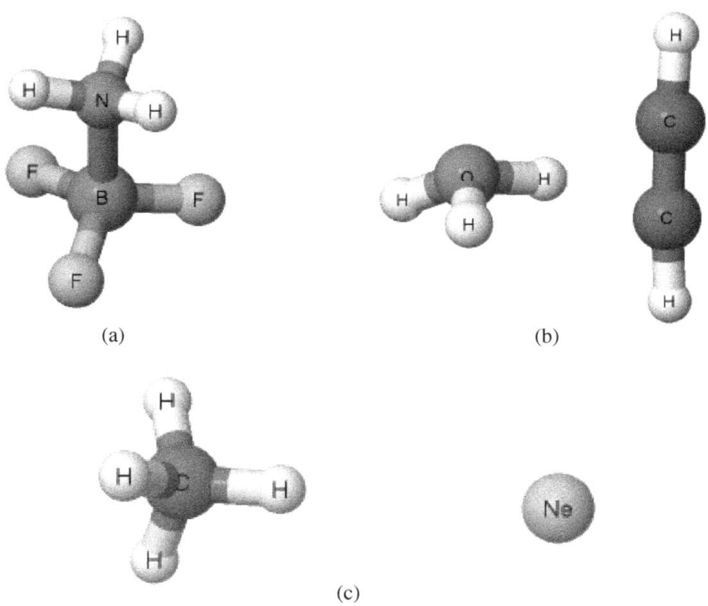

Fig. I.1 Examples of systems linked by different types of interactions: a – BF_3-NH_3, b – H_3O^+-C_2H_2, c – CH_4-Ne.

Lewis acid — Lewis base interaction is very strong and possesses characteristics of a covalent bond.[10] The B···N distance is equal to 1.670 Å, and the interaction energy amounts to –44.8 kcal/mol (corrected for the basis set superposition error, BSSE, MP2/aug-cc-pVTZ calculations).[11]

According to the supermolecular approach the interaction energy in an A-B complex is defined by the difference between the energy of the A-B complex and the sum of energies of A and B monomers (eq. I.1).[12–14]

$$E_{int} = E_{A...B}(A...B) - E_{A...B}(A) - E_{A...B}(B) \quad (I.1)$$

The labels in parentheses refer to species for which energies are calculated and the subscripts indicate which systems were optimised, in other words, from which systems corresponding to the energetic minima the units indicated in parentheses were taken. This definition of the energy of interaction informs that monomers are characterised by geometries that they have in the complex.

Another measure of the strength of interaction that is often presented in various studies is the binding energy expressed by eq. I.2.[12–14]

$$E_{bin} = E_{A...B}(A...B) - E_A(A) - E_A(B) = E_{int} + E_{def} \quad (I.2)$$

One can see that it is the difference between the energy of the complex in its energetic minimum and the sum of energies of monomers in their energetic minima. The latter minima concern isolated monomers, thus their geometries differ from those that they have in the complex. It is usually assumed in a majority of studies that the negative energies correspond to the stabilising interactions and the positive energies to destabilising interactions. The complexation leads to the energetically destabilising change of geometries of interacting species, thus the deformation energy (E_{def} in eq. I.2) that corresponds to this complex formation is positive.[15] However, such destabilisation is overwhelmed usually by the stabilising interaction between monomers. Then the binding energy is negative and consequently the interaction energy is negative.

It is worth mentioning that the binding energy discussed here represents enthalpy difference, ΔH, at 0 K without the zero-point vibrational energy, ZPVE (see next chapter for details concerning ZPVE).[16] The binding energy is related to computed

Born–Oppenheimer potential energy surfaces. The energetic terms are discussed here for the intermolecular interactions; however, more often, both energy and enthalpy terms are applied to consider chemical reaction processes, thus they are differences between the energies (or enthalpies) of products and reactants. The enthalpy difference, ΔH, and the free energy difference, ΔG, are related by the entropy difference, ΔS (eq. I.3).

$$\Delta G = \Delta H - T\Delta S \tag{I.3}$$

For temperatures other than 0 K the enthalpy difference, $\Delta H(T)$, is the energy difference at 0 K, ΔE(0 K), plus the increase in energy terms such as translational, ΔE_{trans}, rotational, ΔE_{rot}, vibrational, ΔE_{vib}, and electronic energy increments, ΔE_{el}, on going from 0 K to the temperature considered. The work of the system related to the change in pressure or volume, $\Delta(PV)$, is also included.

$$\Delta H(T) = \Delta E(0\ \text{K}) + \Delta E_{trans} + \Delta E_{rot} + \Delta E_{vib} + \Delta E_{el} + \Delta(PV) \tag{I.4}$$

Coming back to complexes presented in Fig. I.1, the B···N distance in the BF_3-NH_3 complex of 1.670 Å is much lower than the corresponding sum of van der Waals radii (Table I.1) of the boron and nitrogen atoms; this sum is equal to 3.47 Å! It was stated

Table I.1 Consistent van der Waals radii (in Å) for all main group elements; bold values are from the study of Truhlar and coworkers,[17] value for H is from Rowland and Taylor,[18] and other values are from Bondi.[19]

1	2	13	14	15	16	17	18
H							He
1.10							1.40
Li	Be	B	C	N	O	F	Ne
1.81	**1.53**	**1.92**	1.70	1.55	1.52	1.47	1.54
Na	Mg	Al	Si	P	S	Cl	Ar
2.27	1.73	**1.84**	2.10	1.80	1.80	1.75	1.88
K	Ca	Ga	Ge	As	Se	Br	Kr
2.75	**2.31**	1.87	**2.11**	1.85	1.90	1.83	2.02
Rb	Sr	In	Sn	Sb	Te	I	Xe
3.03	**2.49**	1.93	2.17	**2.06**	2.06	1.98	2.16
Cs	Ba	Tl	Pb	Bi	Po	At	Rn
3.43	**2.68**	1.96	2.02	**2.07**	**1.97**	**2.02**	**2.20**
Fr	Ra						
3.48	**2.83**						

that "One might define the van der Waals radius in terms of that distance r at which this repulsion just balances the attraction forces between the two atoms. [···] this distance corresponds to a very low electron density."[19] In other words, for such interatomic contacts the distance is the sum of corresponding van der Waals radii. The low electron density at the "contact point" of two van der Waals spheres indicates that there is no meaningful electron charge shifts between interacting units. The stabilisation of this system proceeds due to dispersion forces. The interatomic distance is lower than the abovementioned sum if significant electron charge shifts are observed, and it indeed occurs for the Lewis acid-base contacts, in other words for the Lewis acid-base interactions.[20,21] The attribution of charge transfer as the main feature of the Lewis acid — Lewis base interactions such as the hydrogen bond, the halogen bond and others took place in early studies.[20,21] It is also postulated in more recent monographs and articles.[6,22,23]

The interactions such as that for the BF_3-NH_3 complex, between the triel electron-deficient centres (triel atoms are those of the 13th group) and the electron donors, are known as triel bonds.[10,24,25] The large absolute value of the energy of interaction in the BF_3-NH_3 species indicates the covalent character of the triel bond linking this complex, or at least partly covalent character. The formation of this complex leads to enormous electron charge shift from the NH_3 Lewis base unit to the BF_3 species acting as the Lewis acid. This shift is equal to 0.34 e (calculated with the use of the Natural Bond Orbital method).[10,11,26] Such shifts for the weak- and medium-strength interactions are 10–20 times lower; for example, for the trans-linear water dimer, this shift amounts to 0.02 e approximately.[27]

Figure I.1b presents the $H_3O^+\cdots C_2H_2$ complex where the hydronium cation as the Lewis acid unit is linked with the acetylene molecule that is the electron charge donor. The O-H···π charge-assisted hydrogen bond occurs here. The H···π distance is equal to 1.729 Å (distance between the H-centre and mid-point of the CC bond of acetylene), the interaction energy amounts to –19.4 kcal/mol, while the electron charge transfer from the acetylene to the hydronium cation is equal to ~0.31 e (estimated from the CHelpG charges, MP2/6-311++G(3df,3pd) level).[11,28] These results indicate that strong interaction occurs in this complex.

Fig. I.1c shows the $CH_4\cdots Ne$ complex. The H···Ne contact is observed here, and this intermolecular distance is equal to 2.784 Å, which is greater than the corresponding sum of van der Waals radii –2.64 Å (Table I.1); however, these values are close to each other. The interaction energy amounts to –0.04 kcal/mol (BSSE correction included) and the electron charge shift from the Ne-centre into the methane unit is equal to 4×10^{-5} e only, which is below the accuracy of calculations. In other words, there is no electron charge shift practically. These are all MP2/aug-cc-pVTZ results; however, if the Hartree-Fock (HF) energy of interaction in this complex is calculated, it amounts to +0.15 kcal/mol. It means that the correlation energy being the difference between MP2 and HF energies[11] is responsible for the slight stabilisation of this complex.[29] The stabilisation dispersion interaction energy is the main part of the correlation energy.

The following are observations concerning interactions in the above-described complexes. The B···N and H···π interactions in BF_3-NH_3 and $H_3O^+\cdots C_2H_2$ complexes, respectively, are very strong and it seems that they possess numerous characteristics of covalent bonds. The absolute values of energies of interactions are very high, the intermolecular distances are small (close to typical lengths of covalent bonds), and the complexation leads to great electron charge shifts from the Lewis base unit to the Lewis acid. Besides, the centres in direct contact are characterised by opposite charges; for the BF_3-NH_3 complex the boron and nitrogen centres possess positive and negative charges, respectively; for the $H_3O^+\cdots C_2H_2$ complex the positively charged H-centre is in contact with π-electrons of the acetylene. One can say that for these two complexes the Lewis acid and Lewis base centres are well defined because they are characterised by positive and negative charges, respectively. The interactions that stabilise these complexes are characterised by the significant electron charge shifts which are usually observed for the Lewis acid-base interactions;[20,21] Lewis acids are acceptors of the electron charge and Lewis bases donors.

In the case of the $CH_4\cdots Ne$ complex, the H···Ne intermolecular distance is greater than the corresponding sum of van der Waals radii, the interaction energy is close to zero, only negligible stabilisation is observed owing to the dispersion forces, and there is no electron charge shift practically. Additionally, the Lewis acid and Lewis base centres are not clearly defined here. It means that the weak interaction which is observed in this complex is not classified as Lewis acid — Lewis base interaction.

A similar situation is observed in the methane dimer that is a subject of numerous studies. Different conformations of dimers and trimers of methane were analysed early and the theoretical calculations indicate that the dimer conformation of D_{3d} symmetry possesses the lowest energy (it corresponds to the global minimum).[30] It was also found that the dispersion forces are crucial to stabilise the systems.[30] These findings were further confirmed since it was calculated that for this conformation of the dimer possessing the D_{3d} symmetry, the HF energy is positive and the MP2 approach reproduce correctly the slight stabilisation of this dimer.[31] The interaction energy calculated at the MP2/6-311++G(3df,3pd) level amounts to approximately –0.4 kcal/mol; a similar value was calculated at the CCSD(T)/CBS level.[32] The recent SAPT/aug-cc-pVQZ results give the interaction energy of about –0.5 kcal/mol and indicate the dispersion energy as the main attraction interaction energy contribution.[33] In other words, the interactions in the methane dimer are not classified as Lewis acid — Lewis base since this dimer is linked by dispersion forces, the role of the electron charge shifts is not observed and there are no specific Lewis acid-base centres.

Hence, we may define the Lewis acid — Lewis base (La-Lb) interactions as those where La and Lb are well-determined sites in a contact that leads to the electron charge shifts, mainly from the Lb unit to La. This occurs for the BF_3-NH_3 and $H_3O^+\cdots C_2H_2$ complexes but not for the $CH_4\cdots Ne$ complex nor the methane dimer; for the latter two cases one observes weak stabilisation due to dispersion forces. In the water dimer, the O-H···O hydrogen bond is classified as La-Lb interaction because hydrogen and oxygen centres characterised by opposite charges are in intermolecular contact. The electron charge shift is also observed, from a water unit playing the role of the Lewis base into another water molecule that acts as the Lewis acid.

The following question related to the definition given above arises: which centres are well-determined La and Lb sites? It seems that they are characterised by positive and negative charges, respectively. It is explained in detail in the next section of this chapter since there are many legitimate reservations about the physical meaning of atomic charges.[34,35] Besides, it is worth noting that the centres being in contact in La-Lb interactions are not necessarily single atoms; the $H_3O^+\cdots C_2H_2$ complex is an example where π-electrons of the CC bond act as the Lb site.

I.2 The Nature of Intermolecular Interactions

Molecular mechanics (MM) is one of the most often applied groups of methods to calculate geometrical and energetic characteristics of systems; these are usually large systems such as clusters, structures of ions, and molecules in liquids, proteins or crystals. MM concerns species which are linked by covalent bonds, more or less polarised, and which are linked also by weaker interactions. The latter links are often named as non-covalent interactions.[36,37] The set of equations and parameters that allow one to perform MM calculations is called the force field; various force fields are determined for different classes of compounds, different conditions or for different states (gas, liquid, solid). The determination of the parameters is based on their best matching to the experimental data; sometimes such matching to the theoretical results of calculations is considered. This process of the determination of parameters is known as parametrisation.

The force field may be expressed as a function that is applied to model an assemblage of atoms or molecules (eq. I.5).[37]

$$U(r_1,r_2\ldots.r_N)=\sum_{bonds}\frac{k_{li}}{2}(l_i-l_{i,0})+\sum_{angles}\frac{k_{\Theta i}}{2}(\Theta_i-\Theta_{i,0})+\sum_{torsions}\frac{V_n}{2}(1+\cos(n\omega-\gamma))$$
$$+\sum_{i=1}^{N}\sum_{j=i+1}^{N}\left\{4\varepsilon_{ij}\left[\left(\frac{\sigma_{ij}}{r_{ij}}\right)^{12}-\left(\frac{\sigma_{ij}}{r_{ij}}\right)^{6}\right]+\frac{q_iq_j}{4\pi\varepsilon_0 r_{ij}}\right\} \qquad (I.5)$$

The $U(r_1,r_2\ldots.r_N)$ is the potential energy that is a function of r-positions of N atoms constituting the system. The first term of the above expression is the sum of interaction energies between bonded atoms. These energies are approximated by the harmonic oscillator potentials. The energetic term for the single ith bond, $\frac{k_{li}}{2}(l_i-l_{i,0})$, expresses the increase of the energy of interaction if the l_i length of bond considered deviates from the reference $l_{i,0}$ value that corresponds to the potential minimum of the harmonic oscillator. The same concerns valence angles, which are taken into account in the next summation of eq. I.5, where the Θ_i angles are considered and their reference values $\Theta_{i,0}$. The next term is a torsional potential which in some force fields is not included. It concerns the torsion angles, A-B-C-D. The ω and γ values correspond to the A-B-C-D quartet and to the phase factor, respectively. The latter parameter determines where the torsion angle passes through the minimum. The k_{li} and $k_{\Theta i}$ force constants are related to bonds and angles, respectively, while V_n often refers to the barrier height of the torsional potential. However, it is not exactly such a barrier since usually more than one term occurs in the cosine expansion; n refers to the number of points of minima of the torsional potential.[37]

The last term of eq. I.5 that is the double sum concerns interactions of pairs of non-bonded atoms, among them pairs of atom-atom contacts. It is worth mentioning that eq. I.5 for a complex that contains two linked molecules may be related to the binding and interaction energies discussed roughly before (see eqs. I.1 and I.2). It is an approximate value of the binding energy where the first three terms correspond to the deformation energy related to changes of geometries of monomers resulting from complexation, while the last double summation is the approximate interaction energy. In such a case, the Θ_i angles and l_i bond lengths concern the complex in energetic minimum while the $\Theta_{i,0}$ angles and $l_{i,0}$ values are related to isolated monomers in their energetic minima. Equation I.6 shows the relationship between the force fields applied in MM approaches and the energetic measures of the strength of interaction that are often discussed in various studies.

$$U(r_1,r_2,\ldots r_N) \equiv E_{bin} = E_{int} + E_{def} \tag{I.6}$$

The *ab initio* methods that are described in the next chapter are useful to calculate the interaction and binding energies more precisely. According to Coulomb's law, the last part of eq. I.5 expresses the electrostatic interaction (eq. I.7).

$$U_{elec}(r_1, r_2 \ldots r_N) = \sum_{i=1}^{N} \sum_{j=i+1}^{N} \frac{q_i q_j}{4\pi\varepsilon_0 r_{ij}} \tag{I.7}$$

It is worth to note that in eq. I.7, for a system containing N atoms, only non-bonded atom-atom contacts are taken into account; q_i and q_j in the last part of the equation (and in eq. I.5) are atomic charges of interacting atoms, while r_{ij} is the distance between these atoms.

An alternative approach may also be applied to calculate the electrostatic interaction. The interacting species, ions or molecules are treated here as single units. Such approach requires the central multipole expansion, i.e., the consideration of multipoles: charges, dipoles, quadrupoles, octopoles and so on.[12,37,38]

Let us consider a part of the last term of eq. I.5 (the double sum) that does not contain the electrostatic interaction. In the case of two interacting species (molecules, ions, etc.), this term may be modified (eq. I.8).

$$U_{vdW}(r_1, r_2 \ldots r_{N_A \cdot N_B}) = \sum_{i=1}^{N_A}\sum_{j=1}^{N_B} 4\varepsilon_{ij}\left[\left(\frac{\sigma_{ij}}{r_{ij}}\right)^{12} - \left(\frac{\sigma_{ij}}{r_{ij}}\right)^{6}\right] \qquad \text{(I.8)}$$

In the above equation, N_A and N_B label the number of atoms in the A and B interacting species, respectively; σ and ε are the collision diameter and the well depth. This expression (eq. I.8) is often referred to the van der Waals interaction. If one restricts to two interacting atoms, say A and B, then one has the well-known Lennard-Jones 12-6 function, also known as the Lennard-Jones potential (eq. I.9).[39]

$$U_{AB}(r_{AB}) = 4\varepsilon_{AB}\left[\left(\frac{\sigma_{AB}}{r_{AB}}\right)^{12} - \left(\frac{\sigma_{AB}}{r_{AB}}\right)^{6}\right] \qquad \text{(I.9)}$$

The σ parameter corresponds to the distance between atoms A and B if the attraction and repulsion interactions are balanced such that the $U_{AB}(r_{AB})$ energy is equal to zero. The ε-parameter is known as the well depth; it corresponds to the minimum value of the $U_{AB}(r_{AB})$ function. The r_{AB} value for this minimum is the equilibrium interatomic distance; r_0 designation is often applied. The 12-6 potential (eq. I.9) is often expressed in modified form (eq. I.10).

$$U_{AB}(r_{AB}) = \frac{C}{r_{AB}^{12}} - \frac{D}{r_{AB}^{6}} \qquad \text{(I.10)}$$

In such a way the C and D constants are determined independently during parametrisation, while in eq. I.9 the constants of the positive and negative parts, $\sim r_{AB}^{-12}$ and $\sim r_{AB}^{-6}$, respectively, depend on each other.

The Cr_{AB}^{-12} part of the potential is often assigned as the repulsive part, related to the Pauli repulsion interaction, while the $-Dr_{AB}^{-6}$ part is related to the attractive dispersion interaction. There are reservations concerning the repulsive term of the 12-6 potential, thus the Buckingham potential is also applied in various force fields (eq. I.11).[40]

$$U_{AB}(r_{AB}) = Ce^{-\alpha r_{AB}} - \frac{D}{r_{AB}{}^{6}} \tag{I.11}$$

It is worth to note that the latter potential contains three constants. Moreover, it is characterised by a high positive value for very short distances between atoms considered, and the shortening of this distance leads to a sudden reduction in the value of the potential up to negative values; for the distance $r_{AB} \rightarrow 0$, the potential $U_{AB}(r_{AB}) \rightarrow -\infty$.[37] Such small atom-atom distances corresponding to the maximum as well as to the negative potential values are unrealistic. For the equilibrium distance corresponding to the energy minimum, and for other distances that are close to this equilibrium, the potential energy is usually well reproduced by the Buckingham function. However, one should take care that during the simulation process such small distances do not occur randomly.

The other problem is that both potentials, Lennard-Jones as well as Buckingham, take into account a mixture of two components only; the repulsive (Pauli) interactions and the dispersive interactions. This is why the minima of these functions (eqs. I.10 and I.11) are often treated as corresponding to the sum of van der Waals radii of centres being in a contact. One may refer here to the monograph of Desiraju, Vittal and Ramanan where it is stated that "in the context of crystal engineering, the term usually means the sum of all stabilising and destabilising interactions, but excluding hydrogen bonding and other anisotropic interactions (such as dipole-dipole). [...] van der Waals interactions arise from attractive (dispersive) and repulsive forces."[41] This statement is in agreement with the descriptions given above.

Another term of the energy of interaction, the electrostatic component, is described by Coulomb's law, and it is included in the majority of force fields applied (eqs. I.5 and I.7). The great disadvantage of the majority of force fields is that the part of energy of interaction connected with electron charge shifts is not taken into account, at least not directly. However, the latter interaction energy term is often included since the parametrisation is performed in such a way as to take it into account, in eqs. I.10 and I.11, for example.

It is worth mentioning that the interaction related to electron charge shifts is very important for strong interactions that often possess numerous characteristics of covalent bonds.[42,43] Different names of interaction energy connected with electron charge transfers occur in literature; charge transfer, polarisation, delocalisation, induction, etc. The name depends on the interaction energy decomposition scheme applied.[44] It is also worth mentioning that the physical meaning of this energy also depends on the decomposition scheme applied; there are at least slight differences in this meaning in different schemes. However, these energies, in general, refer to electron charge shifts. In this book, the term polarisation energy is applied if any specific decomposition scheme is not discussed. The meaningful charge shifts phenomena often occur in the case of the hydrogen bonds,[42] and in other interactions such as in numerous σ-hole and π-hole bonds.[6,43] Sometimes modifications of potential energy functions are applied in force fields to include specific interaction energy terms. For example, in the case of the hydrogen bond, the 12-6, Buckingham, or any other type of potential applied is replaced by the term related to hydrogen-bonded atoms, and is known as the 12:10 potential (eq. I.12).

$$U_{AB}(r_{AB}) = \frac{C}{r_{AB}^{12}} - \frac{D}{r_{AB}^{10}} \tag{I.12}$$

This function concerns the H and B pair of atoms of the A-H···B hydrogen bond. Numerous other simple models were applied in different force fields, or in general, in different studies to take into account the hydrogen bond or to evaluate the hydrogen bond energy.[37] For example, various potential functions are described in recent studies.[45]

However, the most important is that the interaction or the binding energy may be decomposed to show its terms related to various physical phenomena (eq. I.13).

$$E_{bin} = E_{int} + E_{def} = E_{Pauli} + E_{elstat} + E_{pol} + E_{disp} + E_{def} \tag{I.13}$$

These interaction energy terms are briefly described in this section of the chapter, and will be considered in detail in the next chapter concerning theoretical approaches. E_{Pauli} is the repulsive term, roughly related to the overlap of "electron clouds" of interacting species. The E_{pol} term is related to the electron charge shifts while the E_{def} term that is beyond the interaction energy, E_{int}, describes the increase of the destabilisation of the complex caused by the change of geometries of interacting species resulting from the complexation. The stabilising (attractive) interacting energy terms are negative values (usually E_{elstat}, E_{pol}, and E_{disp}) while the destabilising (repulsive) terms are positive (E_{Pauli}); the E_{def} term of the binding energy is certainly positive.

It is worth mentioning that for La-Lb interactions, all attractive interaction energy terms are very important, particularly the E_{elstat} and E_{pol} terms; the latter expresses the significance of the electron charge shifts resulting from complexation. In the case of van der Waals interactions that are not classified as La-Lb interactions, the E_{disp} term is the most important attractive term, and remaining attractive terms are negligible. Such a situation occurs for interactions between atoms of noble gases and particularly for the methane dimer and the CH_4···Ne complex discussed earlier in this chapter.

I.3 The σ-hole Concept and the Case of the Halogen Bond

The σ-hole concept was proposed by Clark, Murray, Politzer and coworkers to explain the occurrence of regions of electron charge depletion at various atomic centers.[46–49] This concept was firstly and mainly applied to justify the Lewis acid properties of halogen centres despite being electronegative elements.[46,47,50] Moreover, it explains the dual character of halogen atoms which can often act simultaneously as a Lewis acid centre and a Lewis base centre. These properties are related to the electrostatic potential (EP) at their surfaces.[46,47] The EP, $V(r)$, at any r-point of a surface attributed to a molecule or any other species (radical, ion, etc.) is expressed by the following equation:

$$V(r) = \sum_A \frac{Z_A}{|R_A - r|} - \int \frac{\rho(r')dr'}{|r' - r|} \tag{I.14}$$

Z_A is the charge of nucleus A located at R_A while $\rho(r')$ is the electron density at the r'-point. The EP, $V(r)$, is a physical observable which can be determined experimentally by diffraction techniques. It may also be calculated theoretically. The sites of the Lewis acidity and basicity may be indicated approximately if one considers the EP at a molecular surface. The positive EP regions correspond to the sites of nucleophilic attack while the negative EP areas are the sites of reactions with electrophiles.

The surfaces characterised by the electron density of 0.001 au (electron/Bohr^3) are considered most often to determine the values of the EP. This electron density value of 0.001 au was proposed by Bader and coworkers.[51] Such electron density surfaces with maps of the EP are a source of useful information. They show approximate regions of possible nucleophilic or electrophilic attacks, and they show possible arrangements of neighbouring units which are in contacts with the species considered and which are characterised by the electron-donating and electron-accepting sites, respectively. In other words, these regions at the surfaces containing EP maps indicate the Lewis acid and Lewis base sites.

The choice of the electron density surface of $\rho = 0.001$ au for the EP mapping seems to be justified. It was pointed out that such a sphere covers or even agrees with the van der Waals sphere of the species considered.[51,52] One may expect that for two species which have a contact at their 0.001 au surfaces, the electron charge shifts between them are negligible and the interaction is mainly attractive due to the electrostatic, long-range forces. The attractive electrostatic interaction energy tends to zero with the increase of intermolecular distance, r, and such diminution is proportional to r^{-1}. In the case of the dispersion and polarisation energies (short-range), the diminution is proportional mainly to r^{-6}. Thus the attraction due to polarisation and dispersion forces disappears at shorter distances while such disappearance occurs for electrostatic forces at longer distances.

Table I.2 presents the values of atomic charges and of EPs for methane and its halogen derivatives (halogenomethanes); MP2/aug-cc-pVTZ results are presented.[11] Let us discuss firstly the values of atomic charges that concern the carbon and X centres

Table I.2 The EP values (au) at X-centre (H-centre for methane) in the elongations of C-X bonds; the Mulliken (*) and QTAIM (**) charges for C and X centres (marked by qC and qX, respectively) are presented (in au). These are MP2/aug-cc-pVTZ results.

Species	EP	qX*	qC*	qX**	qC**
CH_4	0.0144	0.314	-1.254	0.010	-0.040
CH_3F	-0.0363	-0.356	-0.584	-0.697	0.595
CH_3Cl	0.0045	-0.194	-0.766	-0.278	0.100
CH_3Br	0.0115	-0.058	-0.838	-0.145	-0.044
CH_3I	0.0199	0.062	-0.970	0.021	-0.213
CH_2FI	0.0293	0.056	-0.351	0.049	0.468
CF_3I	0.0491	0.065	0.635	0.170	1.909

(X = H, F, Cl, Br and I). The Mulliken population charges[53] as well as the charges derived from the Quantum Theory of Atoms in Molecules (QTAIM)[54] are presented.[11] The H-centre charge for methane is positive, which explains the weak Lewis acid properties of this species; very weak hydrogen bonds are formed with methane that acts as the proton donor.[55–57] For example, in the case of the $CH_4 \cdots OH_2$ complex, there are conformations where the C-H···O hydrogen bond links the hydrogen of methane with the oxygen centre of water.[55] However, for other conformations of this complex the O-H···C interactions link water and methane units, which may indicate the Lewis base properties of the carbon centre of methane.[55] Table I.2 shows the negative carbon charge for both Mulliken and QTAIM population analyses that confirms these properties. A similar situation is observed for the $CH_4 \cdots SH_2$ complex; there is conformation of this complex linked by the C-H···S hydrogen bond, and there are conformations linked by two hydrogen bonds: C-H···S and S-H···C.[56]

In the case of the CH_3X halogen species (Table I.2), the X-charge increases in the order F < Cl < Br < I, and it is positive only for the CH_3I unit. The latter trend is observed for both types of charges, Mulliken as well as QTAIM. Similar results were obtained at the MP2(full)/6-311++G(d,p) level for the CH_3Cl and CH_3Br species;[58] in general, it was observed that the charges of the halogen centres are negative. However, various population analyses were applied in the latter study and two exceptions occur there; for the chlorine in CH_3Cl the charge is positive for the Mulliken population[53] while the bromine in CH_3Br is positive for the Natural Bond Orbital population.[8] Thus the atomic charges are strongly dependent on the level of calculations as well as on the kind of population analysis applied. Besides it is known that CH_3Br and even CH_3Cl species are characterised by the Lewis acid properties of the halogen centre since their complexes with electron donors are stable and the corresponding energies of interactions are negative.[59]

It was pointed out earlier that for the methane species the negative carbon charge for both population analyses is observed. A similar situation is observed for the majority of halogenomethanes presented in Table I.2. However, the atomic charge of carbon is positive sometimes, particularly in cases with the influence of F-substituents. For example, it is positive for the CF_3I species for both population analyses applied.

Table I.2 shows that the QTAIM positive charge of the iodine centre in the CH_3I species, +0.021 au, increases due to the influence of fluorine electron-withdrawing substituents, while for the CH_2FI species it amounts to +0.049 au and in CF_3I it is equal to +0.170 au. It means that the F-substituents withdraw electron charge that leads to its depletion at the iodine centres. For the Mulliken population the above changes are not so clear; the charge of iodine in the latter three species does not change significantly and it amounts to approximately 0.06–0.07 au.

It is worth noting here that the use of atomic charges to explain properties of centres considered is criticised in numerous studies.[60,61,62] It was pointed out that "most models have their limitations but there is one in particular that leads to false conclusions and cannot rationalise important noncovalent interactions: net atomic charges".[62] It was indicated that there are various definitions of atomic charges (various population analyses), that the atom charge value depends on the level of calculations and that numerous properties cannot be explained by charges, for example, the dual properties of numerous halogen centres that can act at the same time as Lewis acids and Lewis bases.[60–62] On the other hand, it was pointed out that the EP at any site of the centre considered is a better descriptor of properties than the atomic charge. Figure I.2 presents the EP surfaces (for the electron density, $\rho = 0.001$ au) for methane and its halogen derivatives. Let us analyse EP values at the X centre. One can see the C-X (X = H, F, Cl, Br and I) bonds at the right sites of species presented in Fig. I.2. Table I.2 presents the EP values at the abovementioned surfaces in the elongation of C-X bonds. For the methane species the positive EP value at the H-centre is observed, 0.0144 au. For the CH_3F species the EP at the fluorine centre is negative while for the remaining CH_3X halogenometanes the EP at the X halogen centre is positive and it increases with the increase of its atomic number; for the CH_3I species it is equal to 0.0199 au. Thus the positive EP value at the X-centre explains its Lewis acid properties. One should expect Lewis base properties of the fluorine centre in the CH_3F moiety. In the case of CH_2FI and CF_3I species the EP at the iodine centre is equal to 0.0293 au and 0.0491 au, respectively, thus it increases in comparison with the CH_3I unit. This is because of the fluorine substituents that withdraw the electron charge causing its depletion at the iodine centres. The latter trend is in agreement with that concerning the QTAIM iodine charges discussed earlier.

Figure I.2 presents selected species of halogenometanes collected in Table I.2. One can see that for all halogenometanes, except of the CH_3F species, the positive EP at X-halogen in the extension of the C-X bond is accompanied by the "regular ring" of negative EP around the X-centre. This is similar to that occurring in the fluorine

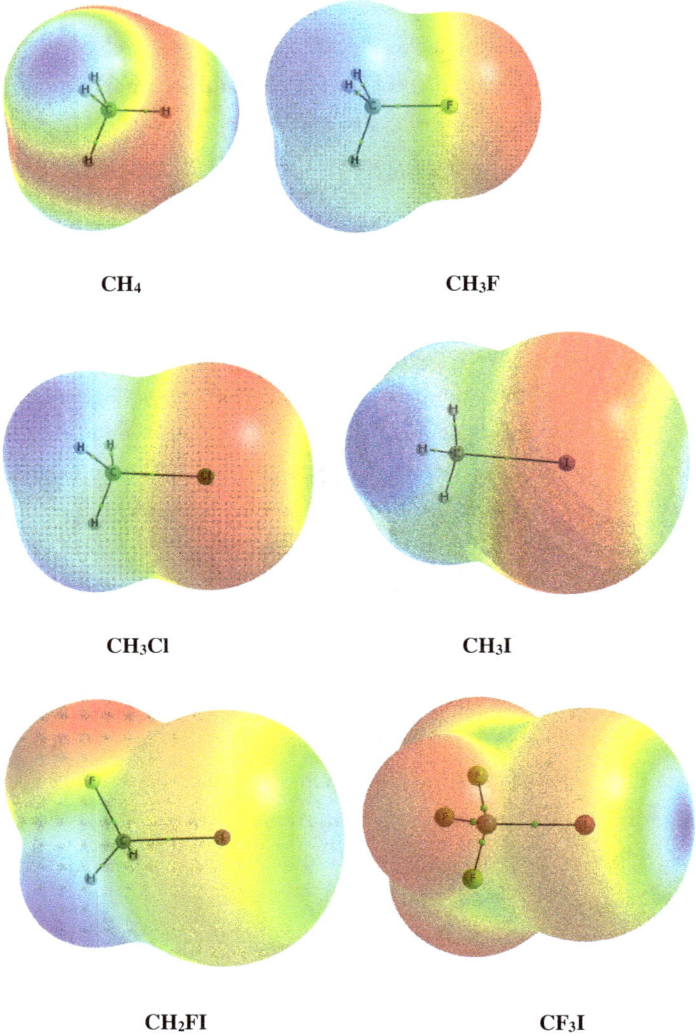

Fig. I.2 The molecular graphs and EP surfaces ($\rho = 0.001$ au) for methane and its halogen derivatives (MP2/aug-cc-pVTZ results). Blue corresponds to maximal positive EP values and red to minimal EP negative values.

centre of the HF molecule.[63] In the halogenometanes considered here, and in numerous species containing C-X bonds, the ring of the electron density around the halogen centre of the C-X bond is formed due to the valence electrons. This results in the negative EP. However, one of the valence electrons is involved in the formation of the C-X bond, which results in the depletion of the electron charge in the extension of the C-X bond at the halogen centre. The latter often leads to the positive EP value in this area and consequently to Lewis acid properties. The abovementioned area is often called the σ-hole.[46,47,50] The phenomenon of the σ-hole occurrence at the halogen centre is well explained in terms of molecular orbitals.[62] Let us discuss a case of the CF_3Br species and the creation of the σ-hole at the bromine centre. The valence electronic structure of bromine that is connected with the carbon centre may be described as $4s^24p_x^24p_y^24p_z^1$ (see Fig. I.3). The single electron of the $4p_z$ atomic orbital is involved in the formation of the σ C-Br bond molecular orbital. This results in the depletion of the electron charge at the outer lobe of the $4p_z$ orbital (so-called σ-hole) and consequently in the increase of the EP at the corresponding area, up to the positive values observed at the elongation of the C-Br bond.[62]

It is known that for the X halogen centre, EP increases with the increase of the atomic number, F < Cl < Br < I, and particularly for the related compounds, for example CH_3X or CF_3X (see Table I.2). It is worth mentioning that the fluorine centre does not reveal regions of positive EP very often. Let us consider the valence electronic structure of fluorine in the CF_4 species (Fig. I.4). Specific characteristics occur here; they relate to the following valence electronic structure $(2s2p_z)^22p_x^22p_y^2(2s2p_z)^1$.

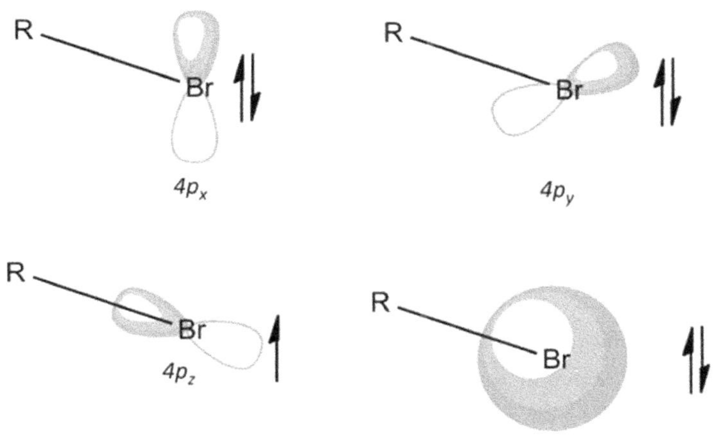

Fig. I.3 Schematic presentation of the bromine atomic orbitals in CF_3Br molecule. Reprinted with permission from ref. 62. Copyright 2013 John Wiley & Sons.

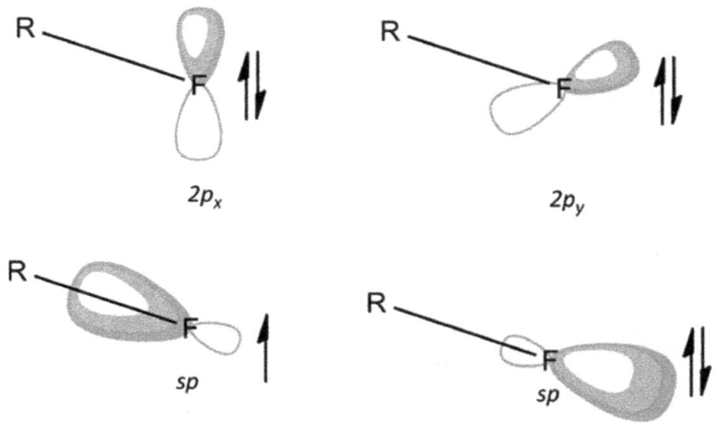

Fig. I.4 Schematic presentation of the fluorine atomic orbitals in CF_4 molecule. Reprinted with permission from ref. 62. Copyright 2013 John Wiley & Sons.

One can see that some degree of sp^n hybridization causes some 2s electronic density to be in the outer lobe of the $2s2p_z$ atomic orbital that is involved in the C-F σ-bond molecular orbital. This counteracts the electron charge depletion to some extent; i.e., it reduces the σ-hole.[62] However, the fluorine can form halogen bonds sometimes, especially if it is linked with electron-withdrawing substituents that may result in positive EP at the fluorine centre (a positive σ-hole).[48,62,64,65]

As mentioned above, the σ-hole concept[60–62,64,65] explains the Lewis acid properties of numerous halogen centres, and particularly their duel character since they can act also as the Lewis base sites; for example, they may be involved at the same time in interactions with nucleophile and electrophile by the σ-hole site and the ring of accumulated electron density, respectively. Fig. I.5 presents the crystal structure of 4-iodobenzoic acid where such a situation occurs.[66] It is shown that the iodine centre of the 4-iodobenzoic acid unit is connected with two other iodine centres. These I···I distances are equal to one another due to symmetry relations and they amount to 2.957 Å. This is within the corresponding sum of van der Waals radii (Table I.1) that may indicate local attractive interactions. In one case the C-I···I halogen bond is formed, approximately in the elongation of the C-I bond since the C-I-I angle amounts to 165.9°. In another case the same C-I bond forms the C-I···I connection with the corresponding angle of 96.1° and the iodine centre connected with the carbon plays

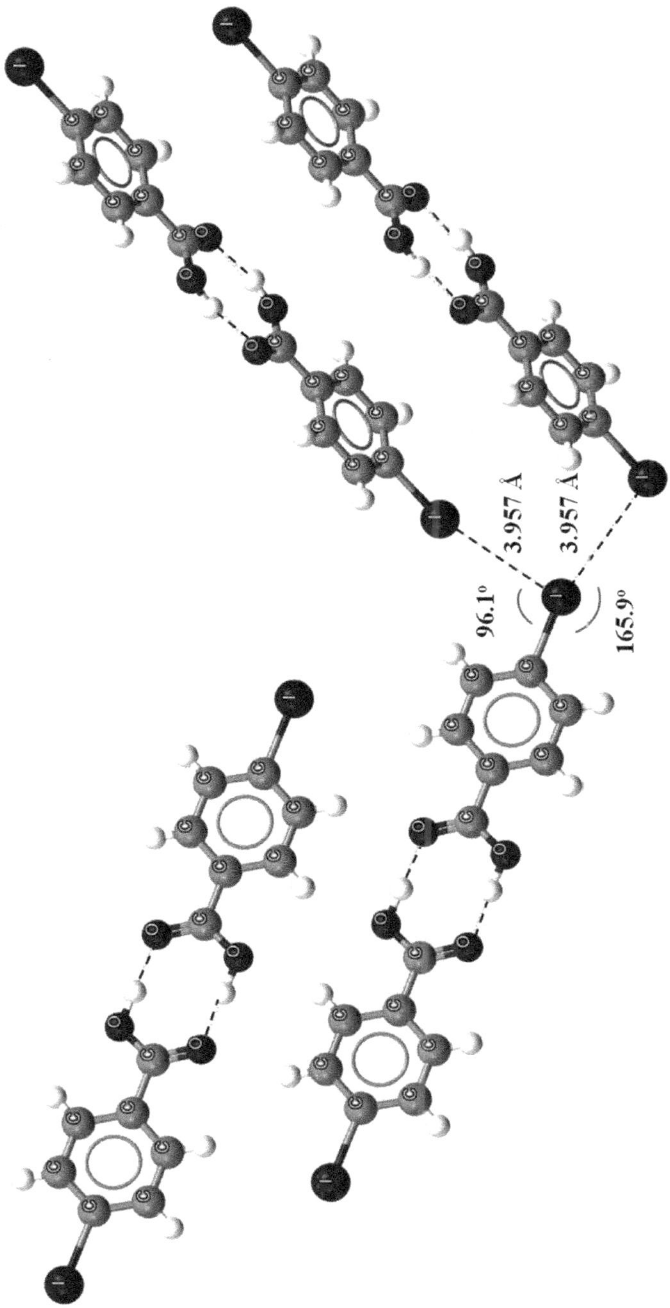

Fig. I.5 Fragment of the crystal structure of 4-iodobenzoic acid (ref. 66). The dual character of the iodine centre is shown.

the role of electron donor; this local interaction may be named as the halide bond. On the other hand, in the latter arrangement, the iodine centre playing the role of the electron acceptor is connected with the carbon atom, thus one may indicate the C-I···I-C arrangement, where two types of interactions occur, the halogen bond and the halide bond.

Figure I.6 presents the crystal structure of dichloro-(2,4-dichloro-6-((3-methylammoniopropylimino)methyl)phenolato-N,O)-zinc(ii) methanol solvate hydrate where the dual role of the chlorine centre may be observed.[67] The C-Cl···O halogen bond is formed with the Cl···O distance equal to 3.095 Å, slightly less than the corresponding sum of van der Waals radii, and a C-Cl-O angle of 163.6°. On the other hand, the C-Cl···H halide bond is formed with Cl···H distance of 2.379 Å, and C-Cl-H angle of 109.1°. The latter distance is lower than the sum of chlorine and hydrogen van der Waals radii by more than 0.4 Å. However, it is worth mentioning that the latter distance is very approximate since the crystal structure was solved by applying X-ray diffraction

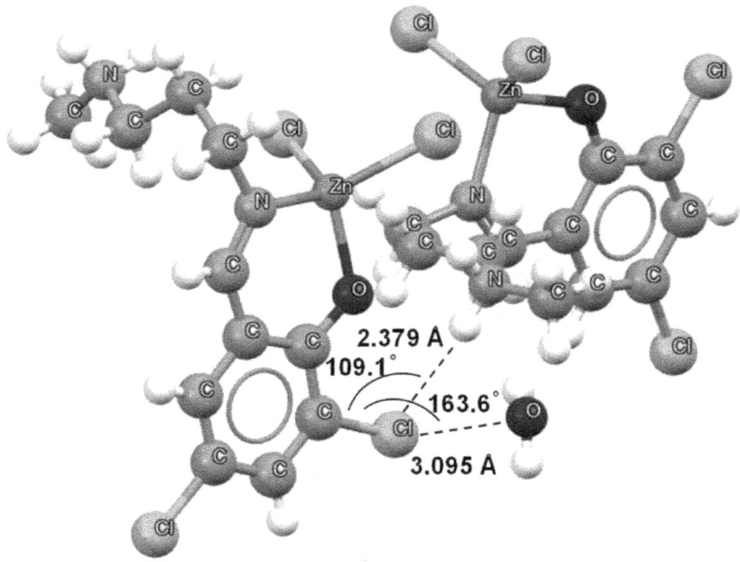

Fig. I.6 Fragment of the crystal structure of dichloro-(2,4-dichloro-6-((3-methylammoniopropylimino)methyl)phenolato-N,O)-zinc(ii) methanol solvate hydrate (ref. 67). The dual character of the chlorine centre is shown.

and a spherical H-atoms' positions refinement (see Chapter III for details concerning experimental techniques). It is worth to note that the latter halide bond belongs to the greater C-Cl···H-N arrangement where a N-H···Cl hydrogen bond is observed that is close to linearity (N-H-Cl angle is equal to 173.0°).

The dual character of halogen centres was discussed in earlier studies, before the proposal of the σ-hole concept. The anisotropy of van der Waals radii was noticed in early studies; particularly, such anisotropies were analysed for halogen centres.[68] Brammer and coworkers have pointed out that the anisotropy of Pauli repulsion and electrostatic interactions occur for halogen atoms (designated as X), particularly for those connected with C carbon centre; the latter leads the halogens to interact with nucleophiles approximately in the elongation of C-X bonds, while in the direction approximately perpendicular to this bond interactions with electrophiles occur (see Scheme I.1).[69] These statements are in agreement with other earlier studies of Desiraju and Parthasarathy where the short halogen···halogen contacts in crystal structures were analysed.[70] It seems that the anisotropy of the electron charge distribution of the majority of halogen centres was first evidenced experimentally for the crystal structure of 1-methylpyrrol-2-yl trichloromethyl ketone.[71] Figure I.7 presents qualitative F_o (structure factor which is related to the electron density — see Chapter III) contour

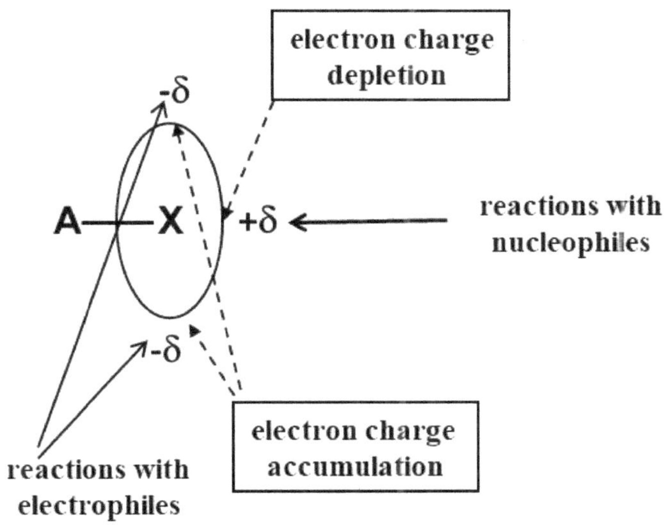

Scheme I.1 The dual character of the X halogen centre.

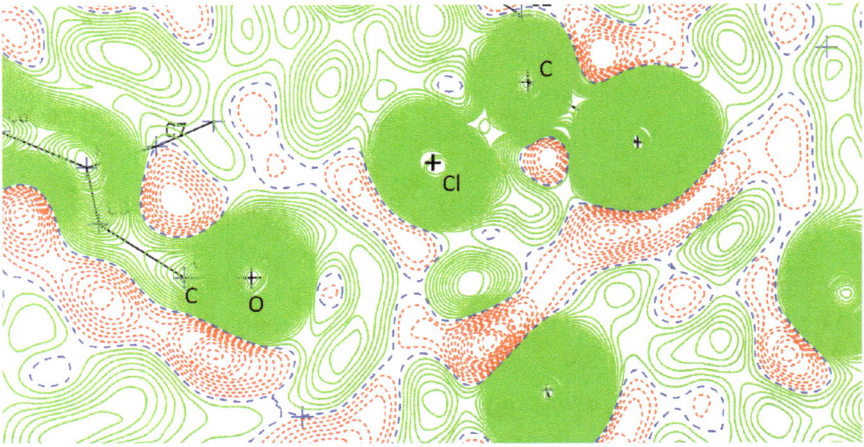

Fig. I.7 Qualitative F_o contour map around the C–Cl···O=C contact. View perpendicular to the mean plane calculated from positional parameters of atoms: C, O, Cl and C which are pointed out by crosses. Positive (continuous, green) and negative (dashed, red) contours are shown at 0.1 e/Å^3 intervals. The cut off for minimum and maximum value are equal to –1.3 Å and 11.6 Å, respectively; axes units in Å. Reprinted with permission from ref. 71. Copyright 2007 Elsevier B.V.

map around the C-Cl···O=C arrangement. One can see the ellipsoidal shape of the structural factor value of Cl centre with shorter semi-axis in the direction of the C-Cl bond. Figure I.8 presents the fragment of the latter crystal structure where the C-Cl···O=C arrangements are shown. These arrangements follow the σ-hole concept since the C-Cl-O angle is equal to 168.6°, thus the σ-hole located in the elongation of the C-Cl bond is in contact with the oxygen centre. The Cl···O contact distance of 3.046 Å is lower than the corresponding sum of van der Waals radii, 3.27 Å. Fig. I.8 shows also the C-H···O hydrogen bonds that are rather weak interactions since the H···O distance of 2.641 Å is slightly greater than the corresponding sum of van der Waals radii of 2.62 Å. However, it was pointed out that the hydrogen bond is mostly electrostatic interaction. The electrostatic attractive forces act far beyond the corresponding sum of van der Waals radii, which may be the case for weak hydrogen bonds.[22]

It is worth mentioning that the anisotropy of electron charge discussed here is well explained in terms of the σ-hole concept and that the A-X···B La-Lb interaction through the σ-hole of the X halogen centre, that is named as the halogen bond, is classified as

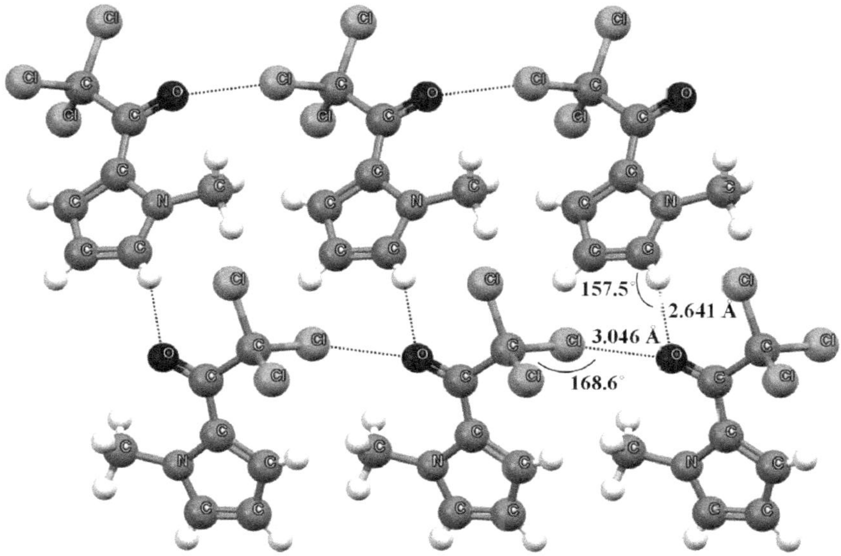

Fig. I.8 Fragment of the crystal structure of 1-methylpyrrol-2-yl trichloromethyl ketone (ref. 71). The geometries of the C-H···O and C-Cl···O arrangements are shown.

the σ-hole bond. A-X and B mark the Lewis acid unit and the Lewis base centre, respectively. These designations as well as X as the halogen centre are applied in this book later. A is most often the carbon centre in cases involving the halogen bond.

The above designation of the halogen bond interaction is similar to that applied in numerous studies as well as in this book for the hydrogen bond, A-H···B. Similar to the halogen X, the H-centre is the Lewis acid site while the A-H and B species are the Lewis acid and Lewis base units, respectively. The hydrogen bond has been classified as a σ-hole bond[62] and it is often compared in various studies with other interactions.[21] The characteristics and sub-classes of the hydrogen bond are described in detail later (see Chapter IV).

I.4 σ-hole Bonds

The σ-hole may occur not only at the halogen centres. This is a phenomenon that occurs also at other elements, often, similar to halogens, electronegative ones. The σ-holes are observed at centres of groups 14, 15, and 16, i.e., at tetrel, pnicogen and chalcogen elements where they occur in elongations of bonds to them. If sufficient depletion of the electron charge at these centres occurs, then positive EP is observed there and these centres may interact with electron donors to form σ-hole bonds. The electron charge depletion and consequently the value of EP depend on the character of substituent attached to the centre considered, as well as on the environment that may influence the electron charge distribution. However, the EP is usually analysed for isolated species to discuss their potential Lewis acid and Lewis base properties.

Figure I.9 presents surfaces of the electron density of 0.001 au with maps of the EP for selected chalcogen moieties. These are the fluorine derivatives since the

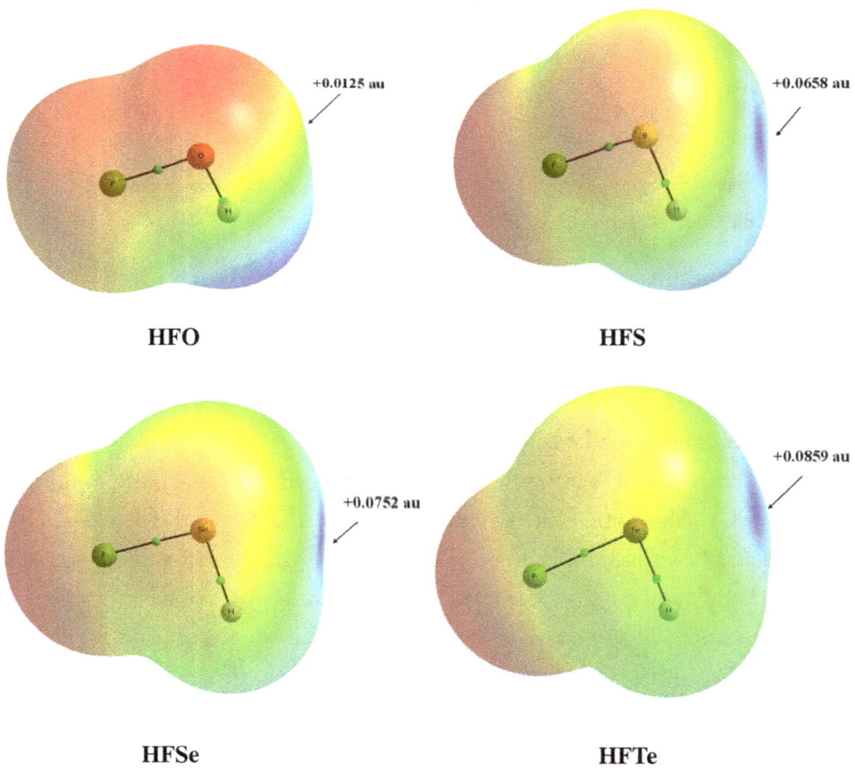

Fig. I.9 The molecular graphs and EP maps for the 0.001 au surfaces for HFZ species (Z = O, S, Se, Te). The EP values in elongations of F-Z bonds are presented. Blue corresponds to maximal positive EP values and red to minimal EP negative values. These are MP2/aug-cc-pVTZ results.

F-substituent strongly withdraws electron charge that leads to the increase of depletion of the electron charge at the attached chalcogen centre. The same influence of substituents was observed and discussed earlier for the halogen centres.

Let us discuss another case that concerns the water molecule. Negative EP is observed here at the oxygen centre for a wide range of electron densities selected to consider molecular surfaces. It is observed at the O-centre for surfaces characterised by values lower than 0.02 au, while for values greater than 0.02 au that occur close to the oxygen nucleus, positive EP is observed.[72] This was mentioned earlier, that the surfaces of 0.001 au are usually analysed in various studies. It is also worth mentioning that the positive EP values usually occur close to the centres considered since the positive nucleus charge overwhelms the negative charge of electrons there. The latter occurs for the oxygen centre in water molecules for greater values of electron densities as mentioned above. The electronegative fluorine substituent in the HFO species causing the σ-hole of the positive EP at the O-centre, approximately in the elongation of the F-O bond, is observed even at the surface characterised by low 0.001 au electron density (Fig. I.9).

A similar situation is observed for other chalcogen species where σ-holes occur approximately in elongations of the fluorine-chalcogen bonds; they are characterised by positive EP values. One can see that the EP value at the chalcogen centre increases with the increase of the atomic number of chalcogen. A similar effect was observed for the halogen centres discussed earlier. It was pointed out that the more polarisable the halogen atom, the more positive the EP at this centre.[47,62] The similar tendency of the increase of EP at the centres with the increase of the atomic number in the same group of the periodic table was observed for various samples of related species.[61,73] It is worth mentioning that the centres of the first row of elements, i.e., carbon, nitrogen, oxygen or fluorine (the period containing H and He atoms is not taken into account here), are usually characterised by negative EP at the corresponding surfaces. It was mentioned above for the fluorine centre that the valence electron configuration has some degree of sp^n hybridisation, which also occurs for other first-row elements. This s-character contribution in valence orbitals leads to the accumulation of the electron charge at elongations of bonds to these centres and to the reduction of potential σ-holes.[62,74]

For the chalcogen species discussed above, additional σ-holes occur in elongations of H-Z bonds (Z designates the chalcogen centre). The positive EP values are much lower than the corresponding values of σ-holes located in elongations of F-Z bonds, even though they are not so clearly visible. One can observe the following tendency for the centres of elements from the 17th to the 14th group, if they are characterised by sp^3 hybridisation: one, two, three and four σ-holes that are accompanied by three, two, one and zero lone electron pairs, respectively, occur. The lone electron pairs may influence the location of σ-holes. One can see that for the chalcogen centres (Fig. I.9) where the σ-holes are situated not exactly in elongations of F-Z bonds.

Figure I.10 shows the molecular graphs with EP surfaces for the $SiFH_3$ and PFH_2 species. One can see that for the former species the σ-hole is located exactly in the extension of the F-Si bond. This species does not have the lone electron pairs which could disturb the location of the σ-hole. However, most importantly this species possesses C_{3v} symmetry where the 3-fold axis overlaps with the Si-F bond. For the second species, PFH_2, the lone electron pair disturbs the location of the σ-holes, particularly that situated in the extension of the F-P bond. Besides, this pnicogen moiety possesses lower symmetry than the $SiFH_3$ unit. Fig. I.10 shows that the σ-hole is not located exactly in the extension of the F-P bond. There are additionally three H-Si and two H-P σ-holes for the $SiFH_3$ and PFH_2 species, respectively. One can see that the symmetrical location of the fluorine σ-hole for the tetrel centres results from the lack of lone electron pairs and from the symmetry of the molecule. A similar situation occurs for the σ-holes located at the halogen centres (Fig. I.2). They are located in extension of C-X bonds due to symmetry of molecules and particularly due to the location of lone electron pairs that form the belt of negative charge around halogen centres in the direction approximately perpendicular to the C-X bond.

The σ-hole bonds discussed so far in this chapter are often responsible for the arrangement of units (ions and molecules) in crystals. Interactions such as halogen, chalcogen, pnicogen and tetrel bonds may be treated as counterparts of the hydrogen bond. It seems that the latter interaction occurs in crystal structures most often. However, other interactions also may play a crucial role in arrangements of species constituting crystals. Figure I.11 presents the fragment of the crystal structure of dicyanoselenide where the C-Se···N chalcogen bonds play the most important role.[75]

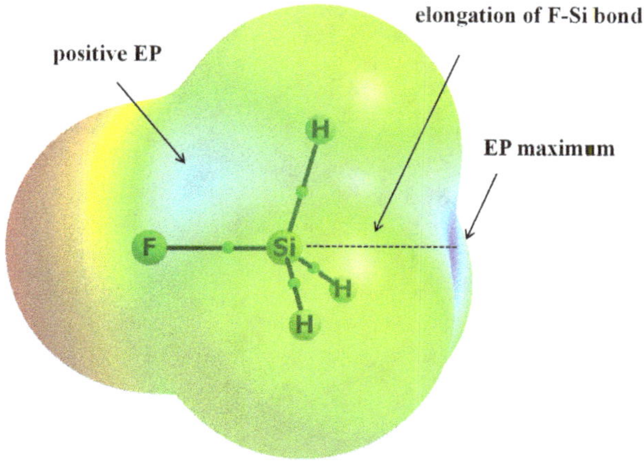

$SiFH_3$

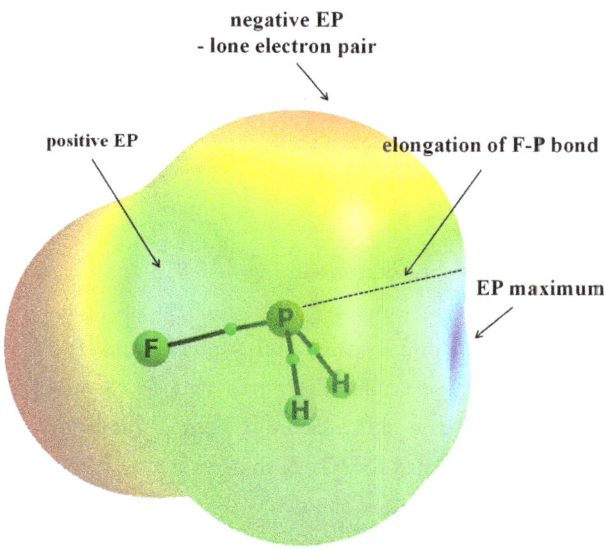

PFH_2

Fig. I.10 Molecular graphs with EP mapped at 0.001 au electron density surfaces for the $SiFH_3$ and PFH_2 species. Blue corresponds to maximal positive EP values and red to minimal EP negative values. These are MP2/aug-cc-pVTZ results.

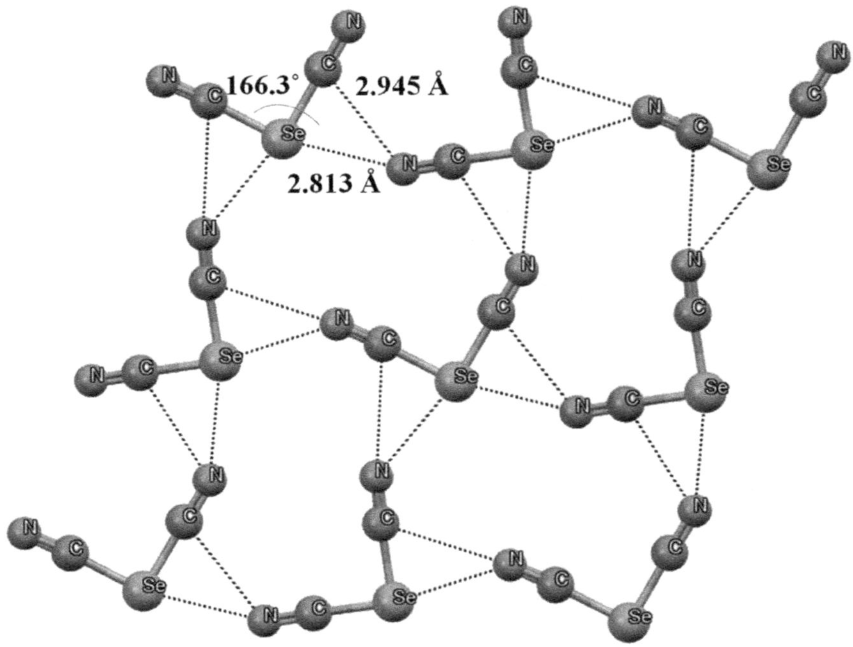

Fig. I.11 Fragment of the crystal structure of dicyanoselenide (ref. 75). The C-Se···N chalcogen bonds and the C···N contacts are shown.

The Se···N intermolecular contacts amounting to 2.813 Å are significantly shorter than the corresponding sum of van der Waals radii, 3.45 Å (Table I.1). The C-Se-N angle of 168.6° indicates the directionality of this interaction, but the arrangement is not linear because of the environmental effects in crystals as well as the factors discussed above, i.e., the lone electron pairs of the selenium centres. One can see for the $Se(CN)_2$ crystal structure the occurrence of the C···N short contacts of 2.945 Å, which is also lower than the corresponding sum of van der Waals radii, 3.25 Å. Thus, it seems the latter contacts are also attractive and stabilising interactions.

The aerogen interactions occur between the noble gas atoms (aerogens), which act as the Lewis acid centres, and the electron-donating sites.[76] Figure I.12 shows, as an example, the fragment of the crystal structure of tris(pyridine) trioxo-xenon where the Xe···N aerogen bonds occur.[77] The latter intermolecular distances are equal to 2.760 Å and 2.820 Å, which are much lower than the sum of Xe and N van der Waals

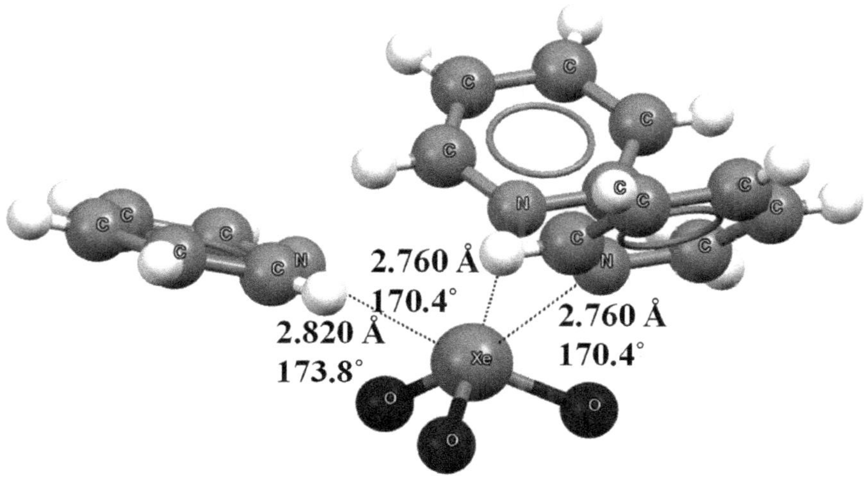

Fig. I.12 Fragment of the crystal structure of tris(pyridine) trioxo-xenon (ref. 77) The Xe···N contacts are shown.

radii, 3.71 Å. The O-Xe-N angles are close to linearity; they are greater than 170°. One can see that the electronegative oxygen centres attached to the xenon withdraw the electron charge causing the σ-holes in elongations of O-Xe bonds and consequently this results in interactions, aerogen bonds which are close to linearity.

I.5 π-hole Bonds, The Triel Bond Interaction

The Lewis acid properties of group 14–18 elements were briefly mentioned in the previous section. The centres, marked henceforth by Z, may interact with electron donors which play the role of Lewis base sites, designated henceforth by B (if B is marked for the boron centre it will be clearly stated). The following interactions may thus be formed: tetrel, pnicogen, chalcogen, halogen and aerogen bonds, respectively. These bonds (interactions) may be marked as A-Z···B, where A is a part of the Lewis acid unit linked with the electron-accepting Z-centre. Similar designations for halogen and hydrogen bonds were proposed earlier.

The Lewis acid properties of Z centres have been explained in terms of the σ-hole concept which states that electron charge shifts from Z to its bond and to the bonded substituent may occur. This may result in electron charge depletion at Z, named as the σ-hole, which is characterised by positive EP if sufficient charge shift (depletion) occurs. These electron charge shifts may be enhanced if the electronegative, electron-withdrawing A substituents are attached to the Z-centres or to further parts of the Lewis acid species. The latter occurs, for example, for the CF_3Z (Z = Cl, Br, I) moieties discussed in previous sections where the fluorine substituents enhance the EP at Z halogen centres in elongations of C-Z bonds.

The question arises regarding whether the electron charge deficiency is related only to the σ-hole occurrence. It was suggested that there are other areas of the depletion of electron charge that are related to molecular planes or to planes of fragments of molecules.[60] Such regions were named as π-holes. The following examples of species where π-holes may occur were mentioned: the boron centre of BF_3, the sulfur of SO_2, the selenium of SeO_2 and silicon centres at silenes.[60] It was stated that "a positive π-hole is a region of positive electrostatic potential that is perpendicular to a portion of a molecular framework."[78] It was also pointed out that π-holes may be related to unoccupied p_π-type orbitals at centres considered.[60] The π-holes that may act as the electron acceptors form π-hole bonds with the Lewis base centres. The latter interactions were a subject of numerous studies.[78,79] For example, it was suggested that π-holes occur in systems containing higher π molecular orbitals, as is observed in the H_2CO and CO molecules which are characterised by empty π^* orbitals.[79,80]

Let us discuss properties of boron and other elements of the 13th group. These centres in numerous compounds are characterised by the occurrence of areas of π-holes, and when these areas are in contact with nucleophiles, such interactions are named as triel bonds.[24,25,81] These interactions may be important in various chemical processes; there are numerous studies on the structure of boron compounds involved in interactions and reactions where they play a key role.[82] For example, regarding the molecular structure of simple boron compounds like trihydridoboron (borane), BH_3, or boron trihalides, BX_3, boron forms three B-H(X) covalent bonds with the use of sp^2 hybrid orbitals. This leads to the planar trigonal molecular structure of such boron compounds.[83,84] The strong Lewis acid properties of the boron centre result from its electron structure — the boron is electron-deficient in borane, boron trihalides and numerous other compounds because it possesses six valence electrons.[85–88]

It was found that the strength of Lewis acid properties of the boron centre in trihalides decreases in the following order: BBr_3 > BCl_3 > BF_3.[89] It seems to be surprising since the most electronegative element, fluorine, is known as a strongly electron-withdrawing substituent. Thus, it should withdraw electron charge from boron more than other halogens. That should lead to the most positive charge of the B-centre in BF_3. It was explained that this observed trend, which is opposite to expected, may be the result of the back bonding effect.[89] The back bonding effect concerns the donation of π-electron density from the halogen centre to the outer 2p orbital of the B-centre; this is characterised by the sp^2-hybridisation. The boron and fluorine possess 2p orbitals of close energies and sizes, thus efficient overlap of orbitals occurs for the BF_3 species. This interaction of orbitals (overlap) is not effective for other boron trihalides or borane molecules because of the greater diversity of their p-orbitals. This back bonding effect that is connected with orbital-orbital overlap leads to the double character of the B-X bonds, which is in agreement with the abovementioned order concerning the Lewis acid strength, where the greatest double character is observed for the B-F bond.

The above findings concerning borane, BH_3, and boron trihalides, BX_3, are in agreement with theoretical calculations on the BF_3-NCH[90] and BH_3-NCH[91] complexes where stronger Lewis acid properties were observed for the BH_3 unit. The binding energies for the BF_3-NCH and BH_3-NCH complexes are −6.3 kcal/mol and −19.9 kcal/mol, respectively (MP2/aug-cc-pVTZ level, BSSE correction not included, see Chapter II for description of theoretical methods). The interaction of the sp^2-hybridised boron centre with electron donors often leads, if this interaction is sufficiently strong, to sp^3 hybridisation of the product since the Lewis base ligand disrupts the π bond in the B trigonal reactant.[25,92] Thus the sp^2-hybridised trivalent electron-deficient B-centre, where the octet rule is not obeyed (so-called hypovalency[8]), changes the electron structure to obey it.[25] One may also mention amine systems, like for example the BF_3-NH_3 complex, where short B···N distances occur (~1.6 Å) and the boron centre possesses near-tetrahedral configuration.[93] The typical tetrahedral configurations are formed as a result of interactions of triel trihalides with strong Lewis base units, like the halide anions.[94] There are also numerous experimental studies on systems containing tetravalent boron atom. It seems that boron and other triel species possessing coordination greater than three may participate in additional interactions that are classified as σ-hole bonds. The planar systems with trigonal triel centres usually form π-hole bonds.

Hence one may state that the triel bond is an interaction between the 13th group element acting as an electron acceptor and the electron-rich site. This interaction may be classified as a π-hole bond if the triel centre possesses trigonal configuration, or as σ-hole bond if the centre possesses tetrahedral configuration or another configuration related to higher coordination.

Figure I.13 presents the 0.001 au electron density isosurfaces with EP maps for simple triel species. One can see the regions of positive EP that are attributed to the π-holes. One can also see that the aluminium species possess greater maximum EPs than their boron counterparts. The same is observed for hydrides; they are characterised by greater

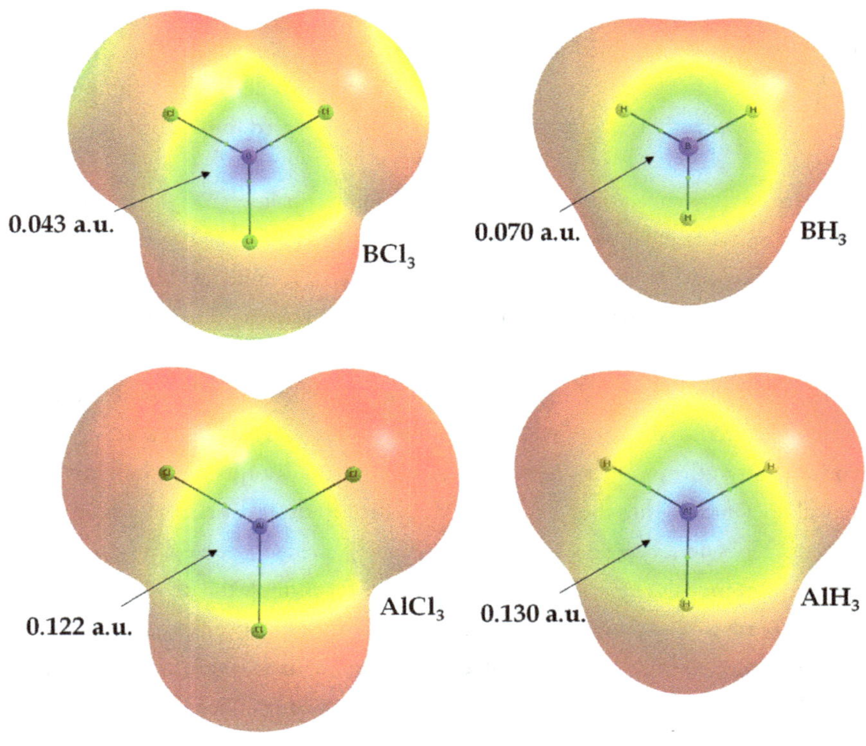

Fig. I.13 The molecular graphs and corresponding 0.001 au electron density isosurfaces with EP maps for simple trigonal triel species. Blue corresponds to maximal positive EP values and red to minimal EP negative values. These are MP2/aug-cc-pVTZ results. Reprinted with permission from ref. 94. Copyright 2020 S.J. Grabowski. Licensee MDPI, Basel, Switzerland.

EP values at triel centres than their chloride counterparts. This may indicate the order of the strength of interactions with Lewis base ligands, or at least the strength regarding the electrostatic term of interaction. However, the situation is not so clear for the triel π-hole bonds. The maximum EP values calculated at the same level (MP2/aug-cc-pVTZ) for the BF_3 and AlF_3 species amount to 0.0895 au and 0.2020 au, respectively.[81] This corresponds to binding energies, E_{bin}'s, of −5.2 kcal/mol and −25.5 kcal/mol, respectively. On the other hand, the EP value corresponding to the π-hole at the boron centre of the BCl_3 species is equal to 0.0432 a.u. and E_{bin} is equal to −3.2 kcal/mol (the E_{bin}'s are corrected for BSSE). Thus EP values are in agreement with binding energies; a more positive EP corresponds to stronger interaction. However, this is not in agreement with the back bonding effect[89] discussed earlier since the greater EP value is observed for the BF_3 species than for the BCl_3 unit; for the former the back bonding effect should be stronger leading to greater reduction of the π-hole.

There are numerous, often contradicting, findings concerning the boron trihalides. However, those concerning their interactions seem to be consistent. For the interactions of boron trihalides with strong Lewis bases, the following order of the Lewis acid strength is observed: $BBr_3 > BCl_3 > BF_3$. This is in line with the back bonding effect despite the EP values for isolated boron trihalide units not being in agreement with this effect. In the case of interactions with weak Lewis bases the reverse order is observed, i.e., the strongest interaction occurs with boron trifluoride.[10]

It is worth mentioning that different triel centres possess unique characteristics.[24,25] For example, boron species form the weakest triel bonds in comparison with their other triel counterparts, aluminium species also form weak interactions which are characterised mainly by their electrostatic nature, while gallium moieties often form covalent links.[24] The differences in coordinations of the triel centres were also discussed.[25] Figure I.14 presents fragments of two crystal structures, complexes of aluminum trichloride-tetrahydrofuran[95] and aluminum trihydride-tetrahydrofuran.[96] Slightly shorter Al···O distances for $AlCl_3$ species than for AlH_3 ones are observed that may suggest stronger interactions (triel bonds) in the former case. In both cases these distances, 1.99 Å and 2.06–2.07 Å, are much shorter than the corresponding sum of van der Waals radii (3.36 Å! Table I.1). This indicates the strong and covalent nature of these interactions.

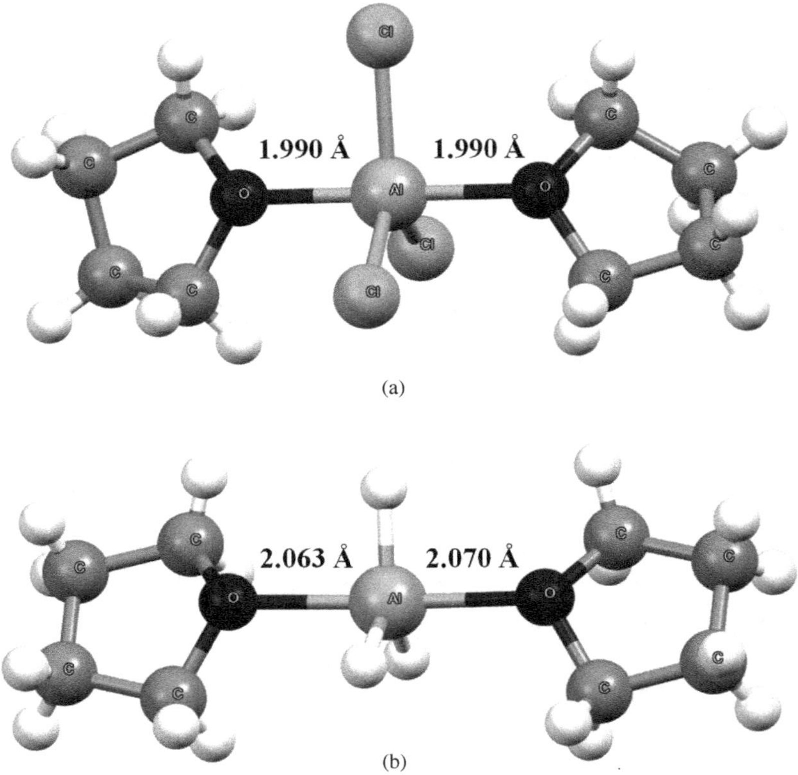

Fig. I.14 Fragments of two crystal structures, complexes of aluminum trichloride-tetrahydrofuran (a, ref. 95) and aluminum trihydride-tetrahydrofuran (b, ref. 96).

The π-holes of triel elements were discussed mainly. However, it is worth mentioning that such electron charge depletions, π-holes, occur for other elements of numerous other species. For example, the nitrogen centre of the nitro group is often characterised by such properties.[97,98] Figure I.15 presents the fragment of the crystal structure of 2-chloro-4-nitro-1H-imidazole where π-hole pnicogen N···Cl bonds occur.[99] The N···Cl distance is equal to 3.264 Å, only slightly less than the corresponding sum of van der Waals radii, 3.30 Å (Table I.1). Thus, it seems these pnicogen bonds are rather weak. The nitrogen plays here the role of the Lewis acid centre through the π-hole while chlorine is the Lewis base site through the belt of the electron charge located around the chlorine atom; note that the C-Cl-N angle amounts to 90.2° (Fig. I.15).

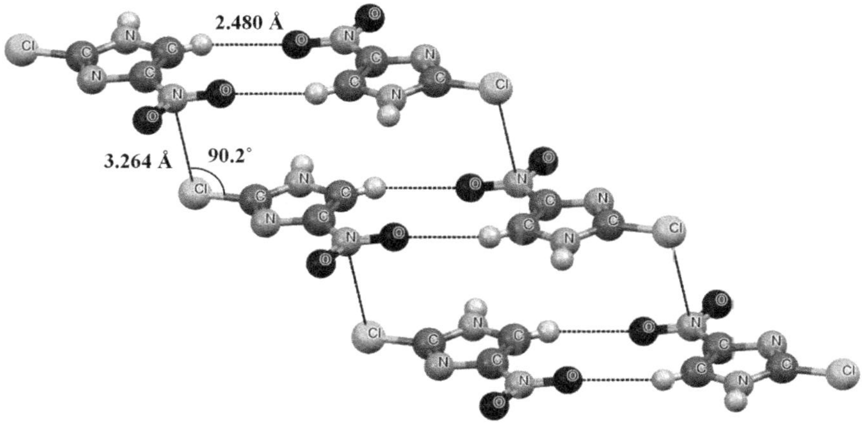

Fig. I.15 Fragment of the crystal structure of 2-chloro-4-nitro-1H-imidazole (ref. 99).

Hence this is in agreement with the σ-hole and π-hole concept. Fig. I.15 also shows the C-H···O hydrogen bonds, which are weak interactions since the H ··O distance of 2.480 Å is close to the van der Waals radii sum, 2.62 Å.

Murray and Politzer discussed in detail the nature of various interactions in terms of the σ-hole and π-hole concept. They indicated various kinds of such interactions. For example, they stated that "another kind of π-hole is above and below a π-region of a molecule having electron withdrawing constituents, e.g., hexafluorobenzene."[100] One can mention other examples of such kind of interactions, and the reader is referred to numerous studies on this topic.[100]

I.6 Centres of Groups 1 and 2 – Corresponding Interactions

There are no doubts that elements of groups 1 and 2 most often act as Lewis acid centres since they are characterised by low values of electronegativity. Electronegativity refers to an attraction of shared electrons when forming a chemical bond; this is a weak attraction in the case of these elements. Let us consider the Pauling scale of electronegativity, the most often applied measure of electronegativity in various studies.[101,102] Regarding the first two groups of the periodic table, the highest electronegativity of 2.2 is observed for hydrogen, which is excluded from the alkali and alkaline-earth elements because of its specific characteristics. Next, beryllium and magnesium possess electronegativities of 1.57 and 1.31, respectively; for the remaining elements the electronegativity does not exceed unity. This is why 3c-4e (three centre — four electron) systems, i.e., A-Z···B arrangements characterised by the A-Z covalent bond (sometimes highly polar), are rather rare for the Z centre of groups 1 and 2. Four electrons concern here the A-Z covalent bond and the lone pair of the B Lewis base unit. However, the occurrence of 3c-4e A-H···B hydrogen bonds was confirmed experimentally and theoretically. Of course, numerous bifurcated systems are known where the hydrogen centre is involved in additional linkages;[103,104] such arrangements are discussed later (in Chapter IV). One may also expect 3c-4e arrangements for the beryllium and magnesium centres since their electronegativities are higher than those of other alkali and alkaline-earth elements. The latter centres easily release electrons and consequently participate in non-directional ionic links (bonds and interactions). There are usually more than two links for such alkali or alkaline-earth elements, so the 3c-4e A-Z···B arrangements are rather rare.

The following definition of the term "coordination number" is recommended by IUPAC: "The coordination number of a specified atom in a chemical species is the number of other atoms directly linked to that specified atom. For example, the coordination number of carbon in methane is four, and it is five in protonated methane, CH_5^+. (The term is used in a different sense in the crystallographic description of ionic crystals.)"[105,106] In one recent study the coordination of the selected centres of groups 1, 2 and 13 was analysed.[107] It was found that only 3.5% of crystal structures containing lithium are characterised by coordination 2 (66 such arrangements were found). This means that the 3c-4e A-Li···B systems which are typical for numerous A-H···B hydrogen bonds, and more generally for numerous A-Z···B σ-hole and π-hole bonds, are not the case for the lithium systems. The most often occurring coordination number for lithium in crystals is four, 53.1% of structures containing lithium. For coordination four the tetrahedral configuration is usually observed for lithium centres. The next most often occurring coordination for the lithium centre is three (trigonal configuration) –17%. In the case of sodium, the occurrence of coordination two is even more rare than in the case of lithium, at 1.5%. Coordination six occurs most often for the sodium structures, 32.6%. Beryllium: coordination two –3.5%, four (most often) –63.1%. Magnesium: coordination two –0.5%, coordination six (most often) –41.4%. Hence one can see that for the centres of the first and second groups of the periodic table, their arrangements and consequently their interactions usually differ from those which occur in the case of hydrogen and other centres of groups 14–18.

Consequently, it is difficult to discuss the 3c-4e A-Z···B interactions in relation to structures of centres of groups 1 and 2. For the latter structures one can usually observe cations surrounded by nucleophilic ligands; such ligand-cation interactions should be dominated by electrostatic forces. However, terms such as lithium bond, sodium bond or beryllium bond appear in various studies from time to time and these are usually theoretical studies where the model, simple systems are discussed.

In various studies the La-Lb interactions were usually compared with the hydrogen bond. That was the case of the lithium bond, LiB; Kollman and co-workers noticed that "The lithium bond is a normal donor-acceptor system, even though it may occur rarely in chemical systems. Dicoordinated lithium is expected only in the gas phase, and perhaps in molten salts."[108] One can see that, according to this definition, the lithium bond may be classified as La-Lb interaction since it is treated here as a contact between acceptor of electron charge and its donor. Besides, it was noticed that dicoordinated lithium, i.e., the A-Li···B arrangement, occurs rarely.[108] The authors considered theoretically simple model 3c-4e A-Li···B systems. They pointed out that the H-bonded linear Li-F···H-F complex formation leads to increase of the positive H-atom charge, while in the case of the H-F···Li-F Li-bonded system such increase of the Li-atom charge is observed only if the *p* orbitals related to the Li-centre are not taken into account in calculations; if these orbitals are included then the charge decrease of the central lithium resulting from the complexation (LiB formation) is observed. The Mulliken population analysis (see Chapter II)[53] was applied in this study for the analysis of changes of atomic charges resulting from the HB and LiB formations.[108] The change of the central H or Li atomic charge as a result of complexation seems to be one of the differences between the hydrogen bond and the lithium bond, respectively, which was further discussed in other studies.[109,110]

It is worth to note that the formation of A-H···B hydrogen bond and of other La-Lb A-Z···B bonds lead to the electron charge shift from the Lewis base unit,

B, to the Lewis acid species, A-H(Z).[6] Further electron charge shifts occur because of the formation of hydrogen bond or other links. This is mainly the electron charge outflow from the H(Z)-centre to the remaining part of the Lewis acid unit.[6] This leads to the increase of the positive charge of the central H(Z)-atom. It is worth mentioning, however, that there are numerous exceptions. For example, in a few hydrogen-bonded systems the complexation leads to decrease of the positive charge of the H-atom.[111] However, these observations are based on theoretical calculations.[111] Similarly, in σ-hole and π-hole bonds, the situation may be more complex and therefore the increase of the Z-centre charge is not always observed. In the case of Li-bond, the complexation usually leads to decrease of the positive charge of the Li-centre.[109,110]

One can mention another early study on the Li-bonded systems.[112] The authors analysed different types of the lithium bond and indicated numerous differences between LiB and HB interactions. They concluded that the binding energy (the absolute value) is greater for the former interaction; the $-E_{bin}$ and $-E_{int}$ energies (see eqs. I.1 and I.2) were discussed in this study.[112] The most important conclusion was that the Li-bond is largely ruled by the electrostatic forces while in the case of H-bond different attractive forces take place. In other words, the electron charge shifts and consequently the polarisation interaction energy (see eq. I.13) crucial for the hydrogen bond are not so important for the lithium bond, which may be classified as cation — nucleophile interaction where the charge shifts are negligible. The authors also stated that dicoordinated lithium moieties are rather rare and that the chemistry of the lithium centre is related to polycoordinated lithium atoms.[112] It is in agreement with other studies discussed earlier.[107–108]

To the knowledge of the author of this book, there is not a clear definition of the lithium bond. In studies related to this topic, even in early studies, at most the main characteristics of this kind of interaction are presented for selected systems, or a rough comparison with the hydrogen bond is given. For example, Hazra and Pal explain that "analogous to hydrogen, lithium can also participate in a three-center interaction leading to the di-coordination of lithium, which is known as lithium bonding."[113] The authors further explain that the Li-bond is stronger than the hydrogen bond because lithium is more electropositive than hydrogen.[113]

It is worth to note that there are no studies on other types of alkali bonds; only sodium bonds were analysed theoretically.[114,115] They were classified as 3c-4e A-Na···B arrangements. However, there are numerous experimental, early and more recent, studies on coordination of sodium cations and more generally on the coordination of alkali and alkaline-earth cations where coordination number is greater than two.[116,117] Similar arrangements as for the hydrogen bond were discussed for the alkali metals; these are three-centre A-Z...B systems (Z may be the central metal element such as Li, Na or K for example).[118] The $AM^{+}...B \Leftrightarrow A...M^{+}B$ (M = Li, Na and K) heteroconjugated bonds were analysed and the processes of M^{-} cations' shifts as analogues to proton transfer were described. Intramolecular $AM^{-}...B$ systems were also discussed. Hence one can see that interactions analogous to the lithium bond and hydrogen bond were described. They possess similar characteristics as the hydrogen bond; it is a shift of a central cation (proton, Li^{+}, Na^{+} etc), for example.

In the case of the second group of elements the beryllium may act as the Lewis acid centre and its contacts with the electron-donating sites are often named as beryllium bonds.[107,119–123] The interactions of the magnesium centre as Lewis acid site with the nucleophiles is named as magnesium bond, and has been analysed in various studies and compared with other interactions, mainly with the beryllium bond.[124–128] The name "magnesium bond" was used for the first time probably in the study on the $LiNH_2$-HMgX complexes (X = H, F, Cl, Br, CH_3, OH, NH_2).[124] However, the term "alkaline-earth bonds" was proposed for interactions where beryllium, magnesium and calcium centres play the role of Lewis acid sites; the complexes of X_2M (X = H, F and Cl; M = Be, Mg and Ca) with various Lewis bases were analysed with the use of theoretical approaches.[128]

It is worth to note that elements of the second group often form divalent and linear moieties. Hence the electropositive alkaline-earth centre is characterised by the "ring" of positive EP that surrounds it. Figure I.16a presents an example of the $Be(CH_3)_2$ species where the ring around the Be-centre is characterised by such positive EP. That is why this centre may interact with nucleophiles. The linear $Be(CH_3)_2$ moiety interacts with HCN through the nitrogen Lewis base site forming the Be···N beryllium bond (Fig. I.16b). The complex created is characterised by the trigonal configuration. It seems that a feature of interactions, especially stronger ones, is that they lead not only to a change of coordination numbers but also to a change of configurations of centres considered.[6]

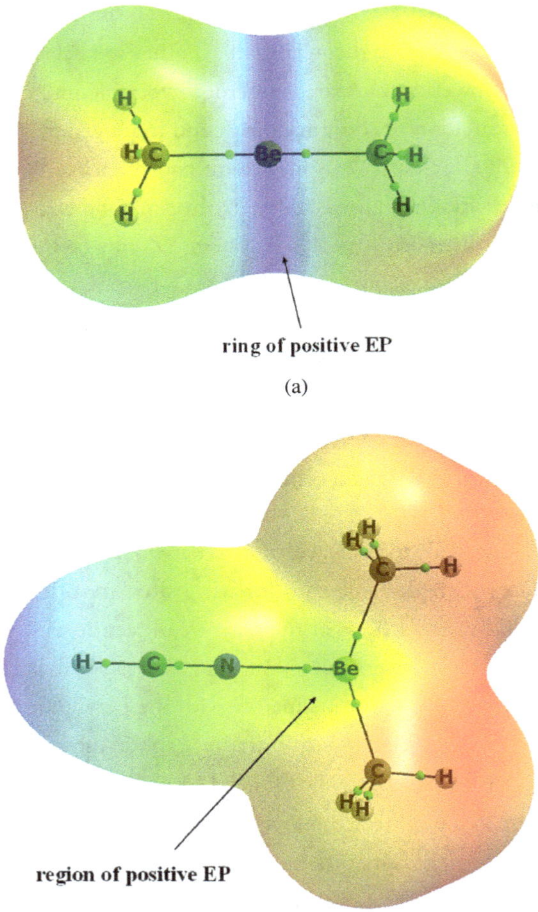

(b)

Fig. I.16 The molecular graphs and corresponding 0.001 au electron density isosurfaces with EP maps for (a) $Be(CH_3)_2$ species, (b) $Be(CH_3)_2(NCH)$ species. Blue corresponds to maximal positive EP values and red to minimal EP negative values. These are ωB97XD/aug-cc-pVTZ results of calculations.

The linear and divalent beryllium and magnesium species, for example, interacting through their Lewis acid centres with electron donors, change into trigonal configurations.[107,128]

Figure I.17 presents changes of configurations of beryllium species as effects of interactions with nucleophilic ligands; the number of ligands (coordination number)

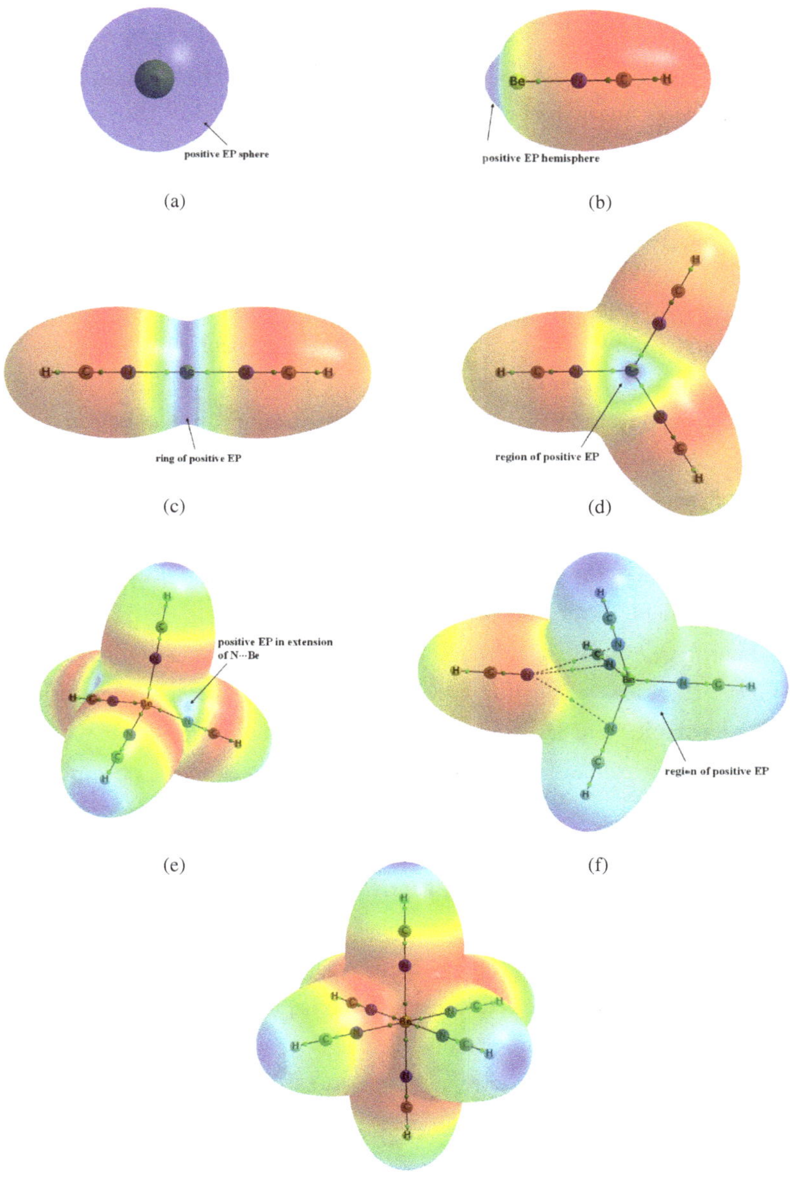

Fig. I.17 The molecular graphs and corresponding 0.001 au electron density isosurfaces with EP maps for Be^{2+} cation and the $Be(NCH)_n^{2+}$ clusters (n = 1–6). Blue corresponds to maximal positive EP values and red to minimal EP negative values. These are ωB97XD/aug-cc-pVTZ results of calculations.

determines the type of configuration. The Be^{2+} cation is characterised by a sphere of positive EP (Fig. I.17a); its interaction with the HCN ligand through the nitrogen centre leads to the formation of the $Be(NCH)^{2+}$ complex linked by the Be···N beryllium bond. In the latter linear species the Be-centre is characterised by the hemisphere of positive EP (Fig. I.17b), thus it still may interact with electron donors. In the next complex, the linear $Be(NCH)_2^{2+}$ species, a situation similar to that of the $Be(CH_3)_2$ moiety occurs. The ring of positive EP around the Be-centre is observed, and this centre is linked by beryllium bonds with two nitrogen centres of ligands (Fig. I.17c). The next $Be(NCH)_3^{2+}$ complex is characterised by trigonal configuration (Fig. I.17d): three beryllium bonds, and the region of positive EP at the Be-centre parallel to the plane of the complex. This region may be treated as a π-hole since the Be···N interactions may be considered as links like covalent bonds according to the σ-hole and π-hole concept. These Be···N interactions are at least partly covalent in nature.[107] Such covalent character of the Be···N contacts is also detected in the $Be(NCH)_4^{2+}$ complex which possesses tetrahedral configuration (Fig. I.17e). One can see here the regions of positive EP at the Be-centre in the extensions of N···Be contacts. Thus one can conclude that interaction of the HCN ligand with the $Be(NCH)_3^{2+}$ complex leads to a change of the trigonal configuration of the latter species into tetrahedral configuration in the $Be(NCH)_4^{2+}$ complex. Interaction with the additional HCN ligand results in the $Be(NCH)_5^{2+}$ complex (Fig. I.17f) which possesses irregular configuration, with regions of positive EP approximately in elongations of the N···Be contacts. However, regular configuration of the magnesium structure with five HCN ligands, $Mg(NCH)_5^{2+}$ complex, is observed. It is the trigonal bipyramid (Fig. I.18a); there is also another irregular configuration (Fig. I.18b), similar to that of the beryllium centre (Fig. I.17f). It seems that in the $Be(NCH)_6^{2+}$ system (Fig. I.17g) the beryllium centre does not possess characteristics of a Lewis acid site, as the regions of positive EP are not observed. This is the regular octahedral structure that is often characteristic for systems containing six ligands. There are similar observations of change of configurations that results from interactions for the magnesium systems.[107]

These findings concerning regions of positive EP, i.e., concerning Lewis acid properties, show that interactions are responsible for the great geometrical changes, which are related to electron charge shifts. They are described in detail in subsequent chapters.

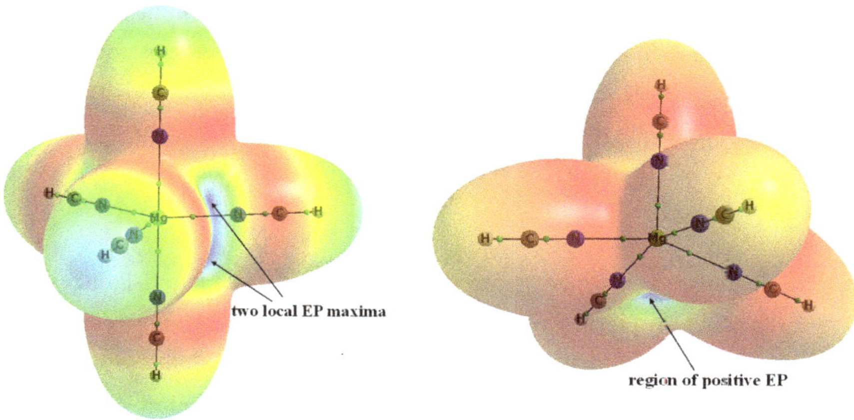

(a) (b)

Fig. I.18 The molecular graphs and corresponding 0.001 au electron density isosurfaces with EP maps for two configurations of the $Mg(NCH)_5^{2+}$ cluster. Blue corresponds to maximal positive EP values and red to minimal EP negative values. These are ωB97XD/aug-cc-pVTZ results of calculations.

Summary

The Lewis acid — Lewis base interactions, La-Lb, are characterised by significant electron charge transfer from the Lewis base unit to the Lewis acid. The complexes characterised by such interactions are usually linked by electron-rich and electron-deficient centres, regions which possess opposite electrostatic potentials. These regions in contact are often marked as the Lewis base and Lewis acid sites.

The interaction and binding energies are measures of the strength of interaction; negative values are attributed to stabilising interactions in most studies, and "more negative" values designate stronger interactions. The binding energy is the sum of interaction energy and deformation energy; the latter term is characterised by positive values, and it informs of the deformation of units which participate in the complex formation. The greater the deformation energy, the greater deformation of units linked in the complex. The interaction energy may be decomposed into electrostatic, polarisation and dispersion attractive interaction energy terms and the Pauli repulsive interaction energy. The polarisation term related to the electron charge shifts resulting from complexation is typical for La-Lb interactions.

In Molecular Mechanics approaches, various force fields are applied to analyse different classes and sub-classes of compounds as well as different systems, for example, clusters, liquids, solids, etc. The force field is a set of equations which describe interactions occurring in the system. The Lennard-Jones and Buckingham potentials accompanying the electrostatic energy terms are functions most often applied in various force fields.

The σ-hole and π-hole concept introduced by Murray, Clark and Politzer and others[46–50] explains the dual Lewis acid and Lewis base character of numerous centres of groups 14–18. It also describes various types of interactions; one can mention triel, tetrel, pnicogen, chalcogen, halogen and aerogen bonds where centres of groups 13, 14, 15, 16, 17 and 18, respectively, act as the Lewis acid sites. The σ-hole is a region of depletion of the electron charge in extension of the bond to the centre considered. If sufficient depletion occurs, it is characterised by positive electrostatic potential, and it may act as the Lewis acid interacting with electron-rich sites. The π-hole is usually the region of depletion of the electron charge that is parallel to the molecule or molecular fragment.

The centres of groups 1 and 2 usually act as Lewis acid centres since they are characterised by low electronegativity values. They interact often with a greater number of nucleophilic ligands, and three centre — four electron A-Z···B interactions are rather rare. A-Z, B and Z are the Lewis acid unit, the Lewis base centre, and the centre considered that acts as the Lewis acid site, respectively.

Appendix

The following tools were applied to construct figures presented in this chapter.

Figures I.2,9,10,16,17,18

AIMAll program — Reference: Keith, T.A. *AIMAll* (*Version 11.08.23*); TK Gristmill Software. Overland Park, KS, USA, **2011**.

Figures I.5,6,8,11,12,14,15

Graphical tools of the Cambridge Structural Database — References: (1) Groom, C.R.; Bruno, I.J.; Lightfoot, M.P.; Ward, S.C. The Cambridge structural database. *Acta Cryst.* **2016**, *B72*, 171–179. (2) Wong, R.; Allen, F.H.; Willett, P. The scientific impact of the Cambridge Structural Database: A citation-based study. *J. Appl. Cryst.* **2010**, *43*, 811–824.

Figure I.1
Jmol program — Reference: Hanson, R.M. Jmol SMILES and Jmol SMARTS: specifications and applications. *J. Cheminform.* **2016**, *8*, 50.

Figures I.3,4,7,13
Reproduced with permission.

Scheme I.1
PowerPoint.

References

1. Vogel, G.C.; Drago, R.S. The ECW model. *J. Chem. Educ.* **1996**, *73*, 701–707.
2. Lewis, G.N. *Valence and the Structure of Atoms and Molecules*, American Chemical Society Monograph Series, New York, **1923**, p. 142.
3. Ingold, C.K. Significance of tautomerism and of the reactions of aromatic compounds in the electronic theory of organic reactions. *J. Chem. Soc.* **1933**, 1120–1127.
4. Brønsted, J.N. Acid and basic catalysis. *Chem. Rev.* **1928**, *5*, 231–338.
5. Savin, K.A. *Writing Reaction Mechanisms in Organic Chemistry: Third Edition* , Academic Press Waltham, San Diego, London, Oxford, **2014**, p. 28.
6. Grabowski, S.J. Hydrogen bonds, and σ-hole and π-hole bonds–mechanisms protecting doublet and octet electron structures. *Phys. Chem. Chem. Phys.* **2017**, *19*, 29742–29759.
7. ref. 2, pp. 98–99.
8. Weinhold, F.; Landis, C. *Valency and Bonding: A Natural Bond Orbital Donor — Acceptor Perspective*, Cambridge University Press, Cambridge, **2005**, pp. 306–353.
9. ref. 8, pp. 275–306.
10. Grabowski, S.J. Two faces of triel bonds in boron trihalide complexes. *J. Comput. Chem.* **2018**, *39*, 472–480.
11. Theoretical methods as well as measures of strength of interactions are described in Chapter II.
12. Piela, L. *Ideas of Quantum Chemistry*, Elsevier, Amsterdam, **2007**, pp. 684–691.
13. Jeziorski, B.; Moszyński, R.; Szalewicz, K. Perturbation theory approach to intermolecular potential energy surfaces of van der Waals complexes. *Chem. Rev.* **1994**, *94*, 1887–1930.

14. Szalewicz, K.; Jeziorski, B. Comment on "On the importance of the fragment relaxation energy terms in the estimation of the basis set superposition error correction to the intermolecular interaction energy" [J. Chem. Phys. 104, 8821 (1996)]. *J. Chem. Phys.* **1998**, *109,* 1198–1200.
15. Grabowski, S.J.; Sokalski, W.A. Different types of hydrogen bonds: correlation analysis of interaction energy components. *J. Phys. Org. Chem.* **2005**, *18*, 779–784.
16. Lewars, E. *Computational Chemistry*, Kluwer Academic Publishers, Dordrecht, **2003**.
17. Mantina, M.; Chamberlin, A.C.; Valero, S.; Cramer, C.J.; Truhlar, D.G. Consistent van der Waals radii for the whole main group. *J. Phys. Chem. A* **2009**, *113*, 5806–5812.
18. Rowland, R.S.; Taylor, R. Intermolecular nonbonded contact distances in organic crystal structures: comparison with distances expected from van der Waals radii. *J. Phys. Chem.* **1996**, *100*, 7384–7391.
19. Bondi, J. Van der Waals volumes and radii. *J. Phys. Chem.* **1964**, *68*, 441–451.
20. Lipkowski P.; Grabowski S.J.; Leszczynski, J. Properties of the halogen-hydride interaction: an ab initio and "atoms in molecules" analysis. *J. Phys. Chem. A* **2006**, *110*, 10296–10302.
21. Grabowski, S.J. Hydrogen Bond and Other Lewis Acid — Lewis Base Interactions — Mechanisms of Formation. In Leszczynski, J.; Shukla, M.K. (Eds.), *Practical Aspects of Computational Chemistry IV*, Springer, New York, **2016**, pp. 245–278.
22. Grabowski, S.J. *Understanding Hydrogen Bonds: Theoretical and Experimental Views*, Royal Society of Chemistry, Cambridge, **2021.**
23. Alkorta, I.; Elguero, J.; Frontera, A. Not only hydrogen bonds: other noncovalent interactions. *Crystals* **2020**, *10*, 180.
24. Grabowski, S.J. Boron and other triel Lewis acid centers: From hypovalency to hypervalency. *ChemPhysChem* **2014**, *15*, 2985–2993.
25. Grabowski, S.J. Triel bond and coordination of triel centres–Comparison with hydrogen bond interaction. *Coord. Chem. Rev.* **2020**, *407*, 213171.
26. Reed, E.; Curtiss, L.A.; Weinhold, F. Intermolecular interactions from a natural bond orbital, donor-acceptor viewpoint. *Chem. Rev.* **1988**, *88*, 899–926.
27. Grabowski, S.J. Ab initio calculations on conventional and unconventional hydrogen bonds — study of the hydrogen bond strength. *J. Phys. Chem. A* **2001**, *105,* 10739–10746.
28. Grabowski, S.J. Hydrogen bonds with π and σ electrons as the multicenter proton acceptors: high level ab initio calculations. *J. Phys. Chem. A* **2007**, *111,* 3387–3393.
29. Grabowski, S.J. Unpublished results.
30. Szczęśniak, M.M.; Chałasiński, G.; Cybulski, S.M.; Scheiner, S. Intermolecular potential of the methane dimer and trimer. *J. Chem. Phys.* **1990**, *93*, 4243–4253.

31. Li, A.H-T.; Chao, S.D. Intermolecular potentials of the methane dimer calculated with Møller-Plesset perturbation theory and density functional theory. *J. Chem. Phys.* **2006**, *125*, 094312.
32. Li, A.H-T.; Chao, S.D. Interaction energies of dispersion-bound methane dimer from coupled cluster method at complete basis set limit. *J. Mol. Struct.* **2009**, *897*, 90–94.
33. Metz, M.M.; Szalewicz, K.; Sarka, J.; Tóbiás, R.; Császár, A.G.; Mátyus, E. Molecular dimers of methane clathrates: ab initio potential energy surfaces and variational vibrational states. *Phys. Chem. Chem. Phys.* **2019**, *21*, 13504–13525.
34. Clark, T.; Murray, J.S.; Politzer, P. A perspective on quantum mechanics and chemical concepts in describing noncovalent interactions. *Phys. Chem. Chem. Phys.* **2018**, *20*, 30076–30082.
35. Murray, J.S.; Politzer, P. Interaction and polarization energy relationships in σ-hole and π–hole bonding. *Crystals* **2020**, *10*, 76.
36. Allinger, N.L. *Molecular Structure: Understanding Steric and Electronic Effects from Molecular Mechanics,* John Wiley & Sons, Hoboken, New Jersey, **2010**.
37. Leach, A.R. *Molecular Modelling: Principles and Applications Second Edition*, Pearson Education Limited, Harlow, **2001**.
38. Hobza, P.; Zahradník, R. *Intermolecular Complexes*, Czechoslovak Academy of Sciences in co-edition with Elsevier, Prague, Amsterdam, **1988**, pp. 65–66.
39. Lennard-Jones, J.E. On the determination of molecular fields.-II. From the equation of state of a gas. *Proc. Roy. Soc. London, Ser. A* **1924**, *106*, 463–477.
40. Buckingham, A.D.; Utting, B.D. Intermolecular forces. *Ann. Rev. Phys. Chem.* **1970**, *21*, 287–316.
41. Desiraju, G.R.; Vittal, J.J.; Ramanan, A. *Crystal Engineering: A Textbook*. World Scientific, Singapore, **2011**, pp. 28–29.
42. Grabowski, S.J. What is the covalency of hydrogen bonding? *Chem. Rev.* **2011**, *11*, 2597–2625.
43. Grabowski, S.J.; Sokalski, W.A. Are various σ-hole bonds steered by the same mechanisms? *ChemPhysChem* **2017**, *18*, 1569–1577.
44. Pendas, A.M.; Blanco, M.A.; Francisco, E. The nature of the hydrogen bond: A synthesis from the interacting quantum atoms picture. *J. Chem. Phys.* **2006**, *125*, 184112.
45. Grabowski, S.J. Intramolecular hydrogen bond energy and its decomposition O-H···O interactions. *Crystals* **2021**, *11*, 5.
46. Clark, T.; Hennemann, M.; Murray, J.S.; Politzer, P. Halogen bonding: the σ-hole. *J. Mol. Model.* **2007**, *13*, 291–296.
47. Politzer, P.; Lane, P.; Concha, M.C.; Ma, Y.; Murray, J.S. An overview of halogen bonding. *J. Mol. Model.* **2007**, *13*, 305–311.

48. Murray, J.S.; Lane, P.; Politzer, P. A predicted new type of directional noncovalent interaction. *Int. J. Quant. Chem.* **2007**, *107*, 2286–2292.
49. Politzer, P.; Murray, J.S.; Lane, P. σ-Hole bonding and hydrogen bonding: competitive interactions. *Int. J. Quant. Chem.* **2007**, *107*, 3046–3052.
50. Politzer, P.; Riley, K.E.; Bulat, F.A.; Murray, J.S. Perspectives on halogen bonding and other σ-hole interactions: Lex parsimoniae (Occam's Razor). *Comput. Theory Chem.* **2012**, *998*, 2–8.
51. Bader, R.F.W.; Carrol, M.T.; Cheeseman, J.R.; Chang, C. Properties of atoms in molecules: atomic volumes. *J. Am. Chem. Soc.* **1987**, *109*, 7968–7979.
52. Gillespie, R.J.; Popelier, P.L.A. *Chemical Bonding and Molecular Geometry*, Oxford University Press, Oxford, **2001**, p. 137.
53. Mulliken, R.S. Electronic population analysis on LCAO–MO molecular wave functions. I. *J. Chem. Phys.* **1955**, *23*, 1833–1840.
54. Bader, R.F.W. *Atoms in Molecules: A Quantum Theory*, Oxford University Press, Oxford, **1990**.
55. Szczęśniak, M.M.; Chałasiński, G.; Cybulski, S.M.; Cieplak, P. *Ab initio* study of the potential energy surface of CH_4-H_2O. *J. Chem. Phys.* **1993**, *98*, 3078–3089.
56. Rovira, C.; Novoa, J.J. Strength and directionality of the $C(sp^3)$-H···$S(sp^3)$ interaction. *Ab initio* study using the H_2S···CH_4 model complex. *Chem. Phys. Lett.* **1997**, *279*, 140–150.
57. Moore, K.B. III.; Sadeghian, K.; Sherrill, C.D.; Ochsenfeld, C.; Schaefer, H.F. III. C–H···O hydrogen bonding. The prototypical methane-formaldehyde system: a critical assessment. *J. Chem. Theory Comput.* **2017**, *13*, 5379–5395.
58. Wang, W.; Hobza, P. Origin of the X-Hal (Hal = Cl, Br) bond-length change in the halogen-bonded complexes. *J. Phys. Chem. A* **2008,** *112,* 4114–4119.
59. Riley, K.E.; Hobza, P. Investigations into the nature of halogen bonding including symmetry adapted perturbation theory analyses. *J. Chem. Theory Comput.* **2008,** *4,* 232–242.
60. Politzer, P.; Murray, J.S.; Clark, T. Halogen bonding: an electrostatically-driven highly directional noncovalent interaction. *Phys. Chem. Chem. Phys.* **2010**, *12*, 7748–7757.
61. Politzer, P.; Murray, J.S.; Clark, T. Halogen bonding and other σ-hole interactions: a perspective. *Phys. Chem. Chem. Phys.* **2013**, *15*, 11178–11189.
62. Clark, T. σ-Holes. *WIREs Comput. Mol. Sci.* **2013**, *3*, 13–20.
63. ref. 52, p. 87.
64. Murray, J.S.; Lane, P.; Clark, T.; Politzer, P. σ-Hole bonding: molecules containing group VI atoms. *J. Mol. Model.* **2007**, *13*, 1033–1038.

65. Murray, J.S.; Lane, P.; Politzer, P. Expansion of the σ-hole concept *J. Mol. Model.* **2009**, *15*, 723–729.
66. Nygren, C.L.; Wilson, C.C.; Turner, J.F.C. On the solid state structure of 4-iodobenzoic acid. *J. Phys. Chem. A* **2005**, *109*, 2586–2593.
67. Hui, R-H.; Zhou, P.; You, Z-L. Synthesis, characterization, and crystal structures of a pair of zinc(ii) complexes derived from 2,4-dichloro-6-[(3-methylaminopropylimino) methyl]phenol. *Synth. React. Inorg., Met.-Org., Nano-Met. Chem.* **2008**, *38*, 464–467.
68. Nyburg, S.C.; Faerman, C.H. A revision of van der Waals atomic radii for molecular crystals: N, O, F, S, CI, Se, Br and I bonded to carbon. *Acta Cryst.* **1985**, *B41*, 274–279.
69. Zordan, F.; Brammer, L.; Sherwood P. Chemistry of halogens: complementary features of inorganic (M-X) and organic (C-X′) halogens applied to M-X···X′-C halogen bond formation. *J. Am. Chem. Soc.* **2005**, *127*, 5979–5989.
70. Desiraju, G.R.; Parthasarathy, R. The nature of halogen…halogen interactions: are short halogen contacts due to specific attractive forces or due to close packing of nonspherical atoms? *J. Am. Chem. Soc.* **1989**, *111*, 8725–8726.
71. Bilewicz, E.; Rybarczyk-Pirek, A.J.; Dubis, A.T.; Grabowski, S.J. Halogen bonding in crystal structure of 1-methylpyrrol-2-yl trichloromethyl ketone. *J. Mol. Struct.* **2007**, *829*, 208–211.
72. ref. 22, p. 58.
73. Bundhun, A.; Ramasami, P.; Murray, J.S.; Politzer, P. Trends in σ-hole strengths and interactions of F_3MX molecules (M = C, Si, Ge and X = F, Cl, Br, I). *J. Mol. Model.* **2013**, *19*, 2739–2746.
74. Kutzelnigg, W. Chemical bonding in higher main group elements. *Angew. Chem. Int. Ed. Engl.* **1984**, *23*, 272–295.
75. Klapötke, T.M.; Krumm, B.; Scherr, M. Homoleptic selenium cyanides: attempted preparation of $Se(CN)_4$ and redetermination of the crystal structure of $Se(CN)_2$. *Inorg. Chem.* **2008**, *47*, 7025–7028.
76. Bauzá, A.; Frontera, A. Aerogen bonding interaction: a new supramolecular force? *Angew. Chem. Int. Ed.* **2015**, *54*, 7340–7343.
77. Goettel, J.T.; Mercier, H.P.A.; Schrobilgen, G.J. XeO_3 adducts of pyridine, 4-dimethylaminopyridine, and their pyridinium salts. *J. Fluor. Chem.* **2018**, *211*, 60–69.
78. Murray, J.S.; Lane, P.; Clark, T.; Riley, K.E.; Politzer, P. σ-Holes, π-holes and electrostatically-driven interactions. *J. Mol. Model.* **2012**, *18*, 541–548.
79. Kozuch, S. Should "anion–π interactions" be called "anion–σ interactions"? A revision of the origin of some hole-bonds and their nomenclature. *Phys. Chem. Chem. Phys.* **2016**, *18*, 30366–30369.

80. Kirshenboim, O.; Kozuch, S. How to twist, split and warp a σ-hole with hypervalent halogens. *J. Phys. Chem. A* **2016**, *120*, 9431–9445.
81. Grabowski, S.J. π-Hole bonds: boron and aluminum lewis acid centers. *ChemPhysChem* **2015**, *16*, 1470–1479.
82. Davidson, M.G.; Hughes, A.K.; Marder T.B.; Wade, K. (Eds.). *Contemporary Boron Chemistry*, Royal Society of Chemistry, Cambridge, **2000**.
83. Hirao, H.; Omoto, K.; Fujimoto, H. Lewis acidity of boron trihalides. *J. Phys. Chem. A* **1999**, *103*, 5807–5811.
84. Jacobsen, H. Kinetic energy density and covalent bonding — a complementary analysis at the border of bond and no bond. *Dalton Trans.* **2009**, 4252–4258.
85. Chen, E.Y.X.; Marks, T.J. Cocatalysts for metal-catalyzed olefin polymerization: activators, activation processes, and structure–activity relationships. *Chem. Rev.* **2000**, *100*, 1391–1434.
86. Erker, G. Tris(pentafluorophenyl)borane: a special boron Lewis acid for special reactions. *Dalton Trans.* **2005**, 1883–1890.
87. Piers, W.E. The chemistry of perfluoroaryl boranes. *Adv. Organomet. Chem.* **2004**, *52*, 1–76.
88. Williams, S.D.; Harper, W.; Mamantov, G.; Tortorelli, L.J.; Shankle, G. *Ab initio* MO study of selected aluminum and boron chlorides and fluorides: Comparison with ^{11}B NMR spectra of a tetrachloroborate melt. *J. Comput. Chem.* **1996**, *17*, 1696–1711.
89. Hiberty, P.C.; Ohanessian, G. Comparison of minimal and extended basis sets in terms of resonant formulas. application to 1,3 dipoles. *J. Am. Chem. Soc.* **1982**, *104*, 66–70.
90. Phillips, J.A.; Cramer, J. Quantum chemical characterization of the structural and energetic properties of HCN–BF_3. *J. Chem. Theory Comput.* **2005**, *1*, 827–833.
91. Smith, E.L.; Sadowsky, D.; Cramer, C.J.; Phillips, J.A. Structure, bonding, and energetic properties of nitrile–borane complexes: RCN–BH_3. *J. Phys. Chem. A* **2011**, *115*, 1955–1963.
92. Cid, J.; Carbó, J.J.; Fernández, E. Disclosing the structure/activity correlation in trivalent boron-containing compounds: a tendency map. *Chem. Eur. J.* **2012**, *18*, 12794–12802.
93. Fujiang, D.; Fowler, P.W.; Legon, A.C. Geometric and electric properties of the donor–acceptor complex H_3N–BF_3. *J. Chem. Soc. Chem. Commun.* **1995**, 113–114.
94. Grabowski, S.J. The nature of triel bonds, a case of B and Al centres bonded with electron rich sites. *Molecules* **2020**, *25*, 2703.
95. Cowley, A.H.; Cushner, M.C.; Davis, R.E.; Riley, P.E. Crystal and molecular structure of the 1:2 aluminum trichloride-tetrahydrofuran complex $AlCl_3$·2THF. *Inorg. Chem.* **1981**, *20*, 1179–1181.

96. Gorrell, I.B.; Hitchcock, P.B.; Smith, J.D. Preparation and crystal structures of aluminium hydride adducts with tetrahydrofuran $[AlH_3{\cdot}OC_4H_8]_2$ and $AlH_3{\cdot}2OC_4H_8$. *J. Chem. Soc. Chem. Commun.* **1993**, 189–190.

97. Bauzá, A.; Ramis, R.; Frontera, A. A combined theoretical and Cambridge Structural Database study of π-hole pnicogen bonding complexes between electron rich molecules and both nitro compounds and inorganic bromides (YO_2Br, Y = N, P, and As). *J. Phys. Chem. A* **2014**, *118*, 2827–2834.

98. Bauzá, A.; Mooibroek, T.J.; Frontera, A. Directionality of π-holes in nitro compounds. *Chem. Commun.* **2015**, *51*, 1491–1493.

99. Fun, H-K.; Goh, J.H.; Chandrakantha, B.; Isloorc, A.M.; Shetty, P. 2-Chloro-4-nitro-1H-imidazole. *Acta Cryst.* **2010**, *E66*, o1828–o1829.

100. Murray, J.S.; Politzer, P. Molecular electrostatic potentials and noncovalent interactions. *WIREs Comput. Mol. Sci.* **2017**, *7*, e1326.

101. Pauling, L. *The Nature of the Chemical Bond: Third Edition.* Cornell University Press, Ithaca, N.Y., **1960**.

102. Mullay, J. Estimation of Atomic and Group Electronegativities. In Sen, K.D., Jørgensen, C.K. (Eds.), *Electronegativity,* **1987**, Springer, Berlin Heidelberg, pp. 1–25.

103. Desiraju, G.R.; Steiner, T. *The Weak Hydrogen Bond in Structural Chemistry and Biology*, Oxford University Press, New York, **1999**.

104. Grabowski, S.J.; Hoffmann, R. Stabilizing H3-: or are we stabilizing a proton? *ChemPhysChem* **2012,** *13*, 2286–2288.

105. https://goldbook.iupac.org/terms/view/C01331

106. Muller, P. Glossary of terms used in physical organic chemistry (IUPAC recommendations 1994). *Pure Appl. Chem.* **1994**, *66*, 1077–1184.

107. Grabowski, S.J. Coordination of Be and Mg centres by HCN ligands — Be…N and Mg…N Interactions. *ChemPhysChem* **2018**, *19*, 1830–1840.

108. Kollman, P.A.; Liebman, J.F.; Allen, L.C. The lithium bond. *J. Am. Chem. Soc.* **1970**, *92*, 1142–1150.

109. Vila, A.; Vila, E.; Mosquera, R.A. Topological characterisation of intermolecular lithium bonding. *Chem. Phys.* **2006**, *326*, 401–408.

110. Lipkowski, P.; Grabowski, S.J. Could the lithium bond be classified as the σ-hole bond? — QTAIM and NBO analysis. *Chem. Phys. Lett.* **2014**, *591*, 113–118.

111. Platts, J.A.; Howard, S.T.; Bracke, B.R.F. Directionality of hydrogen bonds to sulfur and oxygen. *J. Am. Chem. Soc.* **1996**, *118*, 2726–2733.

112. Sannigrahi, A.B.; Kar, T.; Niyogi, B.G.; Hobza, P.; von Schleyer, P.R. The lithium bond reexamined. *Chem. Rev.* **1990**, *90*, 1061–1076.

113. Hazra, A,B.; Pal, S. Weak interaction between HCHY (Y=O, S) and LiCl: a theoretical study. *J. Mol. Struct.* **2000**, *497*, 157–163.

114. Esrafili, M.D.; Mohammadirad, N. Halogen bond interactions enhanced by sodium bonds — Theoretical evidence for cooperative and substitution effects in NCX···NCNa···NCY complexes (X = F, Cl, Br, I; Y = H, F, OH). *Can. J. Chem.* **2014**, *92*, 653–658.
115. Solimannejad, M.; Rabbani, M.; Ahmadi, A.; Esrafili, M.D. Cooperative and diminutive interplay between the sodium bonding with hydrogen and dihydrogen bondings in ternary complexes of NaC_3N with HMgH and HCN (HNC). *Mol. Phys.* **2014**, *112*, 2017–2022.
116. Poonia, N.S.; Bajaj, A.V. Coordination chemistry of alkali and alkaline earth cations. *Chem. Rev.* **1979**, *79*, 389–445.
117. Quezada, G.R.; Toledo, P.G. Complexation of alkali and alkaline-earth metal cations at spodumene-saltwater interfaces by molecular simulation: Impact on oleate adsorption. *Minerals* **2021**, *11*, 12.
118. Zundel, G. *Series of ten lectures on: Proton Polarizability of Hydrogen Bonds and Proton Transfer Processes, Their Role in Electrochemistry and Biology*, Institut für Physikalische Chemie der Universität München, München, **1997**.
119. Yáñez, M.; Sanz, P.; Mó, O.; Alkorta, I.; Elguero, J. Beryllium bonds, do they exist? *J. Chem. Theory Comput.* **2009**, *5*, 2763–2771.
120. Mó, O.; Yáñez, M.; Alkorta, I.; Elguero, J. Modulating the strength of hydrogen bonds through beryllium bonds. *J. Chem. Theory Comput.* **2012**, *8*, 2293–2300.
121. Eskandari, K. Characteristics of beryllium bonds; a QTAIM study. *J. Mol. Model.* **2012**, *18*, 3481–3487.
122. Yáñez, M.; Mó, O.; Alkorta, I.; Elguero, J. Can conventional bases and unsaturated hydrocarbons be converted into gas-phase superacids that are stronger than most of the known oxyacids? The role of beryllium bonds. *Chem. Eur. J.* **2013**, *19*, 11637–11643.
123. Brea, O.; Mó, O.; Yáñez, M.; Alkorta, I.; Elguero, J. Creating σ-holes through the formation of beryllium bonds. *Chem. Eur. J.* **2015**, *21*, 12676–12682.
124. Yang, X.; Li, Q.; Cheng, J.; Li, W. A new interaction mechanism of $LiNH_2$ with MgH_2: magnesium bond. *J. Mol. Model.* **2013**, *19*, 247–253.
125. Li, S-Y.; Wu, D.; Li, Y.; Yu, D.; Liu, J-Y.; Li, Z-R. Insight into structural and π–magnesium bonding characteristics of the X_2Mg···Y (X = H, F; Y = C_2H_2, C_2H_4 and C_6H_6) complexes. *RSC Adv.* **2016**, *6*, 102754–102761.
126. McDowell, S.A.C.; Maynard, J.M. A computational study of model hydrogen-, halogen-, beryllium- and magnesium-bonded complexes of aziridine derivatives. *Mol. Phys.* **2016**, *114*, 1609–1618.
127. Grabowski, S.J. Magnesium bonds: from divalent Mg centres to trigonal and tetrahedral coordination. *ChemistrySelect* **2018**, *3*, 3147–3154.
128. Montero-Campillo, M.M.; Sanz, P.; Mó, O.; Yáñez, M.; Elguero, J. Alkaline-earth (Be, Mg and Ca) bonds at the origin of huge acidity enhancements. *Phys. Chem. Chem. Phys.* **2018**, *20*, 2413–2420.

Chapter II

Theoretical Approaches in Analysis of Interactions

Abstract: Selected theoretical approaches that are often applied to analyse intermolecular interactions, among them Lewis acid — Lewis base interactions, are considered here. These include *ab initio* and Density Functional Theory (DFT) methods, the Quantum Theory of Atoms in Molecules (QTAIM), Energy Decomposition Analysis (EDA), Natural Bond Orbital (NBO) method, Electron Localization Function (ELF) method, Non-Covalent Interactions (NCI) method, Car-Parrinello molecular dynamics and others. Various parameters which could be obtained from these approaches are also presented and the application of these parameters for the description of inter- and intramolecular interactions is described.

Keywords: *ab initio* methods, DFT approach, Quantum Theory of Atoms in Molecules, Natural Bond Orbital method, Energy Decomposition Analysis, Car-Parrinello molecular dynamics

Different methods are applied in theoretical studies to evaluate geometrical and energetic parameters of systems. Among properties that are derived from these approaches are characteristics of various interactions. Hence it is useful to present here approaches that are most often applied. In general, four groups of theoretical methods are used in a majority of studies calculating properties of species: *ab initio*, semiempirical, density functional theory (DFT) and molecular mechanics (MM).[1]

MM approaches[2,3] were described in the previous chapter since they are useful, roughly at first, to define and describe Lewis acid — Lewis base interactions, which are the subject of this book. It is worth mentioning that equations which constitute force fields are often treated as a preliminary stage of other advanced methods like molecular dynamics (MD) or other related approaches.[4] One can mention here approaches that are applications of *ab initio* methods to MD, e.g., Car-Parrinello,[5] Born-Oppenheimer and Ehrenfest,[4] or hybrid QM/MM MD.[4] There are also Monte Carlo methods[6] which are useful in the analysis of large systems such as solvated and particularly hydrated species.[7] Another group of methods concerns solvent effects:[8–10] these are the Polarizable Continuum Model[8] or the Conductor-like Screening Model[9,10] for example.

The semiempirical approaches[11] are related to *ab initio* methods where some theoretical expressions are replaced by experimental values to simplify and make faster calculations. One can mention CNDO (Complete Neglect of Differential Overlap), INDO (Intermediate Neglect of Differential Overlap), NDDO (Neglect of Diatomic Differential Overlap), and MINDO (Modified Intermediate Neglect of Differential Overlap); they were often applied in the seventies and eighties of the last century. However, nowadays computers are much more powerful and such great simplifications that are applied in the latter methods are often not needed. Other methods that are often classified as semiempirical, and used with greater or less modifications until now are, for example, the Hückel method and the so-called extended Hückel method developed by Hoffmann.[12] The semiempirical approaches are not presented here since *ab initio* and DFT methods are applied for calculations on large systems and semiempirical approaches are not adopted nowadays, especially for analyses of Lewis acid — Lewis base interactions.

Results of *ab initio* and DFT methods are presented and discussed in this book mainly. For detailed descriptions of these approaches and their approximations and limitations, the reader is referred to advanced monographs on theoretical chemistry.[1,13–17] Theoretical tools, including *ab initio* and DFT techniques, were briefly explained in an understandable way in the recent monograph on the hydrogen bond interaction.[18] It is worth mentioning that, contrary to the semiempirical approaches that use simplified Hamiltonians with experimental parameters, the *ab initio* methods use the Hamiltonian without any experimental-based simplifications.

II.1 *Ab initio* Methods

The time-independent Schrödinger equation (II.1) is often a starting point of numerous approaches in theoretical chemistry.

$$\hat{H}\Psi = E\Psi \tag{II.1}$$

Ψ is the wave function, and $\hat{H}$ is the Hamiltonian operator that concerns the observable energy, E, of the system. Equation II.1 is related to all particles of a system; these are nuclei and electrons most often. The Hamiltonian operator may be partitioned into operators corresponding to the following energy terms: kinetic energy of nuclei ($\hat{T}_n$), kinetic energy of electrons ($\hat{T}_e$), as well as potential interaction energies — nucleus-nucleus ($\hat{V}_{nm}$), nucleus-electron ($\hat{V}_{ne}$), and electron-electron ($\hat{V}_{ee}$).

$$\hat{H} = \hat{T}_n + \hat{T}_e + \hat{V}_{nm} + \hat{V}_{ne} + \hat{V}_{ee} \tag{II.2}$$

The Born-Oppenheimer approximation establishes that nuclei are "immobile" in relation to electrons. Hence $\hat{T}_n = 0$, and the $\hat{V}_{nm}$ term is constant. The electron energy terms are important in numerous chemical phenomena, thus the electronic Hamiltonian operator containing three terms only is often considered.

$$\hat{H}_{el} = \hat{T}_e + \hat{V}_{ne} + \hat{V}_{ee} \tag{II.3}$$

The Hartree-Fock (HF) method, which is based on eq. II.3, contains numerous approximations. It is often treated as the starting point for more accurate approaches. The following approximations in the HF method, which is often also called SCF method (self-consistent field method), may be mentioned: relativistic effects are neglected, the Born-Oppenheimer approximation is assumed, the energy eigenfunction (in the Schrödinger equation) is described by the single Slater determinant, etc. The Born-Oppenheimer approximation is usually accepted for systems where various interactions are analysed; this is also in force for results which are presented in this book. The relativistic effects are significant for heavier atoms, starting from the third or fourth period of elements (period numeration omits the hydrogen and helium row here and in the majority of theoretical analyses).

There is another HF approximation that is important: HF does not include electron correlation. One may refer to the Löwdin definition of the energy of correlation which states that "The correlation energy for a certain state with respect to a specified Hamiltonian is the difference between the exact eigenvalue of the Hamiltonian and its expectation value in the HF approximation for the state under consideration."[19]

This definition is accepted in many studies and it is also adopted here. One can mention two types of electron correlation which are covered by the Löwdin definition: (1) the dynamical correlation that takes into account the dynamical character of electron-electron interactions, which thus concerns the correlated motion of each electron with each other, and (2) the static (non-dynamical) correlation.[15,16] The methods that take into account more than one determinant allow inclusion of the latter kind of correlation. In the HF method a single Slater-type determinant is applied, thus the static correlation is not considered. Other approaches which consider only the single determinant also cannot cover the static correlation.[15,16] However, the dynamic correlation is included in the single determinant post-HF methods.

The following methods are often applied in various studies that include electron correlation: Configuration Interaction, Coupled Cluster, and Many-Body Perturbation Theory.

The configuration interaction (CI) method, like the HF method, is based on the variational principle. However, in the CI approach the trial wave function is a linear combination of determinants where coefficients are determined in such a way that the wave function corresponds to the energy minimum of the system considered. The singly, doubly, triply, etc., excited determinants are labelled by S, D, T, etc., respectively. The full CI (FCI) approach considers all so-called configuration state functions (CSFs); such a treatment seems to produce the best results that can be obtained for the basis set considered. The FCI approach covers both types of correlations, dynamic and static. This approach is applied only for very small species because of time-consuming calculations; particularly, for larger systems the FCI needs an enormous number of CSFs. It is worth mentioning that the reduction of the number of excited determinants in the CI expansion leads to variants of the CI approach; for example, CID designates CI that includes Doubles only. The other variant, quadratic CI method, QCI, was introduced by Pople and coworkers.[20] Its QCISD, QCISDT and QCISD(T) modifications that are often applied may be classified as situated between CI and coupled cluster methods.

In the coupled cluster (CC) approach, there are different variants where only some excitations are included. The CCD, CCSD, CCSDT and CCSDTQ variants may be mentioned. The last two CC variants are applied only for very small systems because the corresponding calculations are very time consuming. This is why mostly the CCSD variant is used in calculations, or the CCSD(T) approach which is a hybrid method where a contribution of triplets is calculated from the formula of the MP4 method but with the use of CCSD c-coefficients. The CCSD(T) calculations are often called benchmark (or gold) calculations for moieties that may be classified as one-reference systems. The CCSD(TQ) method, that by analogy to CCSD(T) is a hybrid with the perturbation approach, is also known. There are several variants of the CC approach which were introduced in recent years that allow reduction of computational cost; particularly the modifications of the CCSD(T) method are known: Pair Natural Orbital Coupled Cluster Method, PNO-CCSD(T)[21] and Domain-Based Local Pair Natural Orbital Coupled Cluster Method, DLPNO-CCSD(T).[22]

The Møller-Plesset method[23] is the most often applied of the MBPT group of approaches, and the following variants are known: MP2, MP3, MP4, MP4(SDQ) and MP6.[15,16]

Other than the three groups of approaches mentioned above (CI, CC and MBPT), there are others whose main feature is that they consider static correlation. For example, the Multi-Configuration Self-Consistent Field (MCSCF) covers static correlation but not dynamic correlation.[15,16] However, one has to be aware that these correlations cannot be separated. If the number of configurations increases in the MCSCF approach, then dynamic correlation is taken into account to a greater degree, up to the situation corresponding to the FCI approach where the whole correlation is included.

In the Complete Active Space Self-Consistent Field (CASSCF) variant[24] of the MCSCF approach, only configurations necessary to describe the system are selected. The active molecular orbitals chosen in the CASSCF variant are often highest occupied and lowest unoccupied atomic orbitals derived from HF calculations. The CASSCF(*N,M*) label marks the (*N,M*) active space; *N* and *M* correspond to the number of electrons and orbitals, respectively.

The CASPT2 (Complete Active Space Perturbation Theory) method[25,26] includes dynamic correlation which is not taken into account in the CASSCF approach. CASPT2 contains MP2-type expansion that is based on the CASSCF reference. Other modifications of the CASSCF method which include dynamic correlation are

Table II.1 Brief characteristics of post-HF methods. HF is included for comparison. Adapted with permission from ref. 18. Copyright 2021 The Royal Society of Chemistry.

Method	Variants (examples)	Brief characteristics
Hartree-Fock	HF	One-determinant wave function.
Møller-Plesset	MP2, MP3, MP4, MP4(SDQ), MP6	Very popular, commonly used variant of MBPT approach.
Configuration interaction	CIS, CID, CISD, CISD(T), CISD(Q)	Truncated CI variants containing selected excited determinants.
Quadratic configuration interaction	QCISD, QCISD(T)	Methods being intermediated between CI and CC.
Coupled-cluster	CCD, CCSD, CCSD(T), CCSDT, CCSDTQ	The cluster operator T is applied to describe the full-CI wave function; however, truncations are usually applied.
Multi-configurational	MCSCF, CASSCF	Approach where not only the coefficient of determinants are optimised but also the molecular orbitals that construct determinants.
Multi-configurational	CASPT2, NEVPT2	The generalisation of MP theory to the multiconfigurational case.
Full configuration interaction	FCI	Considers all 'configuration state functions', thus it is the best benchmark method. However, it may be applied only for very small systems since calculations are time consuming, thus approximations are needed.

known, for example, the multiconfigurational perturbation theory, based on a CASSCF zeroth-order wave function, called n-electron valence state perturbation theory, NEVPT2.[27–29]

The characteristics of theoretical approaches discussed in this section are given in Table II.1. The following trend in approximate quality of the methods may be observed if the multireference approaches, among them those which include dynamic correlation, are excluded.[15]

HF < MP2 ~ MP3 ~ CCD < CISD < MP4SDQ ~ QCISD ~ CCSD < MP4 < QCISD(T) ~ CCSD(T)

II.2 Density Functional Theory Approximations

There is another class of approaches based on DFT.[1,15,16] DFT is rooted in the proof of Hohenberg and Kohn[30] which states that the ground state electronic energy of the system considered and consequently all its parameters may be determined from its electron density, ρ. It seems that there are numerous advantages of DFT over methods discussed in the previous section which are based on wave functions. First of all, the electron density function depends only on three coordinates, independent of the number of electrons of a system. In the case of *ab initio* methods analysed so far (HF, CC, CI, MBPT and so on), a wave function for an N-electron system depends on 4N coordinates, since for each electron there are three spatial coordinates and one spin coordinate (for time-dependent *ab initio* methods, there is an additional time coordinate). Another advantage of DFT is the reduction of the time of calculations in comparison with acceptable *ab initio* methods (i.e., methods that provide reliable results which reproduce or are at least comparable with experimental data), such as MP2 for example. However, the main problem of DFT is to find the functional that transforms the electron density, ρ, into energy, E.

The energy functional, $E[\rho]$, may be divided into three parts, that of the kinetic energy, $T[\rho]$, the nuclei-electrons attraction, $E_{ne}[\rho]$, and the electron-electron repulsion, $E_{ee}[\rho]$:

$$E[\rho] = T[\rho] + E_{ne}[\rho] + E_{ee}[\rho] \tag{II.4}$$

One may see here the analogy between the energy functional $E[\rho]$ and the electronic Hamiltionan, $\widehat{H}_{el}$, expressed by eq. II.3, where the Hamiltonian contains operators which are analogues to energy terms of eq. II.4.

The last term, $E_{ee}[\rho]$, of eq. II.4 may be divided into Coulomb, $J[\rho]$, and Exchange, $K[\rho]$, parts, respectively (eq. II. 5). This partitioning corresponds to the HF approach.

$$E[\rho] = T[\rho] + E_{ne}[\rho] + J[\rho] + K[\rho] \tag{II.5}$$

The Kohn-Sham equations[31] lead to the modification of eq. II.5 into eq. II.6:

$$E_{DFT}[\rho] = T_S[\rho] + E_{ne}[\rho] + J[\rho] + E_{xc}[\rho] \tag{II.6}$$

The $T_S[\rho]$ part of the functional corresponds to the kinetic energy of electrons under the assumption of non-interacting orbitals. This assumption usually covers almost all

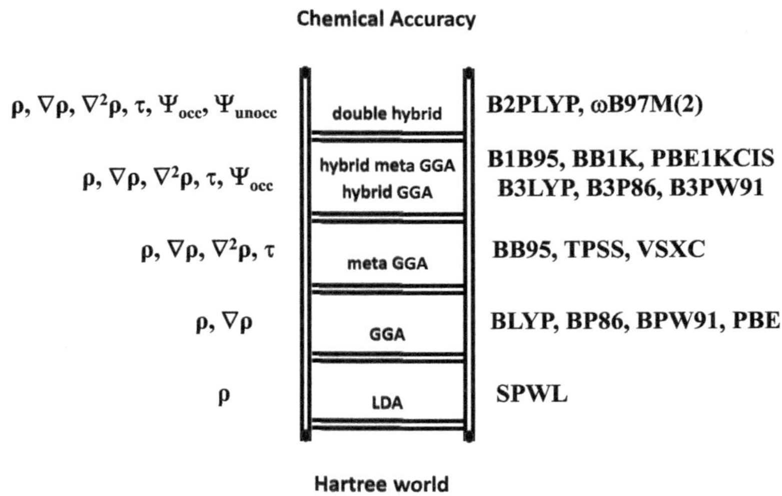

Scheme II.1 The Jacob's ladder of DFT functionals presented as rungs (middle). Information on the electron density are included in functionals of the specific rung (left). The examples of functionals corresponding to the rungs are given (right).

of the kinetic energy. The remaining part of the kinetic energy is included in the exchange-correlation term, $E_{XC}[\rho]$, that is mainly related to the correlation energy.

DFT methods usually differ between themselves in the form of the exchange correlation functional, $E_{XC}[\rho]$; various forms were proposed in literature. This functional is often separated into exchange and correlation parts. It is difficult to compare functionals that are applied in different studies since they often reproduce different parameters. It is worth mentioning that the B3LYP functional is applied most often so far. However, in recent years the ωB97X-D functional that includes dispersion correction is used and recommended in numerous studies; the results with the application of this functional are comparable to the results of post-HF methods which include correlation effects (CCSD or MP2 methods for example). Several selected functionals that are often applied in DFT calculations are presented in Scheme II.1; it is also shown in this scheme to which rung these functionals belong. More extended descriptions of the DFT functionals are given in monographs and review articles.[1,15,16,32,33]

The main source of error in DFT come from the approximations related to the $E_{XC}[\rho]$ functional. The functionals contain various degrees of information of electron density, and the increase of such information may be expressed as climbing Jacob's DFT ladder.[32] The higher rungs of the ladder correspond to more ingredients included (Scheme II.1). The most simple approximation for $E_{XC}[\rho]$ is the local density approximation (LDA), the first rung. In LDA the $E_{XC}[\rho]$ value depends only on the local values (at specific points) of the electron density, ρ. This value may be separated into exchange and correlation terms. The generalised gradient approximation (GGA) applies an energy functional that depends on both the density and its gradient, $\nabla\rho$; this is the second rung. In addition to ρ and $\nabla\rho$, the Laplacian $\nabla^2\rho$ and/or the kinetic energy density τ may be considered. This leads to the next generation of the meta-GGA class of functionals, the third rung. In so-called hybrid-GGA functionals of the fourth rung the exact exchange energy is added, and in the fifth rung the unoccupied (virtual) orbitals are considered.[32,33] It is worth to note that according to hybrid functional approximations, the portion of exact exchange from the HF theory is incorporated while the remaining exchange-correlation energy comes from other sources (theoretical or experimental ones). The double hybrid functionals include a certain amount of HF exchange (similar to "single" hybrid functionals) and PT2 (the second-order perturbation theory) correlation.[34–37]

It is worth to note that the dispersion forces are important to describe properly numerous intermolecular interactions and that the dispersion corrections may be included in DFT approaches.[38] Particularly it concerns the ingredients included into functionals applied. The following dispersion-corrected DFT functionals are often used in numerous studies: B3LYP-3D, B2PLYP-D3, ωB97X-D, TPSS-d3 and others. The ωB97X-D functional is used in numerous studies that are presented in this book.

II.3 Basis Sets

For *ab initio* and DFT methods (as well as for semiempirical methods), the choice of a set of basis functions which express molecular orbitals is very important.[1,15,16] Such a set is usually named in various studies as a basis set. For example, in this book, the MP2/aug-cc-pVTZ and ωB97X-D/aug-cc-pVTZ results of calculations are presented most often. The first terms in the designations, MP2 and ωB97X-D, mark the MP2 perturbation theory method and the DFT functional applied, respectively, while the second term, aug-cc-pVTZ, labels the Dunning-style basis set.

The choice of the basis set is very important since, in connection with the method applied, it determines the quality of results of calculations. MP2/aug-cc-pVTZ and ωB97X-D/aug-cc-pVTZ levels of calculations are a good balance between a method and a basis set applied; the former MP2 method takes into account dynamic correlation, the ωB97X-D hybrid functional takes into account dispersion forces, while the large and sufficiently saturated Dunning-style basis set is often used in calculations and well reproduces experimental parameters.

The basis functions are usually related to atomic orbitals. The atomic orbitals may be represented by so-called Slater-type orbitals[39] or by linear combinations of more Slater-type orbitals. The Slater-type orbitals depend on the $\exp(-\xi r)$ exponential factor, where ξ and r correspond to the orbital exponent and the distance of the electron from the nucleus, respectively. Calculations with basis functions based on Slater-type orbitals are time consuming, thus Gaussian-type functions (GTFs) are usually applied to make calculations faster.[40] This is because the GTF depends on the $\exp(-\xi r^2)$ factor that makes it easier and faster to calculate many-centre integrals which are crucial in a great part of theoretical chemistry methods.

The basis sets based on GTFs are used by most theoretical chemistry programs (Gaussian,[41] Gamess,[42] and ORCA,[43] for example). The ADF program[44] is one of the programs that applies basis sets based on Slater-type orbitals. There are two most often used types of basis sets that are based on GTFs: Pople-type[45] and Dunning-style.[46] In recent years the latter is more often applied, while the former is not as popular as before. There are different reasons; one is that it is possible to achieve the basis set limit thanks to the construction of those basis sets. There are various other groups of basis sets;[47] one may mention the Sadlej-type basis sets[48–50] or Atomic Natural Orbital bases of Roos and coworkers.[51]

Table II.2 shows examples of the basis sets that are often applied in different studies; the classification inserted in this table follows that presented in the monograph of Ramachandran, Deepa and Namboori.[52] Let us discuss briefly various basis sets, starting from the simplest. In minimal basis sets, one basis function is attributed to each atomic orbital. For the STO-3G basis set, for example, the Slater-type orbital, STO, that is attributed to any atomic orbital considered, is replaced by three GTFs. The STO-6G

Table II.2 Different types of basis sets; classification based on that presented in ref. 52.

Type of basis set	Examples
Minimal basis sets	STO-3G, STO-4G, STO-6G, STO-3G*
Pople style	3-21G, 3-21G*, 3-21+G*, 6-31G, 6-311++G**, 6-311++G(3df,3pd)
Correlation consistent (Dunning style)	cc-pVDZ, aug-cc-pVDZ, aug-cc-pVTZ
Other split valence (generic names)	SV(P), SVP, DZV, TZV, TZVPP, QZVPP
Double, triple, and quadruple zeta	D95

is another example of a minimal basis set where six GTFs replace STO. The asterisk is usually applied in names of basis sets to indicate the occurrence of polarisation functions; the STO-3G* basis set is a polarised version of the STO-3G.

The Pople-style basis sets have characteristic designations which inform of their constructions. For example, 6-31G is the split-valence double-ξ (double-zeta), i.e., VDZ basis set where the inner shell atomic orbitals are represented by single basis functions that are linear combinations of six primitive Gaussians functions. For this basis set, each valence orbital is expressed by two basis functions, where one is a linear combination of three primitive Gaussian functions and the second of one primitive Gaussian function.

The diffuse functions are designated in the Pople-style basis sets by "plus", for example, the 6-311++G** basis set that was very popular and often applied several years ago in various studies. The single "+" corresponds to the diffuse functions of non-hydrogen atoms while the double "++" occurs if diffuse functions are also added to hydrogen atoms. The diffuse functions which are introduced to better describe the long-range tails of orbitals are often very useful in calculations concerning excited states, anions, and intermolecular interactions such as van der Waals or hydrogen bonds. In the latter case the diffuse functions added to the hydrogen atoms may be important.

The case of polarisation functions in the Pople-style basis sets should also be clarified. For example, the 6-311++G* basis set, which may also be labelled as 6-311++G(d), is characterised by additional d-type Gaussian polarisation functions for atoms from Li to Ca and additional f-type Gaussian polarisation functions for atoms from Sc to Zn.

In the case of 6-311++G**, which is also labelled as 6-311++G(d,p), the additional three p-type Gaussian polarisation functions are applied for the hydrogen and helium atoms. More extended basis sets are also applied that contain additional polarisation functions. The 6-311++G(3df,3pd) basis set, for example, is characterised by additional d and f polarisation functions at the non-hydrogen centres and p and d polarisation functions at H-atoms.

The Dunning-style basis sets use the correlation functions that allows one to achieve the convergence of the electronic energy for the complete basis set limit (see next section of this chapter). For example, the cc-pVnZ basis sets are often used in calculations (they are also labelled as cc-pVXZ where X is the cardinal number). The basis sets which are often applied are characterised by n equal to 2, 3 and 4, that is D, T and Q, respectively. These are the cc-pVDZ, cc-pVTZ or cc-pVQZ, i.e., the correlation-consistent polarised basis sets that are double-ξ (double-zeta — DZ), triple-ξ (triple-zeta — TZ) and quadruple-ξ (quadruple-zeta — QZ), respectively. The calculations with basis sets of higher n are also performed where the number of GTFs increases in a regular way with the increase of n, i.e., cc-pV5Z, cc-pV6Z basis sets and so on. This regularity is very important to make the complete basis set calculations. The diffuse functions may be added to the Dunning-style basis sets that correspond to augmented sets, aug-cc-pVDZ, aug-cc-pVTZ, aug-cc-pVQZ and so on. The Pople and Dunning basis sets may be compared as concerns the accuracy of calculations; the cc-pVDZ basis set is comparable to the Pople-style 6-31G** basis set,[53] and the aug-cc-pVDZ to 6-311++G(d,p). However, the results of the latter Pople-style basis set are often superior to those of the corresponding Dunning-style, surprisingly.[54]

In general, the use of more functions in basis sets usually improves results of calculations. However, it does not always work as in the example of the latter pair of basis sets above; the aug-cc-pVDZ contains more basis functions than 6-311++G**. Some information about split-valence basis sets is given below. A split-valence (SV) basis set uses two or more basis functions for valence-shell atomic orbitals and one function for each inner-shell atomic orbital. For the valence double-zeta (VDZ) and valence triple-zeta (VTZ) basis sets, the number of basis functions amounts to 9 and 13, respectively, for first-period atoms of the periodic table (H and He atoms are excluded in the period numeration). The D95 class of basis sets (Table II.2) are those introduced by Dunning and coworkers that contain diffuse and polarisation functions.[52,55]

II.4 The Complete Basis Set Limit

The improvement of results is observed with the increase of the basis set size. It occurs also for the series of the Dunning style, i.e., correlation-consistent polarised basis sets. Two such series, cc-pVnZ and aug-cc-pVnZ, were mentioned in the former section. These series are characterised by a regular size increase with the increase of the cardinal number n (X mark is also often applied), and consequently by a regular change of energetic or geometrical parameters resulting from the application of these basis sets for the same approach chosen (HF, MP2 or any functional of the DFT method). Different extrapolations to the so-called complete basis set (CBS) limit ($n \rightarrow \infty$ in the case of the Dunning-style series of basis sets) were proposed in various studies.[56–61] One can mention the exponential expression of extrapolation proposed by Feller:[56]

$$E_R(n) = E_R(\infty) + Ae^{-Bn} \tag{II.7}$$

or that proposed by Helgaker and co-workers:[57]

$$E_R(n) = E_R(\infty) + Cn^{-3} \tag{II.8}$$

In the above equations, n designates the cardinal number in the cc-pVnZ or aug-cc-pVnZ series, and $E_R(\infty)$ stands for energy of a species for the basis set limit ($n \rightarrow \infty$). The $E_R(n)$ is an energy of this species for the basis set characterised by the cardinal number equal to n. A, B and C are constants that are to be determined. $E_R(\infty)$ is also a variable that has to be determined. Hence for expressions II.7 and II.8, $E_R(n)$ values for 3 and 2 cardinal numbers are needed, respectively.

Table II.3 shows MP2 results for various dimers. The Helgaker extrapolation[57] to the CBS limit was applied here. These are results taken from the study concerning interactions in various complexes.[62] However, geometries of these complexes to calculate the interaction energies were taken from earlier calculations of Hobza and co-workers.[63] The CBS results presented in Table II.3 result from extrapolations based on the aug-cc-pVTZ and aug-cc-pVQZ basis sets.[62] The complexes presented in this table are mainly characterised by links that may be classified as hydrogen bonds: the ammonia dimer by N-H···N links, water and formic acid dimers by O-H···O interactions, and formamide dimers by N-H···O hydrogen bonds. The T-shaped benzene dimer is linked by the C-H···π hydrogen bond. Another stacking conformation of the benzene dimer is presented in the table, marked as PD that designates the parallel displaced

Table II.3 The interaction energies for dimers linked by hydrogen bonds (except of the parallel displaced conformation, PD, of benzene dimer where H-bonds do not occur). These are MP2 results taken from ref. 62 where different Dunning-style basis sets were applied. Results for the CBS limit are presented, energies in kcal/mol.

Structure	aug-cc-pVDZ	aug-cc-pVTZ	aug-cc-pVQZ	CBS limit
$(NH_3)_2$	-2.68	-2.99	-3.09	-3.16
$(H_2O)_2$	-4.37	-4.69	-4.86	-4.98
$(HCOOH)_2$	-15.99	-17.55	-18.14	-18.57
$(HCONH_2)_2$	-13.95	-15.03	-15.50	-15.84
$(CH_4)_2$	-0.39	-0.46	-0.48	-0.49
$(C_6H_6)_2$ PD	-4.25	-4.70	-4.85	-4.96
$(C_6H_6)_2$ T	-3.10	-3.46	-3.56	-3.63

PD — parallel displaced conformation
T — T-shaped conformation

conformation. The methane dimer presented in Table II.3 is linked by van der Waals interactions (these interactions were discussed in Chapter I). One can see from this table that the extension of the basis set (choosing a more saturated basis set that follows the increase of the cardinal number) makes the results, i.e., interaction energies in this case, closer to the CBS limit.

It is worth mentioning that there are at least several other schemes of extrapolations to the CBS limit. And there are various treatments of results; for example, the MP2 interaction energy (or any other energy that results from the post-HF method) contains the HF (SCF) interaction energy contribution and the correlation energy (dynamic correlation is taken into account in the case of MP2 method). It was pointed out that the convergence of the SCF and correlation energies differ, thus it is better to separate these energies in the process of the extrapolation to the CBS limit.[64] Hence different schemes are often applied separately for HF and correlation energies.

II.5 Interaction and Binding Energies

The interaction and binding energies were briefly described in Chapter I, eqs. I.1 and I.2, respectively. The deformation energy was also defined by eq. I.2. However, the inconsistency of basis sets used in calculations of the above energies should also be taken into account. According to the supermolecular approach,[65,66] the interaction energy, say between A and B units in the A···B system, that includes the influence of basis sets used, may be defined in the following way.

$$E_{int} = E_{A...B}(A...B)^{A\cup B} - E_{A...B}(A)^{A} - E_{A...B}(B)^{B} \quad \text{(II.9)}$$

The designations in parentheses indicate systems which energies are presented, the superscripts correspond to basis sets used in calculations, and the subscripts inform of systems for which geometries were optimised. One can see that the interaction energy is the difference between the energy of the A···B complex for which the geometry was optimised with the A∪B basis set of the complex and the energies of A and B units calculated with A and B basis sets, respectively. The geometries of A and B are taken from the optimised geometry of the complex. However, the interaction energy does not consider deformations of units which result from the formation of the complex.

The binding energy (eq. II.10) includes the deformation energy, E_{def}, which expresses "an energetic cost" of such geometry deformations.

$$E_{bin} = E_{int} + E_{def} = E_{A...B}(A...B)^{A\cup B} - E_{A}(A)^{A} - E_{B}(B)^{B} \quad \text{(II.10)}$$

The species used in calculation of the interaction energy that have geometries taken from the complex are not in energetic minima, thus the deformation energy is positive. One can see that for the binding energy expressed by eq. II.10, similar to the interaction energy expressed by eq. II.9, there is inconsistency of the basis sets used for the complex and for its interacting units.

This inconsistency may be explained in the following way. In the A···B complex, the B basis functions are available for the A unit and the A basis functions are available for the B unit. It does not occur for the separated A and B units where the basis sets used in calculations are not as extended as in the complex. The basis set inconsistency leads to inaccuracy of the interaction and binding energies expressed by eqs. II.9 and II.10, respectively. To reduce this inaccuracy, all energy terms of the latter equations should be calculated in the same extended basis set (containing both A and B functions). The difference between the interaction (or binding) energy calculated for all units with the consistent (extended) basis sets and the interaction (or binding) energy calculated with inconsistent basis sets described above is known as the Basis Set Superposition Error

(BSSE). A positive BSSE value is observed, and it decreases if the applied basis set is enlarged. It should be equal to zero in the CBS limit. It is worth mentioning that it is difficult or even impossible to apply large basis sets for large moieties like clusters or bio-chemically active systems, for example. Hence the evaluation of the BSSE correction is often needed to achieve results of relatively high accuracy. The Counterpoise Correction is used most often to calculate BSSE (eq. II.11)[67,68] and it exactly corresponds to the definition (energy difference) mentioned above.

$$\Delta E_{CP} = E_{A\ldots B}(A)^{A} + E_{A\ldots B}(B)^{B} - E_{A\ldots B}(A)^{A\cup B} - E_{A\ldots B}(B)^{A\cup B} \qquad \text{(II.11)}$$

Table II.4 shows the energetic parameters discussed above; the BSSE correction is also included for complexes linked by various kinds of σ-hole bonds.[69] These are the tetrel, pnicogen, chalcogen and halogen bonds where acetylene molecule or chloride anion plays the role of the Lewis base unit. Chloride is a stronger Lewis base than acetylene, thus the chloride complexes are linked by stronger interactions than the corresponding acetylene systems. The strongest interactions occur for complexes of the BrF Lewis acid moiety. The interaction energy for the BrF complex with the chloride anion is equal to –47.0 kcal/mol (the binding energy amounts to –38.5 kcal/mol) and the interaction energy for the BrF complex with acetylene is –7.2 kcal/mol (the binding energy is –6.8 kcal/mol).

The acetylene complexes are characterised by lower values of the deformation energy than their chloride counterparts. The greatest value of the deformation energy for the acetylene complexes is observed for the T-shaped F-Br···C_2H_2 complex, 0.4 kcal/mol. It means that the complexation does not change geometries of acetylene and linked Lewis acid units or these changes are negligible. The BSSE correction for acetylene systems does not exceed 2 kcal/mol.

Similar BSSE values are observed for the chloride complexes, also lower than 2 kcal/mol but sometimes these values are slightly greater than those of acetylene complexes. However, the deformation energies for the chloride complexes are large when compared with those of acetylene systems. They are greater than 4 kcal/mol and even in a few cases greater than 10 kcal/mol; the greatest value of 13.1 kcal/mol is observed for the F-Cl···Cl^- complex. This means that the geometries of interacting species, the Lewis acid units in the case of chloride complexes, change significantly because of

Table II.4 The energetic parameters (in kcal/mol): the interaction energy, E_{int}, and the binding energy, E_{bin}, both corrected for BSSE (which is also included in this table separately) and the deformation energy, E_{def}. For the acetylene complexes, T-shaped configurations are considered where the C_2H_2 unit is perpendicular or nearly so to the Lewis acid unit bond. These are MP2/aug-cc-pVTZ results taken from ref. 69.

Lewis acid	E_{bin}	E_{int}	BSSE	E_{def}
	T-shaped C_2H_2 Lewis base			
$SiFH_3$	-2.3	-2.4	0.6	0.0
PFH_2	-3.6	-3.7	0.7	0.0
SFH	-4.7	-4.8	0.7	0.1
ClF	-4.4	-4.5	0.6	0.2
$GeFH_3$	-2.5	-2.6	1.0	0.1
$AsFH_2$	-4.0	-4.1	1.5	0.1
SeFH	-6.1	-6.4	1.8	0.3
BrF	-6.2	-6.6	1.6	0.3
$GeFH_3$*	-2.6	-2.6	1.1	0.1
$AsFH_2$*	-4.1	-4.2	1.5	0.1
SeFH*	-6.5	-6.8	1.9	0.3
BrF*	-6.8	-7.2	1.6	0.4
	Cl^- Lewis base			
$SiFH_3$	-19.8	-32.6	1.3	12.8
PFH_2	-19.6	-23.6	1.1	4.0
SFH	-24.2	-31.3	1.1	7.2
ClF	-31.9	-45.0	1.2	13.1
$GeFH_3$	-22.5	-32.6	1.7	10.1
$AsFH_2$	-24.9	-29.4	1.8	4.5
SeFH	-30.5	-36.7	1.7	6.2
BrF	-37.6	-46.3	1.7	8.7
$GeFH_3$*	-22.3	-31.7	1.7	9.4
$AsFH_2$*	-25.1	-29.6	1.7	4.5
SeFH*	-31.0	-37.2	1.7	6.2
BrF*	-38.5	-47.0	1.7	8.5

*relativistic effects taken into account for heavier Ge, As, Se and Br elements.

complexation. These changes concern mainly elongations of bonds to the halogen Lewis acid centres in halogen-bonded complexes or the deformation of tetrahedral conformation of the Lewis acid units that becomes closer to the trigonal bipyramid structure (these changes are discussed in detail in Chapter VI). Table II.4 includes, for comparison, results where relativistic effects were taken into account for the heavier Lewis acid centres. Notable differences occur between these results and those where relativistic effects were not taken into account; however, they are not significant.

Table II.5 presents the binding energies for two conformations of the benzene dimer,[70] the T-shaped conformation possessing C_{2v} symmetry and the conformation characterised by C_s symmetry; the latter may be treated as a slightly distorted T-shaped structure. For the former conformation the C-H···π hydrogen bond links benzene molecules; for the latter conformation the same type C-H···π hydrogen bond or C-H···C interaction may be considered since the C-H proton donating bond is closer to one of the carbon centres of the Lewis base unit. The MP2 results with different Dunning-style aug-cc-pVnZ basis sets are presented in this table; the cardinal number n for basis sets changes from 2 to 5. One can see that the BSSE correction decreases with the increase of the size of basis set, i.e., the increase of cardinal number. It is about 4 kcal/mol for the aug-cc-pVDZ basis set and only 0.4 kcal/mol for the aug-cc-pV5Z basis set. The CBS limit results for the binding energies of both conformations are presented. The CBS limit binding energy was evaluated by fitting binding energies at increasing basis set size to the double exponential function expressed by eq. II.12.[71]

$$E_R(n) = E_R(\infty) + Ae^{-(n-1)} + Be^{-(n-1)^2} \tag{II.12}$$

Table II.5 The binding energy, E_{bin}, and this energy corrected for BSSE, E_{bin}BSSE, taken from ref. 70. The BSSE values and CBS limit values are included, all in kcal/mol. Results for C_{2v} and C_S conformations of the benzene dimer are given.

	C_{2v}			C_S		
Basis set	E_{int}	E_{int}**BSSE**	**BSSE**	E_{int}	E_{int}**BSSE**	**BSSE**
aug-cc-pVDZ	-6.80	-2.85	3.95	-7.00	-2.99	4.01
aug-cc-pVTZ	-4.96	-3.40	1.56	-5.11	-3.55	1.56
aug-cc-pVQZ	-4.08	-3.55	0.53	-4.25	-3.70	0.55
aug-cc-pV5Z	-4.08	-3.69	0.39	-4.32	-3.89	0.43
CBS	-3.77			-3.98		

n designates the cardinal number in the aug-cc-pVnZ series, $E_R(\infty)$ expresses energy of a species for the basis set limit (n → ∞), while $E_R(n)$ is energy of this species for the basis set with n cardinal number.

Table II.5 shows that the BSSE corrections are needed to improve accuracy of results, especially for non-extended basis sets. However, despite such corrections they are often still far from accurate. For example, for the T-shaped dimer of benzene, for the MP2/aug-cc-pVDZ calculations, the binding energy corrected for BSSE is equal to –2.85 kcal/mol while the CBS energy is equal to –3.77 kcal/mol.[70]

It is worth mentioning that in the Boys and Bernardi approach, the system is optimised and then the CP correction (eq. II.11) is calculated.[67] This may lead to intermolecular distances which are too short. In the other approach proposed several years ago the counterpoise correction is considered during optimisations.[72–75]

One can see from descriptions of the binding and interaction energies presented so far that there are two effects which influence these energies. One of the effects, the deformation energy, E_{def}, is of physical nature since it expresses the "lost" stabilisation energy due to the complexation process. E_{def} is positive and it increases with the increase of geometrical changes accompanying complexation. This is why it increases for the stronger interactions, which may be treated as preliminary stages of chemical reactions. The other effect, BSSE, may be treated as of technical nature since it is related to the choice of the level of calculations — the method and basis set.

It is worth mentioning that methods of calculations lead to energies that refer to molecules in their ground states and in 0 K. This approach does not take into account vibrations of a system. This is why another correction should be included in calculations, the so-called zero-point vibrational energy (ZPVE).[76] The electronic energy of a molecule of n vibrational degrees of freedom is corrected by the following expression:

$$E_{ZPVE} = \frac{1}{2}\sum_{i=1}^{n} h\nu_i \qquad \text{(II.13)}$$

The ZPVE should be considered for all species where it is needed to calculate the interaction or binding energy; ν_i is the frequency of ith vibration mode.

More complicated expressions to calculate binding and interaction energies as well as the BSSE are needed in the case of larger systems containing more than two units, like clusters, hydration shells of cations, or in an analysis of cooperativity effects, for example. The corresponding relationships are described in detail in other studies.[18] For example the binding energy of the system that contains three monomers, i.e., a triad, may be expressed by eq. II.14.

$$E_{bin} = E_{int} + E_{def} = E_{A\ldots B\ldots C}(A\ldots B\ldots C)^{A\cup B\cup C} - E_{A}(A)^{A} - E_{B}(B)^{B} - E_{C}(C)^{C} \qquad \text{(II.14)}$$

The above expression on binding energy includes the deformation energy which expresses the change of geometries of monomers that results from the process of complexation. The Counterpoise Correction for this binding energy as well as for the corresponding interaction energy which does not contain the deformation energy term may be expressed by eq. II.15, by analogy to eq. II.11.

$$\begin{aligned}\Delta E_{CP} = {} & E_{A\ldots B\ldots C}(A)^{A} + E_{A\ldots B\ldots C}(B)^{B} + E_{A\ldots B\ldots C}(C)^{C} \\ & - E_{A\ldots B\ldots C}(A)^{A\cup B\cup C} - E_{A\ldots B\ldots C}(B)^{A\cup B\cup C} - E_{A\ldots B\ldots C}(C)^{A\cup B\cup C}\end{aligned} \qquad \text{(II.15)}$$

In general, for a system that consists of n units (designated as U-units: U_1, $U_2 \cdots U_n$) that may be marked as the $U_1 \cdots U_n$ system, the binding energy and its BSSE correction may be expressed by eqs. II.16 and II.17, respectively.

$$E_{bin} = E_{int} + E_{def} = E_{U_1\cdots U_n}(U_1 \cdots U_n)^{U_1\cdots U_n} - \sum_{i=1}^{n} E_{U_i}(U_i)^{U_i} \qquad \text{(II.16)}$$

$$\Delta E_{CP} = \sum_{i=1}^{n} \left[E_{U_1\cdots U_n}(U_i)^{U_i} - (E_{U_1\cdots U_n}(U_i)^{U_1\cdots U_n} \right] \qquad \text{(II.17)}$$

It is worth mentioning that other expressions for the energies discussed above for systems containing more than two monomers are used in various studies. Sometimes, dyad properties of more complex systems are considered in calculations of energetic parameters since the large system is divided into two units depending on the intermolecular contact that is analysed.

There are also other energetic properties analysed in various studies; for example, an energetic cooperativity parameter, E_{coop}, may be considered. For a system containing three units, this parameter is expressed by eq. II.18.[77]

$$E_{coop} = E(A\ldots B\ldots C) - E(A\ldots B) - E(B\ldots C) - E(A\ldots C) \qquad \text{(II.18)}$$

If one considers this triad as a "chain system" where A is in contact with B and B with C, but A and C are not connected directly, one can understand the terms of eq. II.18 in the following way: *E*(A…B…C) is the interaction energy of this triad, *E*(A…B) and *E*(B…C) are the interaction energies of the isolated dyads within their corresponding energetic minima configurations, and *E*(A…C) is the interaction energy of A and C units in the geometry they have in the triad.

Let us consider an example of dimers and complexes of dichlorogermylene and dibromogermylene with one and with two molecules of 1,4-dioxide.[78] Table II.6 presents the binding and interaction energies, E_{bin} and E_{int}, the deformation energy, E_{def}, and the BSSE correction. The E_{bin} and E_{int} terms are corrected for BSSE. The BSSE correction is large, approximately between 5 kcal/mol and 10 kcal/mol, despite a saturated basis set being applied, the triple-ξ (triple-zeta — TZ) aug-cc-pVTZ basis set.

The absolute E_{int} values for the $(GeCl_2)_2$ and $(GeBr_2)_2$ dimers exceed 30 kcal/mol, thus these are strong interactions. These interactions may be classified as π-hole tetrel bonds. Scheme II.2 shows the probable Lewis acid and Lewis base sites for dihalometallylenes. The σ-holes and π-holes of the tetrel centre may act as the Lewis acid sites while the valence electrons not involved in bond formation may be the Lewis base regions, particularly in the case of singlet configurations of tetrel centres where these are lone electron pairs. However, it was found that the heavier divalent tetrel centres in dihalides possess mainly Lewis acid properties thanks to the π-holes (see Scheme II.2).[78] The

Table II.6 MP2/aug-cc-pVTZ results of calculations (in kcal/mol) taken from ref. 78. E_{int} – interaction energy corrected for BSSE, E_{bin} – binding energy corrected for BSSE, E_{def} – deformation energy, BSSE – basis set superposition error correction.

Species	E_{int}	E_{bin}	E_{def}	BSSE
$(GeCl_2)_2$	-32.5	-15.8	16.7	6.7
$(GeBr_2)_2$	-32.0	-18.2	13.8	10.0
$GeCl_2$-$C_4H_8O_2$	-19.5	-17.4	2.0	6.8
$GeBr_2$- $C_4H_8O_2$	-19.3	-17.2	2.1	9.4
$GeCl_2$- $(C_4H_8O_2)_2$	-11.9	-9.2	2.7	4.9
$GeBr_2$- $(C_4H_8O_2)_2$	-11.9	-9.1	2.8	7.1

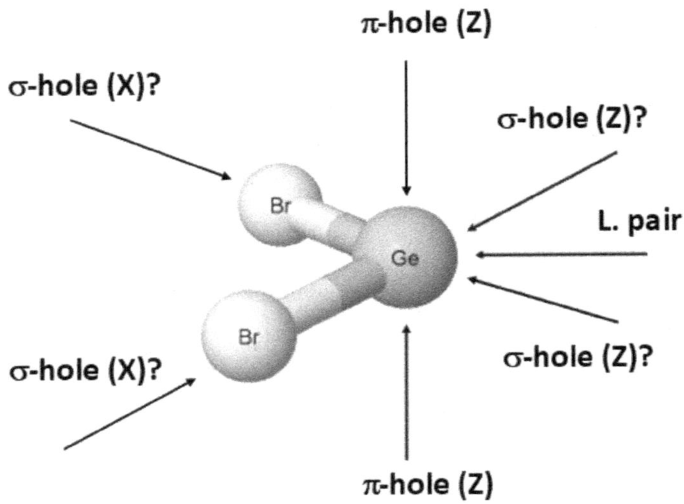

Scheme II.2 The Lewis acid and Lewis base sites of the $GeBr_2$ species. Reprinted with permission from ref. 78. Copyright 2022 S.J. Grabowski. Licensee MDPI, Basel, Switzerland.

Lewis base properties due to lone electron pairs appear for lighter carbon and silicon tetrels in divalent species.

It is interesting that the bromine centres in $GeBr_2$ species presented here, similar to other bromine atoms discussed in Chapter I, possess dual character. They may act as Lewis acid sites thanks to the σ-holes located in elongations of Ge-Br bonds, while rings around Br centres are characterised by negative electrostatic potential, thus they may act as Lewis base sites. Hence the $(GeCl_2)_2$ and $(GeBr_2)_2$ presented in Table II.6 are linked through two Ge···Br contacts (see Fig. II.1a), where bromine acts as the Lewis base through the negative electrostatic potential region while germanium acts as the Lewis acid through the π-hole. The absolute values of binding energies for these dimers are about two times smaller than corresponding interaction energies, because of large deformation energies.

Figure II.1b presents the complex of dibromogermylene with one 1,4-dioxide molecule while Figure II.1c presents the complex of dichlorogermylene with two 1,4-dioxide molecules. These complexes are linked by the Ge···O π-hole tetrel bonds since similar Ge···O links are observed for the complexes of dichlorogermylene and dibromogermylene.

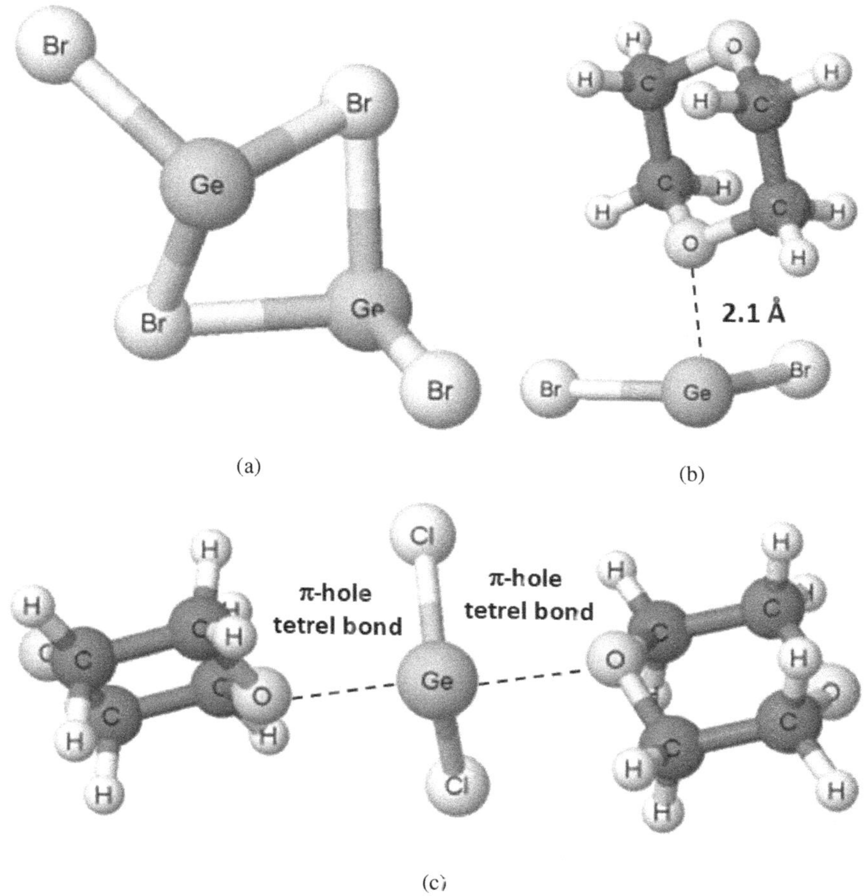

Fig. II.1 (a) The $(GeBr_2)_2$ dimer. (b) The complex of dibromogermylene with one 1,4-dioxide molecule. (c) The complex of dichlorogermylene with two 1,4-dioxide molecules.

The results in Table II.6 show that the absolute values of interaction energies for complexes containing one 1,4-dioxide molecule are close to 20 kcal/mol while those containing two 1,4-dioxide molecules are about 12 kcal/mol. The GeX_2-$C_4H_8O_2$ (X = Cl, Br) complexes are treated as consisting of GeX_2 and $C_4H_8O_2$ units; hence eqs. II.9 and II.10 are applied here to calculate the interaction and binding energies, respectively. In the GeX_2-$(C_4H_8O_2)_2$ complexes two units are also considered: GeX_2-$C_4H_8O_2$ and $C_4H_8O_2$. Thus, the same eqs. II.9 and II.10 are applied here to calculate these energies, and not eq. II.14 which is applied for a triad. The O···Ge interactions are weaker in the GeX_2-$(C_4H_8O_2)_2$ complexes than in those of GeX_2-$C_4H_8O_2$. This is because the Lewis acid properties of the germanium centre in GeX_2-$C_4H_8O_2$ are weaker than in the GeX_2 species. In the former system the electron charge was transferred from $C_4H_8O_2$ into the germanium dihalide that weakens electrophilic properties of the Ge-centre. The BSSE values are still large for the complexes of 1,4-dioxide molecules and the deformation energies are 2–3 kcal/mol.

II.6 The Interaction Energy Partitioning for Hydrogen Bonds and Other Interactions

The partitioning of energy of interaction is a subject of various analyses. It allows us to deepen understanding of the nature of various intermolecular links. Such partitioning was briefly described in the former chapter, and it was expressed by eq. I.13. Roughly speaking, according to this equation, the decomposition schemes usually contain the following "attractive" interaction energy terms: electrostatic, polarisation, and dispersion as well as the "repulsive" exchange interaction energy term. These names of the interaction energy are used throughout this book, unless specific decomposition schemes are referred to. The "attractive" interaction energy terms are referred to as stabilisation of the system considered, and it is most often assumed in expressions on interaction energies that they have negative signs. Hence, consequently, the "repulsive" exchange energy term has a positive sign.

The variational approach of Kitaura and Morokuma[79,80] is one of the first proposed decomposition schemes. It seems that this scheme was also one of the most often applied in the last century. In the Kitaura-Morokuma approach the HF interaction energy is decomposed into the following components (eq. II.19).

$$\Delta E_{HF} = \mathrm{EX} + \mathrm{ES} + \mathrm{PL} + \mathrm{CT} + \mathrm{MIX} \qquad \text{(II.19)}$$

The ES term is the electrostatic interaction energy, PL designates the polarisation interaction energy that is related in this decomposition scheme to electron charge shifts being results of complexation within monomers, and the CT term is the energy of electron charge transfer between two units constituting the complex. Most often, the ES, CT and PL terms are attractive, while the exchange energy, EX, is the repulsive term that results from antisymmetrisation of the wave function. Physically, EX may be related to electron cloud repulsion. The MIX term is the energy difference between SCF interaction energy, ΔE_{HF}, and the sum of ES, PL, CT and EX contributions. This term results from the incomplete partitioning of the total HF energy.

The partitioning expressed by eq. II.19 concerns the HF energy, thus it does not contain the correlation energy term. However, the correlation energy may be calculated as a difference between the energy calculated by the post-HF method, where correlation is taken into account, and the SCF energy. The accuracy of the correlation energy depends on the post-HF method and the basis set applied. The dispersion energy often related to van der Waals interaction in different studies (see explanations in section I.2 of Chapter I) is usually the most important attractive interaction energy that contributes to the correlation energy.

One of the variants of energy decomposition analysis (EDA) which follows the Kitaura-Morokuma scheme is that implemented in the ADF program.[44,81] In this variant,[82,83] the electron charge shifts resulting from complexation are not separated into those related to transfers between interacting units and those within units. Thus there is no separation into the related CT and PL terms. Besides, it was discussed in other studies that these two terms cannot be well separated[84] and that their percentage contributions to the total attractive interaction energy depend strongly on the level of calculation, particularly on the basis set used.[85–87]

The deficiency of the Kitaura-Morokuma approach, that it does not contain the dispersion energy term, may be cancelled in the ADF version of EDA. However, in the ADF set of codes it is possible to perform DFT calculations, not *ab initio* ones. Hence the dispersion energy term may be included if the proper functional is applied. The interaction energy, ΔE_{int}, in the EDA-ADF scheme is divided into three main components and the additional dispersion term, ΔE_{disp} (eq. II. 20).

$$\Delta E_{int} = \Delta E_{Pauli} + \Delta E_{elstat} + \Delta E_{orb} + \Delta E_{disp} \qquad \text{(II.20)}$$

The term ΔE_{Pauli} labels the Pauli repulsion interaction energy. ΔE_{elstat} corresponds to the electrostatic interaction between the unperturbed charge distributions of atoms. The orbital interaction, ΔE_{orb}, concerns charge transfer and polarisation effects, thus it roughly corresponds to the sum of CT and PL terms in the Kitaura-Morokuma approach.

The Symmetry-Adapted Perturbation Theory (SAPT) is one of the most often applied theoretical approaches in recent decades.[88,89] It is very important that the SAPT approach is applied to perform calculations on interaction energy of two closed-shell moieties that is calculated directly as a sum of defined terms. Thus, it differs from other approaches, including those which are classified as decomposition schemes, where the interaction energy is calculated as a difference between the energy of the complex and the sum of energies of monomers.

In the SAPT approach, the energy of interaction is calculated as the sum of the following terms: the first-order electrostatic interaction energy ($E_{elst}^{(1)}$), second-order induction ($E_{ind}^{(2)}$), dispersion contribution ($E_{disp}^{(2)}$), and their exchange counterparts: first-order exchange ($E_{exch}^{(1)}$), second-order exchange-induction ($E_{exch\text{-}ind}^{(2)}$) and exchange-dispersion ($E_{exch\text{-}disp}^{(2)}$). The SAPT approach contains the energy of interaction up to the second order. In order to take into account higher-order induction and exchange-induction energies, the HF "delta" correction term δE_{HF} is included. The so-called SAPT2 interaction energy is calculated according to eq. II.21.

$$E_{int}^{SAPT2} = E_{exch}^{(1)} + E_{elst}^{(1)} + E_{ind}^{(2)} + E_{disp}^{(2)} + E_{exch-ind}^{(2)} + E_{exch-disp}^{(2)} + \delta E_{HF} \qquad \text{(II.21)}$$

In the case of the truncation of the SAPT energy expressed by eq. II.21, the following grouping of energy contributions are performed in numerous studies:

$$E_{int}^{SAPT2} = E_{exch} + E_{elst} + E_{ind} + E_{disp} \qquad \text{(II.22)}$$

The following sums occur in the above equation: $E_{exch} = E_{exch}^{(1)}$, $E_{elst} = E_{elst}^{(1)}$, $E_{ind} = E_{ind}^{(2)} + E_{exch-ind}^{(2)} + \delta E_{HF}$, $E_{disp} = E_{disp}^{(2)} + E_{exch-disp}^{(2)}$. The grouping allows us to obtain energetic terms that are closer in interpretation to terms expressed by other approaches (see eqs. II.19 and II.20).

Table II.7 presents the results of the SAPT2 calculations for selected carboxylic and thiocarboxylic acids whose geometries were optimised at the ωB97X-D/aug-cc-pVTZ level.[90] The abovementioned grouping of energetic terms (eq. II.22) was applied here. These are centrosymmetric dimers linked by a pair of equivalent O-H···O or S-H···O hydrogen bonds. Figure II.2 shows the structures of dimers of chlorodifluoroacetic acid and 2-hydroxythiobenzoic acid as examples; the H···O intermolecular distances

Table II.7 The interaction energy terms (in kcal/mol) that are defined by eq. II.22 for carboxylic and thiocarboxylic acid dimers. Results are based on those of ref. 90.

Dimer	E_{exch}	E_{elst}	E_{ind}	E_{disp}	E_{int}^{SAPT2}
chlorodifluorothioacetic acid	15.0	-11.1	-5.0	-5.6	-6.7
2-hydroxythiobenzoic acid	15.3	-10.6	-4.7	-5.0	-6.0
4-methylthiobenzoic acid	17.3	-12.2	-5.6	-6.2	-6.8
thioacetic acid	16.9	-12.3	-5.5	-5.8	-6.7
thiobenzoic acid	17.2	-12.2	-5.6	-6.2	-6.9
chlorodifluoroacetic acid	41.2	-30.5	-19.8	-8.0	-17.0
acetic acid	45.1	-35.1	-20.0	-8.2	-18.2
2-hydroxybenzoic acid	43.1	-31.9	-19.1	-8.6	-16.6
benzoic acid	46.3	-35.5	-20.7	-8.8	-18.8
4-methylbenzoic acid	46.7	-35.8	-20.9	-8.8	-18.8

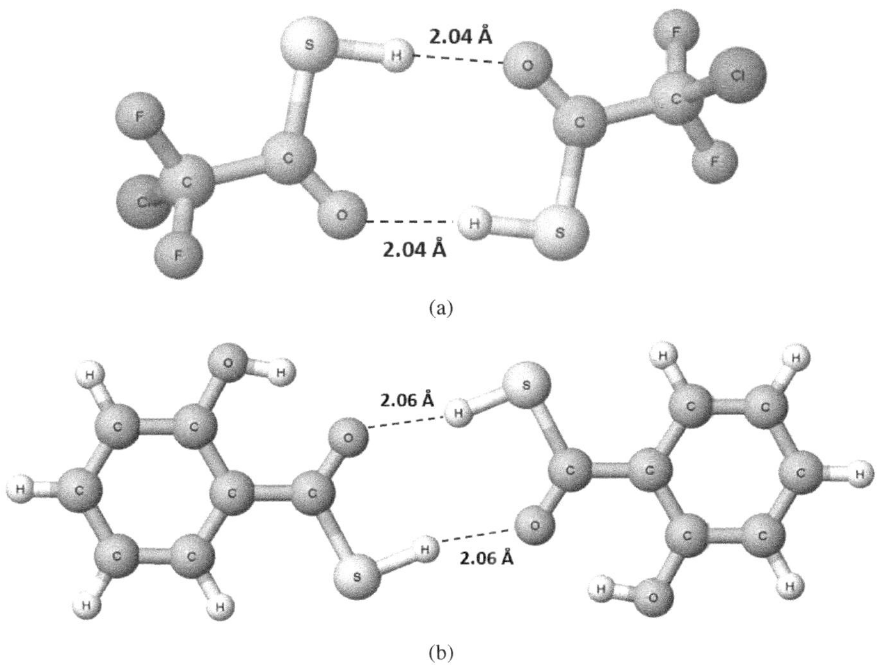

Fig. II.2 The structures of dimers of (a) chlorodifluoroacetic acid and (b) 2-hydroxythiobenzoic acid. They are discussed in ref. 90.

are stated. The results presented in the table are related to interaction energies that correspond to two hydrogen bonds in all structures. These results indicate that the O-H···O hydrogen bonds in carboxylic acids are more than two times stronger than S-H···O interactions in their thiocarboxylic acid counterparts since the absolute values of E_{int}^{SAPT2} are equal to 17–19 kcal/mol for the former species while for the latter dimers they are equal to 6–7 kcal/mol. In the case of thiocarboxylic acids the absolute values of electrostatic energy terms are approximately two times greater than the absolute values of induction terms, which are almost equivalent to the dispersion terms. In the case of stronger hydrogen bonds in carboxylic acids, the induction energy is more important (Table II.7); its absolute values are clearly greater (at least two times) than absolute values of the dispersion energy values, and the induction energy values are closer to the values of electrostatic energies than for the thiocarboxylic acids counterparts.

Table II.8 presents the BP86-D3/TZ2P//ωB97XD/aug-cc-pVTZ results of the decomposition of the energies of interaction (decomposition scheme expressed by eq. II.20) for complexes of imidazol-2-ylidene, NHC.[91] This means that for the complexes optimised at the ωB97XD/aug-cc-pVTZ level the interaction energies partitioning was performed at the BP86-D3/TZ2P level; relativistic effects for aug-cc-pVTZ and TZ2P basis sets were taken into account for iodine, aug-cc-pVTZ-PP and ZORA-

Table II.8 The total interaction energy, E_{int}, and their terms: the Pauli repulsion, E_{Pauli}, the electrostatic energy, E_{elstat}, the orbital-orbital interaction energy, E_{orb}, and the dispersion energy, E_{disp} for the NHC complexes (all in kcal/mol). The partitioning of energies of interactions was performed according to eq. II.20. The percentage contributions of the E_{elstat} and E_{orb} attractive interaction energy terms to the sum of all attractive terms are presented while the E_{disp} percentage contribution is calculated as $\%E_{disp} = 100\% - \%E_{elstat} - \%E_{orb}$. These are BP86-D3/TZ2P//ωB97XD/aug-cc-pVTZ results taken from ref. 91.

Lewis acid	E_{int}	E_{Pauli}	E_{elstat}	E_{orb}	E_{disp}	$\%E_{elstat}$	$\%E_{orb}$
ClCN	-7.12	10.24	-10.96	-5.44	-0.95	63.2	31.4
BrCN	-12.86	21.72	-21.06	-12.26	-1.27	60.9	35.4
ICN	-21.57	55.13	-46.50	-28.16	-2.05	60.6	36.7
Cl_2	-61.58	404.85	-224.66	-240.14	-1.63	48.2	51.5
Br_2	-45.61	260.50	-165.56	-138.46	-2.09	54.1	45.2
I_2	-32.85	146.30	-103.47	-73.10	-2.59	57.8	40.8

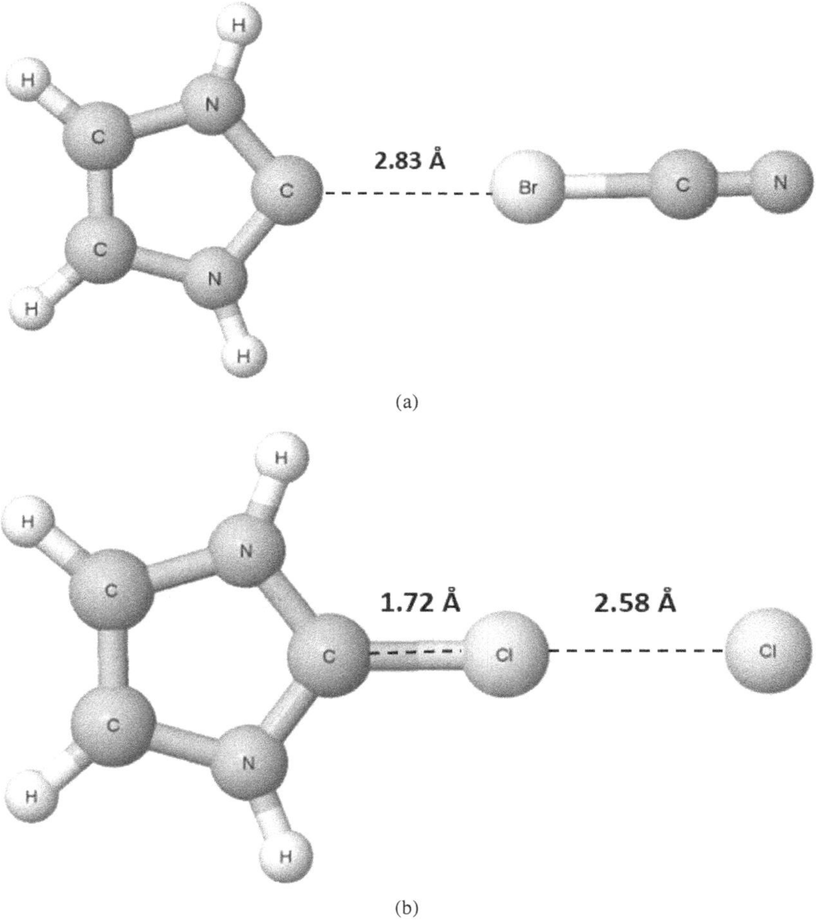

Fig. II.3 The complexes of imidazole-2-ylidene: (a) with BrCN (b) with Cl_2.

TZ2P, respectively. The complexes linked by halogen bonds have been analysed[91] and part of the results for these complexes, those where XCN and X_2 (X = Cl, Br and I) act as the Lewis acid units, is presented in Table II.8. Figure II.3 shows the NHC-BrCN and NHC-Cl_2 complexes.

Table II.8 shows that, for the complexes of XCN, the interaction is stronger for heavier halogen atoms. For example, the interaction energy for the complex of ICN amounts to –21.6 kcal/mol. This was discussed in other studies[84,87] (and Chapter 1), that for the series of related complexes linked by the same type of interaction (chalcogen bond, pnicogen bond, halogen bond, etc.), the strength of interaction increases with the increase of the atomic number of the Lewis acid centre. However, in the case of complexes of X_2, the strength of interaction decreases with the increase of the halogen atomic number (Table II.8). This is connected with the nature of C···X contacts for such complexes. Since they possess characteristics of covalent bonds, enormous interactions are observed for them; E_{int} is equal to -61.6 kcal/mol for the NHC-Cl_2 complex, for example. The NHC-Cl^+···Cl^- system should be considered for the latter since the partitioning into NHC and Cl_2 units leads to the interaction in the C-Cl bond. Hence for all X_2 complexes the orbital interaction energy, E_{orb}, related to covalency of interactions, is comparable, or even more important, than the electrostatic energy of interaction, E_{elstat}, because the partitioning into X_2 and NHC species leads to the calculation of the interaction energy for the C-X covalent bond. Table II.8 presents the percentage contributions of attractive interaction energy terms (related to sums of all attractive interactions, $E_{elstat} + E_{orb} + E_{disp}$). One can see that for the XCN complexes the electrostatic interaction is dominant, while for the X_2 complexes the electrostatic and orbital interactions are approximately equivalent.

II.7 The Natural Bond Orbital Method

The Natural Bond Orbital (NBO) method[92,93] is described in this section since this approach is often used to analyse different types of interactions. For example, the NBO of a σ bond between two A and B atoms, σ_{AB}, is a linear combination of directed orthogonal hybrids (natural hybrid orbitals) h_A and h_B.[92,93]

$$\sigma_{AB} = c_A h_A + c_B h_B \qquad \text{(II.23)}$$

These natural hybrid orbitals, h_A and h_B, consist of effective valence-shell atomic orbitals optimised for the corresponding wave function. The antibonding orbital, σ_{AB}^*, is constructed in the way expressed by eq. II.24.

$$\sigma_{AB}^* = c_B h_A - c_A h_B \qquad \text{(II.24)}$$

The percentage of the A (or B) hybrid orbital in the σ_{AB} bond orbital is expressed as $100|c_A|^2$ (or $100|c_B|^2$ for B hybrid orbital) and it is defined as the polarisation of the A-B bond. It may also be treated as the percentage of the electron density at the

A (or B) atom. $c_A{}^2 + c_B{}^2 = 1$ is in force, thus the percentage of the electron density at the B-atom in the A-B σ-bond is equal to $100|c_B|^2 = 100\% - 100|c_A|^2$.

The analysis of orbital-orbital overlaps resulting from the complexation, i.e., from interactions, is often performed in various studies. This is discussed later but now another issue is discussed that concerns the change of the form of orbitals due to interactions. This issue is not often discussed in studies on interactions. For example, BP86-D3/TZ2P calculations were performed for the water molecule and its simple complexes.[94] The NBO 6.0 program[95] implemented in the ADF set of codes[44,81] was used to calculate the NBO characteristics of these systems. For the water molecule, its O-H NBO is expressed by eq. II.25.

$$\sigma_{OH} = c_H h_H + c_O h_O = 0.5069\ s(H) + 0.8620\ sp^{3.35}(O) \qquad (II.25)$$

One can observe that the hydrogen centre natural hybrid orbital possesses purely s-character having the s(H) form. The oxygen $sp^{3.35}(O)$ hybrid orbital is mainly of the p-type. The coefficient of this hybrid, equal to 3.35, shows the 76.7% p-orbital contribution while 22.9% is of the s-orbital. The contributions from orbitals characterised by higher angular momentum quantum numbers, such as d or f, are meaningless; d-orbital contribution is equal to 0.4%. The values of c_H and c_O coefficients for the σ-orbital of this isolated water molecule inform of the strong O-H bond polarisation; the percentage of electron density at the oxygen centre is equal to 74.3% ($|c_O|^2 \cdot 100\%$, eq. II.25).

Let us consider changes of the O-H bond if it is involved in the O-H···O hydrogen bond in the dimer of water (Scheme II.3). It is very well known from early[96–99] and

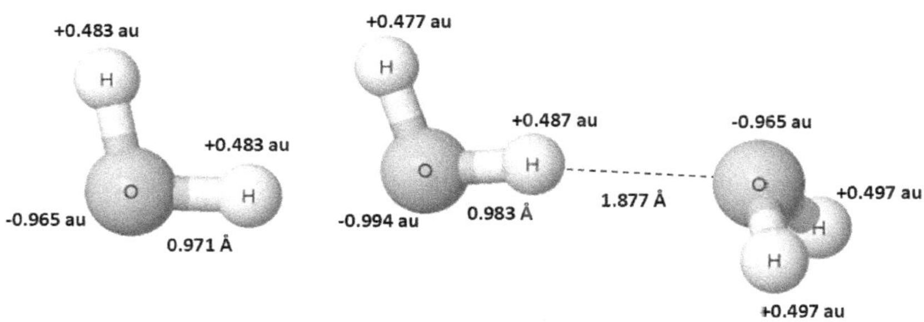

Scheme II.3 Water monomer and dimer, distances (Å) and NBO charges (au).

current studies[18] that the A-H···B hydrogen bond formation usually leads to A-H bond elongation and its weakening; the elongation relates to a shift of the corresponding stretching A-H bond frequency to lower values. Hence it is often named as a red-shift hydrogen bond. However, much less common cases of the A-H bond shortening also occur[100] where the strengthening of this bond is observed with the stretching frequency shift to higher values; this is known as a blue-shift hydrogen bond.

BP86-D3/TZ2P results of calculations show that the O-H bond length for isolated water is equal to 0.971 Å (both OH bonds are equivalent since the water molecule possesses C_{2v} symmetry). For the water dimer, for the same level of calculations, the O-H bond is elongated to 0.983 Å (Scheme II.3). The O-H NBO of this elongated bond is expressed by eq. II.26.

$$\sigma_{OH} = c_H h_H + c_O h_O = 0.4938\ s(H) + 0.8696\ sp^{2.87}(O) \qquad (II.26)$$

It shows that the hydrogen bond formation leads to the increase of s-character of the h_O oxygen hybrid orbital from 22.9% in the monomer to 25.7% in the dimer and the increase of polarisation of the O-H bond, from 74.3% to 75.6% (percentage of the electron density at the oxygen centre). In general, this was justified that the increase of s-character of the h_A hybrid orbital of the σ-orbital of the proton-donating A-H bond that results from the formation of the A-H···B hydrogen bond[101] agrees with the Bent rule.[102]

One can also see that the formation of the water dimer relates to the shift of the electron density from the Lewis base unit (water molecule at the right of the image in Scheme II.3) to the Lewis acid unit that contains the proton-donating O-H bond. This shift amounts to –0.029 au. The next arrangements of the electron density occur because of complexation, since in the isolated water monomer the net atomic charges of +0.483 and –0.965 at the hydrogen and oxygen centres, respectively, are observed for the O-H bond while in the water dimer for the proton-donating O-H bond the charges of +0.487 au and –0.994 au at these centres are observed. Thus, the charge at the H-centre being in contact with the Lewis base unit increases (becomes more positive). This is one of the common features of the hydrogen bond interactions connected with the rehybridisation process.[93]

The electron charge redistributions being a result of hydrogen bond formation were analysed in detail by Alabugin and co-workers.[101] Two factors controlling hydrogen bond linkage were indicated. The hyperconjugative interaction (charge transfer) from a lone pair of the proton acceptor (B) to the antibonding σ^* orbital of the A-H proton-donating bond is the first factor. This orbital-orbital interaction (overlap) is often designated as $n(B) \rightarrow \sigma_{A\text{-}H}{}^*$ and it is often treated as a signature of the occurrence of hydrogen bond interaction.[92,93]

This is also a key part of the Weinhold and Klein definition of the hydrogen bond:[103] "**Hydrogen bond:** A fractional chemical bond of coordinative A-H···B Lewis acid-base type, associated with the partial intermolecular A-H···:B ↔ A:···H-B resonance (3-center/4-electron proton-sharing) commonly originating in the $n_B \rightarrow \sigma_{AH}{}^*$ donor-acceptor interaction between the lone pair n_B of the Lewis base and the hydride antibonding $\sigma_{AH}{}^*$ of the Lewis acid."[103,104]

Summarising, $n_B \rightarrow \sigma_{AH}{}^*$ overlap leads to A-H bond weakening, its elongation and contaminant red shift for the corresponding stretching frequency. The second factor controlling hydrogen bond formation is related to the increase in s-character of A-atom hybrid orbital[101] which leads to strengthening of the A-H bond, its shortening and the corresponding blue shift of stretching frequency. This factor agrees with the Bent rule.[102] The Bent rule states that in the rehybridisation process, atoms maximise their s-character in hybrid orbitals directed towards electropositive substituents and they maximise their p-character towards electronegative substituents. It was described earlier and in other studies[18] that for most hydrogen bonds the redistribution of the electron charge density leads to increase of the positive charge of the hydrogen centre of the A-H proton-donating bond. Hence the hydrogen centre becomes more electropositive, which relates to increase of the s-character of A hybrid orbital and this is exactly in line with the Bent rule. It was stated that the red-shifted hydrogen bonds are observed if the hyperconjugation is superior; if the rehybridisation effects outweigh the former factor then blue-shifted hydrogen bonds occur. However, it was found that both factors are unseparable since, for example, the values of parameters related to both factors increase with the increase of the strength of hydrogen bond.[105]

Joseph and Jemmis[106] have also analysed two effects as main factors steering the hydrogen bond. These factors are also considered in terms of the electron charge distribution being the result of hydrogen bond formation. In the A-H···B system, for the A-H bond, A is more negative, and H is more positive than for the A-H bond of the isolated Lewis acid unit. One of the factors relates to the attraction of H-atom of the A-H bond by the electronegative B-centre that leads to the A-H bond lengthening. The second factor relates to A-H bond contraction being the result of the attraction of H-atom by the A negative centre.[106] Hence these factors are based on electrostatic forces and orbital-orbital overlaps, and orbital changes are not considered to explain geometrical changes of units involved in the hydrogen bond.

The changes of bond orbitals are observed not only for the hydrogen bond but also for other types of interactions, like σ-hole bonds,[107] for example. Tables II.9 and II.10 show the changes of different parameters, among them NBO ones, of the F_3CCl unit if it is involved in various types of interactions (see Scheme II.4). Since the chlorine centre possesses dual character, it may act simultaneously as the Lewis acid in the extension of the C-Cl bond through the σ-hole and as the Lewis base centre in the direction perpendicular to the C-Cl bond or nearly so, through the accumulation of the electron density in this direction at the Cl-centre. Hence the changes of NBO and other parameters of the F_3CCl unit due to complexation may differ significantly in various complexes since chlorine may act as the Lewis base or acid.

Table II.9 The chlorine — Lewis acid/base centre distance (Cl···A/B, in Å), C-Cl bond length (Å), C-Cl···B/A angle (degrees) as well as binding energy, E_{bin} (corrected for BSSE, in kcal/mol). Results taken from ref. 107.

Moiety	**Cl···B/A**	**C-Cl**	**C-Cl···B/A**	E_{bin}
OCH_2	3.059	1.740	179.8	-1.42
Cl^-	3.098	1.721	180.0	-7.30
OH^-	2.460	1.734	175.5	-13.76
FCl	2.990	1.744	175.6	-0.43
$ClCH_3$	3.467	1.745	168.8	-0.69
MgH_2	2.964	1.742	169.4	-1.10
HF	2.592	1.758	96.6	-0.13
H_3O^+	1.979	1.795	94.1	-7.53
HCl_2O^+	1.787	1.823	95.6	-9.02
Li^+	2.482	1.773	78.7	-12.13
F_3CCl*	–	1.746	–	–

*monomer

Table II.10 NBO parameters for the F_3CCl complexes. El_{trans} (in millielectrons) is the electron charge transferred from the Lewis base to the Lewis acid (positive values if F_3CCl is the Lewis acid, negative values if it acts as the Lewis base). C-charge and Cl-charge are NBO charges (in au). Pol(%C) is the C-Cl bond polarisation (% at C-atom), s(%C) is the percentage s-character in C orbital of the C-Cl bond, and s(%Cl) is the percentage s-character in Cl orbital of the C-Cl bond. E_{NBO} (in kcal/mol) is the energy connected with the n → σ* charge transfer (overlap). Results taken from ref. 107.

Moiety	El_{trans}	C-charge	Cl-charge	Pol(%C)	s(%C)	s(%Cl)	E_{NBO}
OCH_2	1.5	0.9844	0.0318	47.76	27.27	12.47	0.58
Cl^-	54.2	0.9426	0.0967	50.18	29.96	10.96	5.53
OH^-	123.4	0.9162	0.1060	51.88	31.54	8.43	15.65
FCl	2.2	0.9904	0.0154	47.29	26.90	12.73	0.70
$ClCH_3$	6.0	0.9906	0.0116	47.28	26.89	12.64	0.88
MgH_2	2.6	0.9882	0.0234	47.53	27.16	12.40	0.72
HF	-5.6	1.0021	-0.0129	46.38	25.95	12.60	1.50
H_3O^+	-122.7	1.0318	0.0241	44.28	23.84	11.46	36.27
HCl_2O^+	-221.7	1.0507	0.0652	42.23	22.25	13.11	82.44
Li^+	-69.2	1.0122	0.0110	45.99	25.4	11.63	22.27
F_3CCl*	–	0.9942	0.0073	47.0	26.62	12.91	–

*monomer

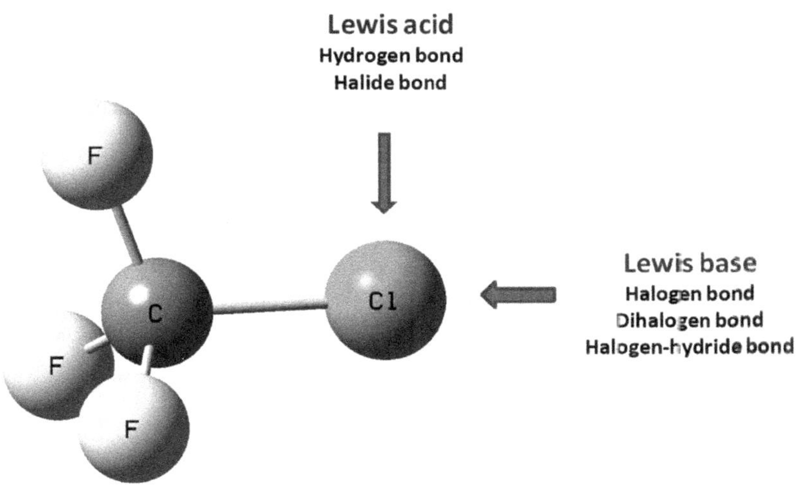

Scheme II.4 The Lewis acid-base properties of the chlorine centre in CF_3Cl species. Reprinted with permission from ref. 107. Copyright 2011 American Chemical Society.

Various variants of the halogen bond are formed if, for the F_3CCl unit, the interaction with the chlorine centre occurs through its σ-hole. It is observed for complexes with the following Lewis base species: OCH_2, Cl^-, OH^-, FCl, $ClCH_3$ and MgH_2. Some of these complexes may be classified as linked by dihalogen bonds; the complex with the MgH_2 species is connected through the so-called halogen-hydride bond (different kinds of interactions are analysed in Chapters I and V).[107,108] Table II.9 shows that the C-Cl···B angle (B is the Lewis base centre) for this group of complexes is situated in the 168.8–180.0° range. Thus, the latter complexes are linked through the σ-hole. The absolute value of the binding energy amounts to 0.4–13.8 kcal/mol here. The strongest interactions are charge-assisted, for complexes with Cl^- and OH^- anions. One can see that all C-Cl bonds in complexes are shorter than this bond for the isolated F_3CCl unit, thus the corresponding interactions may be classified as blue-shifted halogen bonds. The Cl···B distances are shorter than the corresponding sum of van der Waals radii (see Table I.1 in Chapter I for these radii values). The greatest distance shortening regarding this sum occurs for charge-assisted interactions, especially for the F_3CCl···OH^- complex where the strongest interaction is observed. The Cl···H distance for the MgH_2 complex is slightly longer than the van der Waals radii sum. For the F_3CCl complexes with FCl and $ClCH_3$, the Cl···F and Cl···Cl distances are only slightly shorter than the corresponding van der Waals sums. These are complexes linked by the weakest interactions.

Let us consider complexes where the F_3CCl unit acts as the Lewis base through the chlorine centre. The C-Cl···A angle (A designates here the Lewis acid centre, it is the hydrogen atom or lithium cation in this group of complexes) amounts to 78.7–96.6°, which means the F_3CCl species interacts with the Lewis acid units through the region of accumulation of electron density at the chlorine centre. This region is located approximately in the direction perpendicular to the C-Cl bond. The hydrogen bonds link complexes in this group, except for the complex with the lithium cation where the interaction may be classified as lithium bond or halide bond (see Chapter V). These are relatively strong interactions since binding energies between –7.5 and –12.1 kcal/mol are observed. The enhancement of the strength of these interactions may result from charge assistance since ions play the role of Lewis acid units here. The complex with hydrogen fluoride is an exception since it is not charge-assisted hydrogen bonding, and the binding energy amounts to only –0.1 kcal/mol. The elongation of the C-Cl bond because of complexation is observed here, thus the links in this group of complexes are also classified as red-shifted interactions (if they are classified as halide bonds according to the chlorine centre). The Cl···H distances for hydrogen-bonded (halide-bonded) complexes are shorter than the corresponding sum of van der Waals radii (Table I.1 in Chapter I), 2.85 Å; extremely short distances are observed for the charge-assisted hydrogen bonds. In the case of the Cl···Li link, the corresponding distance is equal to 2.482 Å. However, here the van der Waals radius of lithium should probably not be considered (1.81 Å — Table I.1) but rather the ionic radius that is 0.59 Å to 0.92 Å,[109] depending on the coordination of the lithium centre.

Table II.10 presents the NBO parameters for two groups of complexes based on results of Table II.9. The electron charge transfer is presented (in millielectrons) in Table II.10 showing that the division into two sub-groups of complexes of the F_3CCl species was correct. The electron charge transfer from the Lewis base unit to the Lewis acid species is observed. In the case of halogen-bonded systems, it is the transfer to the F_3CCl species where the halogen bond is formed through the σ-hole of the chlorine centre. The greatest transfer for this sub-group of complexes occurs for the charge-assisted halogen bonded systems; for complexes with Cl^- and OH^- anions, the transfer of 54.2 and 123.4 millielectrons is observed, respectively. In the case of hydrogen-bonded and halide-bonded complexes, the transfer of electron charge from the F_3CCl species to the Lewis acid unit is observed; this transfer is rather great since charge-assisted interactions occur here, up to 221.7 millielectrons for the complex with the HCl_2O^+ unit. The neutral hydrogen-bonded F_3CCl···HF system is an exception since only a charge of 5.6 millielectrons is transferred to the hydrogen fluoride unit.

Let us consider the NBO atomic charges of the C-Cl bond of the F_3CCl unit. If the latter species act as the Lewis acid, the charge of chlorine increases, from +0.0073 au in the isolated species to the +0.1060 au in the complex with the hydroxide anion. The carbon centre charge decreases for this sub-group of complexes; it is "less positive". This is an analogous situation to that of the hydrogen bond, thus similar processes of redistribution of the electron charge occur for hydrogen and halogen bonds. One can see that for these complexes, where the F_3CCl species acts as the Lewis acid unit, the polarisation of the C-Cl bond increases (the percentage of the electron density at the carbon centre increases) and the increase of s-character of the carbon hybrid orbital is observed, if compared with the isolated F_3CCl species. For the chlorine hybrid orbital, a decrease of the s-character occurs. However, the chlorine centre is connected here with the carbon centre, whose electronegativity increases because of complexation. The carbon centre is less positive in complexes with Lewis base units than in the isolated F_3CCl unit. These changes are also like those which occur in the hydrogen-bonded systems.

Let us consider a few NBO parameters for complexes with the F_3CCl species acting as the Lewis base unit. Both carbon and chlorine charges of the C-Cl bond increase because of complexation. The complex with HF unit is an exception since the chlorine charge decreases here and becomes slightly negative. The polarisation of the C-Cl bond decreases for this sub-group. Similarly, the s-character for both hybrid orbitals of carbon and chlorine decreases if compared with the isolated F_3CCl species. However, the C-Cl bond in this sub-group of complexes does not belong to the Lewis acid unit, thus the changes of parameters here are not analogous to those occurring in the hydrogen-bonded or halogen-bonded systems.

It is worth to note that NBO atomic charges are calculated as sums of occupations of atomic orbitals considered; the positive charge of the nucleus is also considered. Table II.11 shows the atomic charges of the hydrogen and oxygen centres of the O-H proton-donating bond in the water dimer (see Scheme II.3). The atomic charges in the isolated water monomer are given for comparison as well as the corresponding Mulliken charges. For the NBO charges the orbital contributions are specified, i.e., occupancies of core, valence and Rydberg orbitals are presented. It is worth to note that the Rydberg orbital for an atom is defined as “an orbital with principal quantum number greater than that of any occupied orbital of the ground state”.[110] Hence the low occupancy of Rydberg orbitals is usually observed for the species in energetic minima. That is also observed for atomic orbitals of water (Table II.11). The complexation leads to increase of the positive charge of the H-centre as discussed above; however, this charge change concerns mainly the valence electrons, and the occupancy of Rydberg orbitals practically does not change. In the case of oxygen, its charge decreases (negative charge increases) and here the change is related to valence electrons; for the Rydberg orbitals the slight, meaningless decrease of occupancy is observed as a result of complexation. The Mulliken charges differ significantly from those derived

Table II.11 The NBO and Mulliken atomic charges (in au) for atoms of the O-H proton-donating bond in water dimer are given. The occupancies of core, valence and Rydberg orbitals are presented. The corresponding atomic charges and Mulliken charges for the isolated water molecule are presented as well for comparison. The Mulliken charges are marked by an asterisk.

Atom	Core	Valence	Rydberg	Charge	Charge*
			Water dimer		
H	0.0	0.5099	0.0034	0.4867	0.2867
O	2.0	6.9668	0.0267	-0.9935	-0.5064
			Water monomer		
H	0.0	0.5138	0.0035	0.4827	0.2564
O	2.0	6.9375	0.0279	-0.9655	-0.5128

from the NBO approach. In general, the electron charge shift is detected here to the unit acting as the Lewis acid (the second hydrogen atom of water containing the proton-donating bond is characterised by a charge equal to 0.1943 au). However, only the increase of H-atom charge of the proton-donating bond that results from complexation is observed here; the negative charge of oxygen decreases here in opposition to the NBO results and other observations related to the formation of the hydrogen bond.

It is worth to note here that the Mulliken charge is defined based on orbitals. For an atom, all electron charge contributions from orbitals centred at this atom are summed up, and electron overlaps between two neighbouring atoms are divided equally.[111] The latter assumption is considered as the weakness of Mulliken population analysis. Hence other population analyses are usually applied in recent studies.[18]

Table II.10 also presents the E_{NBO} energies that correspond to orbital-orbital overlaps (interactions). The interactions between "filled" (donor) Lewis-type NBOs and "empty" (acceptor) non-Lewis NBOs are considered and often are treated as contributions to the charge transfer and polarisation interaction energies since they lead to electron charge shifts from orbitals of the localised Lewis-type structure to the empty non-Lewis orbitals. Hence these shifts are also treated as a delocalisation correction to the idealised

Lewis structure. The $n_B \rightarrow \sigma_{AH}^*$ overlap, which is treated as a signature of the hydrogen bond, was mentioned earlier. n_B marks the lone electron pair of the B-centre of the Lewis base unit while σ_{AH}^* is an antibonding orbital of the A-H proton-donating bond. The $n_B \rightarrow \sigma_{AH}^*$ overlap is calculated as the second-order perturbation theory energy (eq. II.27).

$$\Delta E\ (n_B \rightarrow \sigma_{AH}^*) = q_i\ \langle n_B | F | \sigma_{AH}^* \rangle^2 / (\varepsilon(\sigma_{AH}^*) - \varepsilon(n_B)) \qquad \text{(II.27)}$$

$\langle n_B | F | \sigma_{AH}^* \rangle$ is the Fock matrix element, $(\varepsilon\,(\sigma_{AH}^*) - \varepsilon\,(n_B))$ is the orbital energy difference and q_i is the electron donor orbital occupancy (it is usually equal to 2).

The E_{NBO} energies for complexes of HF, H_3O^+ and HCl_2O^+ concern the hydrogen-bonded systems and they exactly correspond to eq. II.27. They correspond to the $n_{Cl} \rightarrow \sigma_{FH}^*$ and $n_{Cl} \rightarrow \sigma_{OH}^*$ overlaps for the HF complex and for two other H-bonded (charge-assisted) systems. In the case of the system with Li^+ cation, the $n_{Cl} \rightarrow n_{Li}^*$ overlap is the main orbital-orbital interaction energy. For the other complexes where the F_3CCl species acts as the Lewis acid unit through the σ-hole, all interactions are classified as halogen bonds (including sub-classes of the halogen bond interaction: the dihalogen bond and the halogen-hydride bond) and the $n_B \rightarrow \sigma_{CCl}^*$ overlap is mainly observed here.

Table II.12 presents the main overlaps which occur for different types of hydrogen bonds and for different types of other interactions (σ-hole bonds and π-hole bonds). Such overlaps were discussed in detail in other studies.[18,112] The following kinds of overlaps are shown in this table; $n \rightarrow \sigma^*$, $\sigma \rightarrow \sigma^*$, $\pi \rightarrow \sigma^*$, $n \rightarrow n^*$, $\sigma \rightarrow n^*$ and $\pi \rightarrow n^*$; the star occurs here for the antibonding orbitals that correspond to the Rydberg states. There are also subscripts in designations in the table to specify kinds of bonds and centres to which orbitals considered are referred. First of all, various types of hydrogen bonds are presented in this table (types of interactions are discussed in detail in Chapter V); one can mention three main sub-classes of hydrogen bonds: A-H···B, A-H···σ and A-H···π where there is the one-centre Lewis base site (proton acceptor) and multi-centre Lewis base sites σ-electrons or π-electrons, respectively. For the A-H···σ interactions, the case of molecular hydrogen as the Lewis base unit is considered here. However, in several recent studies the A-H···σ hydrogen bonds with the C-C bonds acting as proton acceptors are considered.[113] In A-H···π interactions, the

Table II.12 The most important orbital-orbital interactions for different types of hydrogen bonds and σ-hole bonds; the π-hole bond is included for comparison. A similar table was presented in ref. 112, but only hydrogen bonds were included there.

Interaction designation	Overlap supreme	Energy expression	Examples
		hydrogen bonds	
A-H···B	$n_B \rightarrow \sigma_{AH}^*$	$q_i \langle n_B \vert F \vert \sigma_{AH}^* \rangle^2 / (\varepsilon(\sigma_{AH}^*) - \varepsilon(n_B))$	O-H…O, N-H…O, N-H…N
A-H···H-B	$\sigma_{BH} \rightarrow \sigma_{AH}^*$	$q_i \langle \sigma_{BH} \vert F \vert \sigma_{AH}^* \rangle^2 / (\varepsilon(\sigma_{AH}^*) - \varepsilon(\sigma_{BH}))$	N-H…H-Ru, O-H…H-Be, C-H ..H-C
A-H···σ(H_2)	$\sigma_{H2} \rightarrow \sigma_{AH}^*$	$q_i \langle \sigma_{H2} \vert F \vert \sigma_{AH}^* \rangle^2 / (\varepsilon(\sigma_{AH}^*) - \varepsilon(\sigma_{H2}))$	O-H…σ, N-H…σ
A-H···π	$\pi \rightarrow \sigma_{AH}^*$	$q_i \langle \pi \vert F \vert \sigma_{AH}^* \rangle^2 / (\varepsilon(\sigma_{AH}^*) - \varepsilon(\pi))$	O-H…π, N-H…π
		σ-hole bonds	
A-Z···B	$n_B \rightarrow \sigma_{AZ}^*$	$q_i \langle n_B \vert F \vert \sigma_{AZ}^* \rangle^2 / (\varepsilon(\sigma_{AZ}^*) - \varepsilon(n_B))$	H-Si…N, F-S…O
A-Z···σ(H_2)	$\sigma_{H2} \rightarrow \sigma_{AZ}^*$	$q_i \langle \sigma_{H2} \vert F \vert \sigma_{AZ}^* \rangle^2 / (\varepsilon(\sigma_{AZ}^*) - \varepsilon(\sigma_{H2}))$	H-N...σ, F-N ..σ, H-P...σ
A-Z···π	$\pi \rightarrow \sigma_{AZ}^*$	$q_i \langle \pi \vert F \vert \sigma_{AZ}^* \rangle^2 / (\varepsilon(\sigma_{AZ}^*) - \varepsilon(\pi))$	H-N...π, F-N ..π, H-P...π
		π-hole bonds; the triel bond is presented as an example	
Z···B	$n_B \rightarrow n_Z^*$	$q_i \langle n_B \vert F \vert n_Z^* \rangle^2 / (\varepsilon(n_Z^*) - \varepsilon(n_B))$	X_3Z...B; X = F, Cl, H; Z = 13th Group; B = N, Cl
Z…σ(H_2)	$\sigma_{H2} \rightarrow n_Z^*$	$q_i \langle \sigma_{H2} \vert F \vert n_Z^* \rangle^2 / (\varepsilon(n_Z^*) - \varepsilon(\sigma_{H2}))$	X_3Z...σ; X = F, Cl, H; Z = 13th Group
Z…π	$\pi \rightarrow n_Z^*$	$q_i \langle \pi \vert F \vert n_Z^* \rangle^2 / (\varepsilon(n_Z^*) - \varepsilon(\pi))$	X_3Z...π; X = F, Cl, H; Z = 13th Group

π-electron systems belong to the Lewis base units; ethylene and acetylene and their derivatives or aromatic systems may be given here as examples of such units. Thus, one can see that not only lone electron pairs of single-centre sites are involved in hydrogen bonds, but also multi-centre σ- and π-electrons may act as Lewis base sites in these interactions. Another sub-class of the hydrogen bonds, the dihydrogen bonds, are similar to the A-H···σ(H_2) interactions; this was discussed in another study.[112] For the dihydrogen bond the occupied orbital which participates in the overlap is the single σ-bond orbital, as it occurs for the A-H···σ interaction (Table II.12). However, there are also significant differences between A-H···H-B and A-H···σ interactions. For the former interaction the electrostatic term is usually the most important attractive interaction energy term and the polarisation term becomes more important only for strong dihydrogen bonds. For the A-H···σ interactions the polarisation term is very important, and often outweighs the electrostatic term.[112] There are other differences expressed by other theoretical methods, as for example by the Quantum Theory of Atoms in Molecules (QTAIM) approach.[114]

Table II.12 presents several characteristics of the σ-hole bonds as a broader class of interactions than hydrogen bonds. Also, here three types of Lewis base units and three main sub-classes of interactions are indicated, where the orbital-orbital overlaps are very similar to those occurring in the analogous hydrogen bond sub-classes. Finally, three types of overlap are shown in the table for the π-hole bond. Table II.12 presents also several examples of specific interactions (right column of the table). Several characteristics of this table are further discussed in subsequent chapters of this book.

It is worth to note here that the HOMO and LUMO terms are related to orbital-orbital overlaps discussed above. The Highest Occupied Molecular Orbital (HOMO) and the Lowest Unoccupied Molecular Orbital (LUMO) are considered in numerous studies as those which steer chemical reactivity, and which are involved in various chemical processes. HOMO and LUMO are referred to as the energy of the highest valence and lowest virtual orbitals, respectively. HOMO and LUMO are not always involved in chemical reactions because the symmetry of orbitals also plays an important role.[115,116] This is why other orbitals that are energetically close to HOMO and LUMO are considered sometimes as being involved in reactions. For example, the HOMO-1 or LUMO+1 designations are applied for orbitals being lower in energy to the HOMO and being higher in energy to LUMO, respectively. For example, the $n_B \rightarrow \sigma_{AH}^*$ overlap for the A-H···B hydrogen-bonded systems concerns the n_B (lone electron pair orbital) and σ_{AH}^* (A-H antibonding orbital) which may correspond to HOMO and LUMO, respectively, or to orbitals close in energies to this pair.

The HOMO-LUMO gap, which is the energy difference for this pair, is also discussed in various studies; the lower the difference, the less stable the system. One may consider formation of the O-H···O hydrogen bonds between two water molecules. It was shown that the HOMO-LUMO gap increases if the water dimer is formed, hence the stabilisation of the system increases. For the monomer two n_O and σ_{OH}^* orbitals are considered, while for the dimer these are the same orbitals which participate in the $n_O \rightarrow \sigma_{OH}^*$ interaction (see the dimer in Scheme II.3: n_O is an orbital of the -molecule on the right with oxygen possessing a charge equal to –0.965 au, while the σ_{OH}^* orbital corresponds to the molecule on the left, whose bond length is equal to 0.983 Å). The overlap discussed above is presented in Table II.12. One may expect that other overlaps presented in this table and corresponding to the other interactions may be related to the HOMO-LUMO pairs or to orbitals close in energies to them.

It is important that in the NBO approach the natural orbitals that are related to well-known chemical terms such as lone electron pairs, covalent bonds, etc., are considered. On the other hand, the canonical molecular orbitals usually applied in theoretical calculations are often far from "chemical intuition" since they only contain NBO contributions. Hence the canonical orbitals are more difficult in interpretations, particularly in HOMO-LUMO analyses.

II.8 The Quantum Theory of Atoms in Molecules

One can say that QTAIM[114,117,118] is related to an analysis of electron density of a system considered; this may be a small system like atom, ion or molecule but it may also be a greater structure such as a complex, cluster, or even crystal structure. The properties of electron density depend on the corresponding structure, thus positions of atoms and interrelations between them determine areas of electron density accumulation and depletion. Types of interactions between atoms are reflected in characteristics of the electron density that is a function of spatial coordinates. Hence the QTAIM approach is a useful tool to analyse various types of interactions through electron density analysis. The analysis of electron density is one of the main applications of QTAIM.

It is worth to note that the QTAIM approach is also a useful tool to perform the partition of electron density space of a structure into fragments, most often into atomic basins.[114,117,118] It was assumed that some properties of atoms or molecular fragments are transferable from one structure to another.[119,120] This space partitioning is related to other properties often described in various studies, to the integrated QTAIM atomic charges, to atomic volumes, etc; these properties are discussed later in this section.

However, let's go back now to relationships between the electron density and the corresponding molecular structure which are often referred to as characteristics of interactions. The local maxima of electron density correspond to atomic positions if such positions are understood as those of nuclei. Noticeable disagreements between positions of electron density maxima and positions of nuclei may appear for hydrogen atoms — this is also discussed later.

The views of electron density cannot be directly presented since it is a function of three space coordinates (x, y, z); however, its intersections can be presented and they are often discussed in studies on molecular systems and their mutual arrangements. For example, Figure II.4 presents the relief map of electron density for the plane containing the benzoic acid dimer.[121] The whole dimer is close to planarity, thus all maxima of electron density that correspond to positions of atoms are clearly visible. The maxima of electron density corresponding to non-hydrogen atoms are characterised by much greater values than those of hydrogen atoms. Hence the former are cut in Fig. II.4 at 4 au. One can see in the central part of this figure the maxima corresponding to two linked carboxylic groups. They are connected through two equivalent O-H···O hydrogen bonds since these bonds are related by inversion symmetry; the inversion centre is located between these –COOH groups.

There are special points, i.e., critical points (CPs), for the electron density of any molecular system discussed; for these points the gradient of electron density, $\nabla\rho$, is equal to zero (eq. II.28).

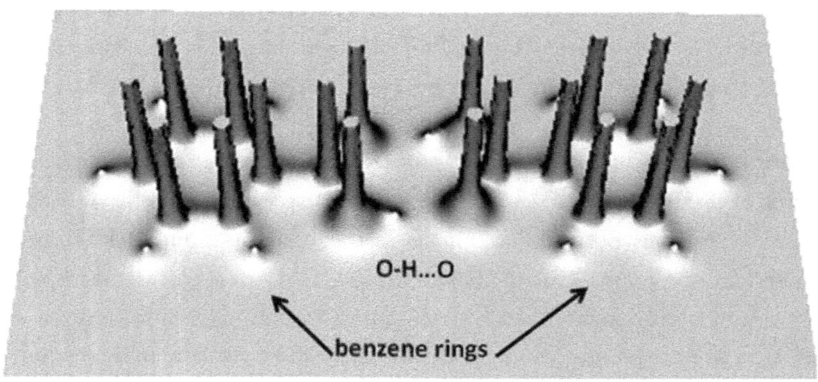

Fig. II.4 The relief map of the electron density of the benzoic acid dimer expressed in the plane of this system. In front of the picture the O-H...O hydrogen bond is presented; benzene rings are also designated. Reprinted with permission from ref. 121. Copyright 2011 American Chemical Society.

$$\nabla\rho = \boldsymbol{i}\frac{d\rho}{dx} + \boldsymbol{j}\frac{d\rho}{dy} + \boldsymbol{k}\frac{d\rho}{dz} = \vec{0} \text{ (at critical points and at } \infty) \tag{II.28}$$

The CPs may be categorised according to the abovementioned local maxima of electron density, ρ, i.e., the positions of nuclei, the saddle points and the local minima. The consideration of second derivatives of electron density allow us to perform such categorisation of CPs. The Hessian matrix of second derivatives that concerns CP, where r_C is its position, is expressed by eq. II.29.

$$A(r_c) = \begin{pmatrix} \frac{\partial^2\rho}{\partial x^2} & \frac{\partial^2\rho}{\partial x\partial y} & \frac{\partial^2\rho}{\partial x\partial z} \\ \frac{\partial^2\rho}{\partial y\partial x} & \frac{\partial^2\rho}{\partial y^2} & \frac{\partial^2\rho}{\partial y\partial z} \\ \frac{\partial^2\rho}{\partial z\partial x} & \frac{\partial^2\rho}{\partial z\partial y} & \frac{\partial^2\rho}{\partial z^2} \end{pmatrix}_{r=r_c} \tag{II.29}$$

The Hessian matrix may be diagonalised through the unitary transformation that leads to eq. II.30.

$$\Lambda(r_c) = \begin{pmatrix} \frac{\partial^2\rho}{\partial x'^2} & 0 & 0 \\ 0 & \frac{\partial^2\rho}{\partial y'^2} & 0 \\ 0 & 0 & \frac{\partial^2\rho}{\partial z'^2} \end{pmatrix}_{r'=r_c} = \begin{pmatrix} \lambda_1 & 0 & 0 \\ 0 & \lambda_2 & 0 \\ 0 & 0 & \lambda_3 \end{pmatrix} \tag{II.30}$$

The above unitary transformation is related to the rotation of the coordinate system, thus x', y' and z' correspond to new coordinates. The sum of eigenvalues (diagonal elements) of the latter Λ matrix, i.e., the trace of this matrix, is equal to the Laplacian of the electron density at the CP (eq. II.31).

$$\nabla^2\rho(r) = \frac{\partial^2\rho}{\partial x'^2} + \frac{\partial^2\rho}{\partial y'^2} + \frac{\partial^2\rho}{\partial z'^2} = \lambda_1 + \lambda_2 + \lambda_3 \tag{II.31}$$

The designation (ω, σ) is often used, where ω is the rank of CP and σ is its signature.[117] The rank informs of the number of non-zero eigenvalues of the electron density of the CP. The CPs with rank lower than 3 do not occur for structures in equilibrium. The σ signature is the sum of signs of eigenvalues. There are the following CPs defined by the above designations: (3,–3) that marks the local electron density maximum and is often named as the attractor, (3,–1) corresponds to the bond critical point (BCP), (3,+1) is attributed to the ring critical point (RCP) and (3,+3) corresponds to the cage critical point (CCP) that is the local minimum of the electron density.

The analysis of BCPs is most often performed in studies on interactions. The BCP is located on the bond path (BP). BP may be defined for the structure in the energetic minimum as the line of maximum electron density connecting two attractors understood often as atoms (it was mentioned earlier that this is an approximate agreement between positions of attractors and nuclei).[118] The BCP corresponds to the lowest electron density on this BP. The BPs are often straight lines, but structures with curved BPs are known. It has been pointed out that the meaning of the BP differs from the meaning of the covalent bond.[122,123] BPs correspond to preferable interactions in structures considered; these interactions may also be covalent bonds.[122–124]

The structure expressed by all CPs, including BCPs and attractors, as well as by BPs is often named as the molecular graph. Figure II.5 presents molecular graphs of complexes linked by various types of interactions.[69] There are complexes of ammonia, $FBr\cdots NH_3$, $HFSe\cdots NH_3$, $H_2FAs\cdots NH_3$, and $H_3FGe\cdots NH_3$ which are linked by halogen, chalcogen, pnicogen and tetrel bonds, respectively. The above links concern those between the electron-rich site of ammonia (lone electron pair) and the σ-hole of the Lewis acid unit. The σ-holes for NH_3 complexes are located in elongations of F-Br, F-Se, F-As and F-Ge bonds approximately. Lewis acid units containing F-substituents were chosen because this electronegative atom withdraws electrons making the σ-hole in the F-Z bond elongation stronger (Z = Br, Se, As and Ge), i.e., more electron density depleted. One can see that BPs correspond here not only to bonds but also to the $Z\cdots N$ intermolecular contacts. The BPs are designated by continuous or broken lines; the former occurs for bonds and strong interactions and the latter to remaining interactions.

Figure II.5 also presents complexes of acetylene that act as the Lewis base unit through π-electrons of the CC bond. Such interactions with π-electrons as the Lewis base region

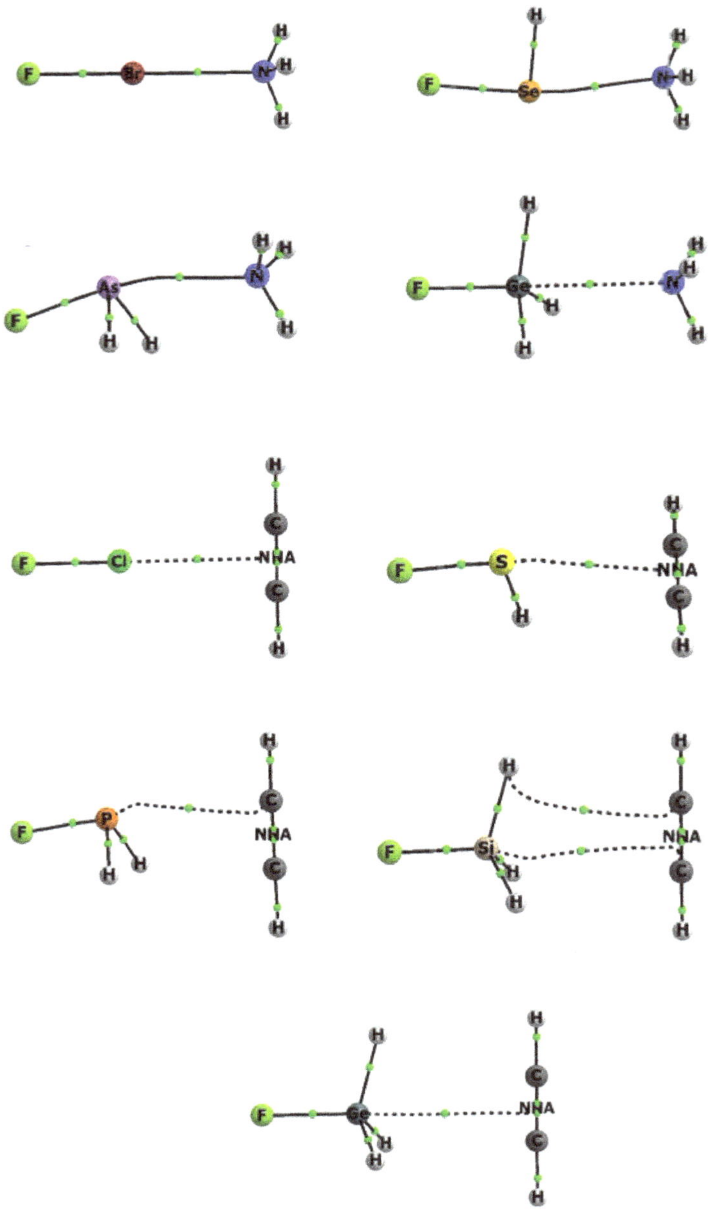

Fig. II.5 Molecular graphs of structures linked by various types of interactions. The attractors are marked by big circles and the CPs by small ones.

were analysed extensively for hydrogen bonds, and they were often designated as A-H···π (A-H is the proton-donating bond).[18,125] Special attention is paid in numerous studies on the C-H···π hydrogen bonds which are often crucial in arrangements of molecules and ions in crystals.[126] The complexes of acetylene presented in Fig. II.5 are characterised by specific BPs corresponding to the Z···π interactions. In the case of FBr···C_2H_2, HFS···C_2H_2, and H_3FGe···C_2H_2 complexes the Z···NNA BPs are observed. NNA is the non-nuclear attractor, which is the accumulation of the electron density not attributed to the nucleus. Such NNAs were analysed theoretically for links between alkali metals,[127] while the first experimental confirmation of the existence of such electron density accumulation concerns the crystal structure where such NNA occurs on the Mg-Mg BP, i.e., in the Mg-Mg bond region.[128]

It was found for the complexes of acetylene with FBr, HFS, and H_3FGe···C_2H_2 that the NNA is situated between C-centres of acetylene, and between two BCPs. These CPs (BCP, NNA, BCP) are located in the middle part of the CC bond, at a flat region of the electron density. However, the NNA is characterised by meaninglessly greater value of electron density than the surrounding two BCPs; this occurs for some T-shaped complexes of acetylene, particularly those linked by hydrogen bonds.[129]

Let us analyse remaining molecular graphs of Fig. II.5. In H_2FP···C_2H_2 and H_3FSi···C_2H_2 complexes, similar to three other complexes of acetylene discussed earlier, the NNA occurs between carbon centres. However, there is no link with the latter CP, i.e., there is no Z···NNA BP (Z = P, Si). In the H_2FP···C_2H_2 complex the BP, corresponding to the pnicogen bond, connects the P-centre with one of the carbon centres of acetylene. In the case of the H_3FSi···C_2H_2 complex, there is the similar Si···C BP corresponding to the tetrel bond and another H···C BP that links the Lewis acid and Lewis base units.

Figure II.6 presents the molecular graph with contour map of the Laplacian of electron density of the F-Cl···C_2H_2 complex discussed earlier; its molecular graph is also presented in Fig. II.5. The complex is planar, thus the contour map concerns the plane of the complex. The big circles correspond to attractors and the small circles to BCPs and NNA. The continuous straight lines are attributed to BPs that correspond to covalent bonds while the broken line corresponds to the BP related to the Cl···NNA intermolecular contact. The curved lines, continuous and broken, correspond to the

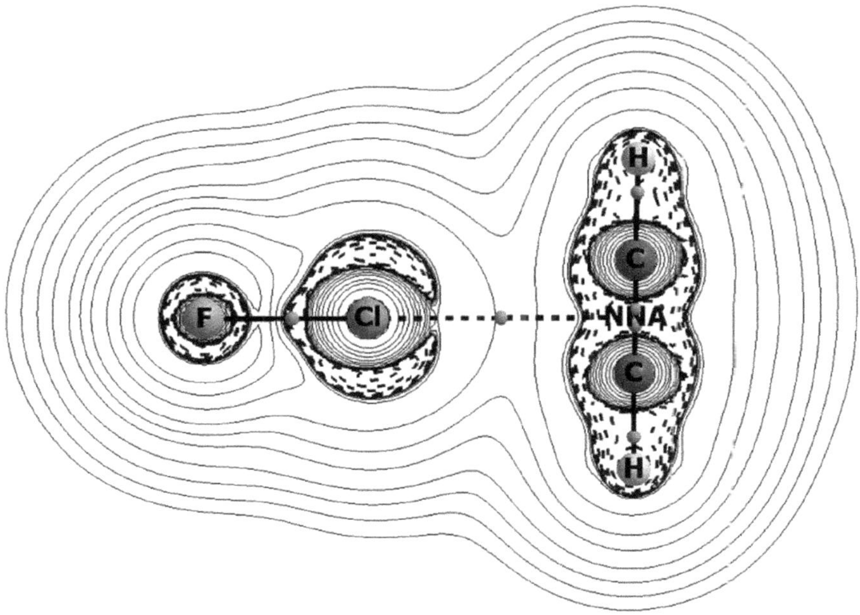

Fig. II.6 The contour map of the Laplacian of electron density of the F-Cl···C_2H_2 complex that concerns its plane. The big circles correspond to attractors and the small circles to BCPs and to NNA. The continuous straight lines are attributed to BPs that correspond to covalent bonds while the broken line corresponds to the BP related to Cl···NNA intermolecular contact. The curved lines, continuous and broken, correspond to positive and negative values of the Laplacian of electron density and thus to the electron density-poor regions and regions characterised by its accumulation, respectively.

positive and negative values of the Laplacian of electron density and thus to the electron density-poor regions and regions characterised by its accumulation, respectively. One can see that the F-centre is surrounded by a sphere of accumulation of the electron density; the same concerns the Cl-centre approximately. However, in the latter case there is an area of positive Laplacian in the extension of the F-Cl bond. This area corresponds to the σ-hole that is responsible for the positive electrostatic potential and consequently for the Lewis acid properties of the Cl-centre in the elongation of the F-Cl bond.

Let us describe the characteristics of BCPs that are often analysed in studies concerning various types of interactions. The broad spectrum of interactions may be analysed with the QTAIM approach, from strong covalent bonds to weak van der Waals interactions.[130] For covalent and polarised covalent bonds, the Laplacian of electron density at BCP located at the BP linking centres corresponding to the bond considered, $\nabla^2\rho_{BCP}$, is negative, indicating the concentration of electron density within the atom-atom region. The negative value of $\nabla^2\rho_{BCP}$ is often treated as confirmation of the covalent character of interaction. For interactions between closed-shell species, i.e., for van der Waals interactions, ionic interactions, the main part of hydrogen bonds, and Lewis acid — Lewis base interactions, among them the σ-hole bonds and π-hole bonds, the $\nabla^2\rho_{BCP}$ value is positive. The latter is connected with the depletion of electron density within the atom-atom region considered. One may refer here to Fig. II.6 where chlorine and fluorine centres are separated by the area of the positive Laplacian. Hence one may suppose that the ionic character of interaction between "cationic chlorine" and "anionic fluorine" occurs here. The MP2/aug-cc-pVTZ NBO atomic charges of chlorine and fluorine in this complex are equal to +0.394 au and -0.429 au, respectively.[94] The total charge of the FCl species in this FCl···C_2H_2 complex linked by the Cl···π halogen bond is equal to -0.035 au since there is electron charge transfer from the acetylene to this unit that results from complexation. This is the typical Lewis base-acid electron charge transfer discussed in the former chapter. However, the BCP located at the BP between Cl and F centres is situated in the region of negative Laplacian, closer to the Cl-nucleus than to the F-nucleus. It confirms that F-Cl should be treated as a covalent bond. It is a polarised bond since 71.1% of the σ-bond orbital electron density is located at the fluorine centre according to MP2/aug-cc-pVTZ NBO results.[94] Another situation is observed for the Cl···π BP (Fig. II.6). The BCP at this BP is located in the region of positive Laplacian values. This means that the F-Cl···π halogen bond is not covalent in nature, rather, and that the interactions in the corresponding F-Cl···C_2H_2 complex are classified as those between closed-shell systems.

One can see that characteristics of CPs inform of the nature of interactions. It is worth mentioning that the occurrence of BPs is often treated as a signature of the existence of stabilising interactions. We mention here the criteria for the existence of A-H···B hydrogen bonds that are based on the QTAIM approach and that were proposed by Koch and Popelier:[117,131]

— The H···B BP, with the corresponding BCP, is observed between the H-atom and the proton acceptor centre, B.

- The electron density at the H···B BCP, $\rho_{H...B}$, should be characterised by a value between 0.002 a.u. and 0.034 a.u.
- The Laplacian of the electron density at the H···B BCP, $\nabla^2\rho_{H...B}$, should be within the 0.024–0.139 a.u. range.
- The mutual penetration of the hydrogen, H, and the acceptor, B, centres occurs. This penetration may be separated into parts of the hydrogen and acceptor atoms: $\Delta r_H^0 = r_H^0 - r_H$ and $\Delta r_B^0 = r_B^0 - r_B$. The nonbonded radius r^0 (r_H^0 or r_B^0) is defined as the distance between the nucleus and the fixed electron density contour that is usually assumed to be equal to 0.001 a.u. for separated A-H and B units. The r_B and r_H radii in the A-H···B system are defined as the distances between the corresponding positions of nuclei and H···B BCP.
- The loss of electron charge of the hydrogen atom (increase of its positive charge) resulting from the A-H···B formation is observed.
- The energetic destabilisation of the hydrogen atom is observed, which is expressed as the positive difference between the total energy of the hydrogen atom within the complex and the energy of this atom not involved in any interaction.
- The decrease of dipolar polarisation of the hydrogen atom and the decrease of its volume as a result of hydrogen bond formation.

The above criteria concern typical and charge-neutral A-H···B hydrogen bonds, and in other cases the properties of these interactions may not fit. For example, for very strong hydrogen bonds, the values of the electron density and of its Laplacian at the H...B BCP, $\rho_{H...B}$ and $\nabla^2\rho_{H...B}$, respectively, are often outside the range of values proposed. The $\rho_{H...B}$ value may be of the order of ~0.1 au for such hydrogen bonds and the Laplacian may be negative.[121,132–135]

It is important to note that the characteristics of BCP are often treated as measures of the strength of interactions. Particularly, the electron density at BCP, ρ_{BCP}. Numerous correlations were found between the latter parameter and others, especially those related to the energy of interaction.[136–144] These relationships mainly concern hydrogen-bonded systems; however, similar dependencies often occur for other interactions, particularly for the σ-hole bonds.[69] It is very important that the QTAIM parameters related to the strength of interaction can also be considered for intramolecular interactions. It is often a problem to evaluate interaction energies for such systems since there are no reference monomers that are easily defined for systems linked by intermolecular interactions.[145]

The ωB97-XD/6-311++G(d,p) calculations performed on dimers and monomers of salicylic acid and salicylamide as well as on their thiol analogues are examples of possibilities to apply QTAIM parameters.[146] Figure II.7a presents one of these structures, the molecular graph of the dimer of the *trans* conformer of thiosalicylic acid; this is a centrosymmetric dimer since it contains the centre of inversion between carboxylic groups that is located at the same place as the RCP. This is why the two intermolecular O-H···O hydrogen bonds observed here are equivalent. However, there are also two equivalent intramolecular H-S···O chalcogen bonds in this dimer. The BPs and corresponding BCPs occur here between sulfur and oxygen centres. Figure II.7b shows the same molecular graph as that presented in Fig. II.7a, with the electron density isolines in black and the gradient paths in blue; red lines are BPs connecting attractors, red circles — attractors attributed to atoms, blue circles — BCPs, yellow circles — RCPs.

There are other characteristics of BCP besides the electron density and its Laplacian that are applied in various studies to describe interactions. These are, for example, the kinetic electron energy density at CP, G_C, and its potential electron energy density analogue, V_C. These terms are related to the Laplacian of the electron density by a virial equation (eq. II.32).

$$\frac{1}{4}\nabla^2\rho = 2G_C + V_C \qquad \text{(II.32)}$$

These terms related to BCP, i.e., G_{BCP} and V_{BCP}, are analysed most often since they are directly related to centre-centre contacts. The first term is always positive while the second is always negative. Thus, the absolute value of the second term must be two times greater than the kinetic electron energy density to have a negative value of the Laplacian. Another parameter is often applied; this is the total electron energy density at the CP, H_C, that is expressed by eq. II.33.

$$H_C = V_C + G_C \qquad \text{(II.33)}$$

Classification of the A-H···B hydrogen bonds that is based on the above-mentioned parameters (eqs. II.32 and II.33) concerning the H···B contact was proposed.[147] It was stated that hydrogen bonds which are weak and medium in strength are characterised by both positive H_{BCP} and $\nabla^2\rho_{BCP}$ values, for strong hydrogen bonds $\nabla^2\rho_{BCP}$ is positive

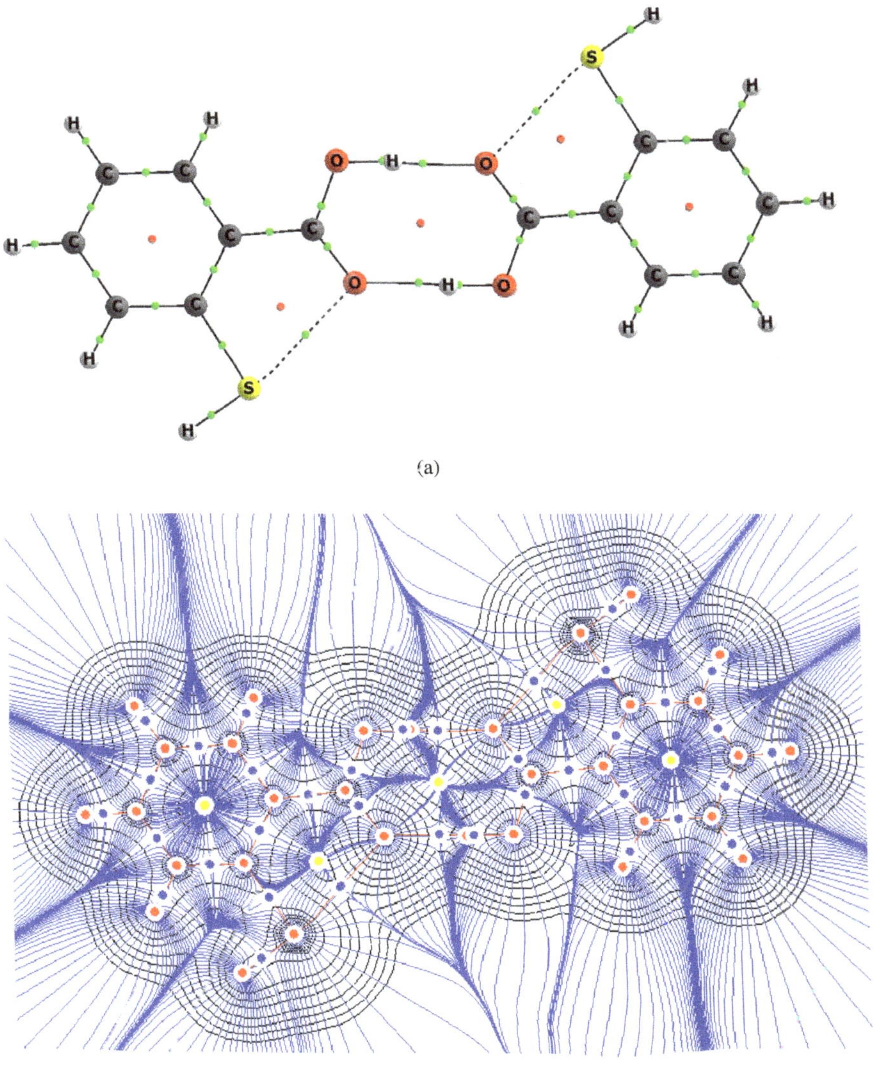

Fig. II.7 (a) The molecular graph of the dimer of the *trans* conformer of thiosalicylic acid. Big circles correspond to attractors, small ones to BCPs and RCPs. (b) Molecular graph with electron density isolines (black) and the gradient paths (blue). Red lines — BPs connecting attractors, red circles — attractors attributed to atoms, blue circles — BCPs, yellow circles — RCPs. Reprinted with permission from ref. 146. Copyright 2021 American Chemical Society.

and H_{BCP} is negative, and for very strong hydrogen bonds $\nabla^2\rho_{BCP}$, thus consequently H_{BCP}, are negative.[147] Similar classification may be extended for other types of interactions.[69] Table II.13 presents selected QTAIM characteristics of complexes linked by halogen, chalcogen, pnicogen and tetrel bonds, such as characteristics of the Z···B BCP. Z is the halogen, chalcogen, pnicogen or tetrel Lewis acid centre while B is the Lewis base centre, i.e., nitrogen, NNA (or carbon) and Cl^- in the case of ammonia, acetylene and chloride anion, respectively. If the electron density at the Z···B BCP, ρ_{BCP}, expresses the strength of interaction, one can see that the chloride anion is the strongest Lewis base centre, followed by ammonia and then acetylene for the same Lewis acid unit. The following ranges of ρ_{BCP} values are observed for Cl^-, NH_3 and C_2H_2 complexes: (0.04–0.09 au), (0.02–0.07 au) and (0.01–0.03 au), respectively. The Laplacian of electron density at BCP, $\nabla^2\rho_{BCP}$, is positive for all complexes of this table but the H_{BCP} value is negative for complexes of chloride anion and ammonia. In the case of acetylene complexes, the H_{BCP} value is positive except for the FBr···C_2H_2 and HFSe···C_2H_2 complexes where it is negative. It is worth mentioning that the negative value of H_{BCP} is often treated as confirmation of partial covalent character of interaction.[148–150] One can also see that the electron charge transfer (charges presented in Table II.13 were calculated by the NBO approach) follow the strength of interaction; the greatest electron charge shifts occur for chloride anion complexes, the smallest for complexes of acetylene.

Figure II.8 presents relationships between the Z···B distance and the DEL/EL ratio, χ, for complexes of chloride anion, ammonia and acetylene. In the case of acetylene complexes, the Z···B distances between Z-centres and mid-points of CC bonds of acetylene (corresponding at least approximately to positions of NNAs) are considered. These relationships are expressed by second-order polynomial functions. The DEL and EL in the χ ratios express the delocalisation and electrostatic interaction energy terms resulting from the decomposition analysis, respectively. The DEL term that was calculated in the decomposition scheme applied[85,86] is related to the electron charge density shifts resulting from the complexation, thus it is related to the covalent character of interaction, similar to other terms corresponding to such charge shifts that appear in various decomposition schemes: polarisation, charge transfer or induction contributions, for example.[121] It was found that the covalency of interaction increases with the increase of the strength of interaction, up to the formation of covalent bond.[121] The changes of covalent character of interactions are clearly shown in Fig. II.8 where

Table II.13 The QTAIM characteristics (in au) of the Z...B BCP: electron density at BCP, ρ_{BCP}, its Laplacian, $\nabla^2\rho_{BCP}$, the total electron energy density at BCP, H_{BCP}, the potential electron energy density at BCP, V_{BCP}, and the kinetic electron energy density at BCP, G_{BCP}. For the acetylene complexes, B denotes the NNA or C-centre of acetylene (see Fig. II.5); Tr is the electron charge transfer from the Lewis base to the Lewis acid (au units, NBO calculations). Results related to ref. 69.

Lewis acid	ρ_{BCP}	$\nabla^2\rho_{BCP}$	V_{BCP}	G_{BCP}	H_{BCP}	Tr
			NH_3 Lewis base			
$SiFH_3$	0.0281	0.0522	−0.0244	0.0187	−0.0057	−0.071
PFH_2	0.0285	0.0622	−0.0209	0.0182	−0.0027	−0.050
SFH	0.0408	0.0976	−0.0349	0.0296	−0.0052	−0.076
ClF	0.0651	0.1396	−0.0637	0.0493	−0.0144	−0.144
$GeFH_3$	0.0247	0.0678	−0.0187	0.0178	−0.0009	−0.055
$AsFH_2$	0.0313	0.0708	−0.0237	0.0207	−0.003	−0.061
SeFH	0.0464	0.0977	−0.0412	0.0328	−0.0084	−0.100
BrF	0.0642	0.1232	−0.0633	0.0471	−0.0163	−0.153
			C_2H_2 Lewis base			
$SiFH_3$*	0.0071	0.0223	−0.0039	0.0047	0.0008	−0.010
PFH_2*	0.0141	0.0354	−0.0084	0.0086	0.0002	−0.010
SFH	0.0196	0.0607	−0.0143	0.0147	0.0004	−0.018
ClF	0.0221	0.0715	−0.0164	0.0171	0.0007	−0.034
$GeFH_3$	0.0083	0.0255	−0.0045	0.0054	0.0009	−0.013
$AsFH_2$*	0.0157	0.0378	−0.0093	0.0094	0.0001	−0.016
SeFH	0.0260	0.0679	−0.0196	0.0183	−0.0013	−0.031
BrF	0.0292	0.0788	−0.0226	0.0211	−0.0014	−0.058
			Cl^- Lewis base			
$SiFH_3$	0.0634	0.0668	−0.0521	0.0344	−0.0177	−0.228
PFH_2	0.0436	0.0433	−0.0323	0.0216	−0.0107	−0.178
SFH	0.0634	0.0668	−0.0521	0.0344	−0.0177	−0.270
ClF	0.0883	0.0946	−0.0789	0.0513	−0.0276	−0.444
$GeFH_3$	0.0417	0.0729	−0.0369	0.0276	−0.0093	−0.200
$AsFH_2$	0.0453	0.0575	−0.0356	0.025	−0.0106	−0.196
SeFH	0.0601	0.0663	−0.0506	0.0336	−0.017	−0.273
BrF	0.0757	0.08	−0.0665	0.0433	−0.0233	−0.378

*concerns the Z...C BP; for remaining complexes with C_2H_2 these are the Z...NNA BPs (NNA is situated between carbon atoms of acetylene).

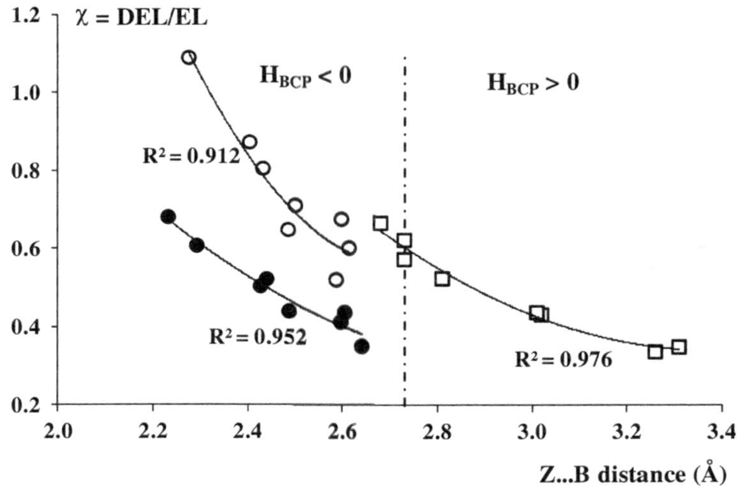

Fig. II.8 The relationships between the Z···B length (Å) and the DEL/El ratio (designated as χ, these are the second-order polynomial regressions); the subgroups of complexes were considered separately. Black circles: ammonia complexes, open circles: chloride anion complexes, white squares: acetylene complexes. The areas of positive and negative H_{BCP} values are shown. Reprinted with permission from ref. 69. Copyright 2017 Wiley-VCH.

one can observe the increase of importance of the attractive delocalisation interaction, DEL, with the shortening of Z···B distance. The latter distance shortening corresponds approximately to the increase of strength of interaction.[151] Figure II.8 shows that interactions between closed-shell systems are observed for longer Z···B distances that is indicated by appropriate EDA and QTAIM parameters ($H_{BCP} > 0$, for example); these parameters change for shorter Z···B distances where interactions covalent in nature occur ($H_{BCP} < 0$).

It is worth mentioning that various correlations between geometrical, energetic and topological parameters were analysed in numerous studies. The intermolecular A-H···B resonance-assisted hydrogen bonds were discussed[152] and various correlations were found, for example the correlation between delocalisation energy, DEL, and energy corresponding to the $n_B \rightarrow \sigma_{A\text{-}H}^*$ overlap , and the correlation between the DEL/EL ratio and the $-V_{BCP}/G_{BCP}$ ratio. The DEL/EL (χ) and $-V_{BCP}/G_{BCP}$ parameters are related directly to the covalent character of interaction. One can see from eqs. II.32 and II.33 that $-V_{BCP}/G_{BCP}$ ratio greater than 2 is equivalent to the negative value of Laplacian of electron density at BCP which, according to the QTAIM approach, indicates covalent interaction. For this ratio greater than 1, the total electron energy density at BCP, H_{BCP}, is negative and corresponds to partial covalent character of interaction. For ratio lower than 1, typical interaction between closed-shell systems occurs.

One can also refer to a recent study on halogen bonds where carbene systems play the role of the Lewis base units.[91] ωB97XD/aug-cc-pVDZ and ωB97XD/aug-cc-pVTZ calculations were performed in this study on complexes of imidazol-2-ylidene with XCCH, XCN and X_2 units (X = Cl, Br and I) that play the role of the Lewis acid. The lone electron pair at the carbon centre of imidazol-2-ylidene acts as the Lewis base site. The results of DFT calculations were supported in this study by parameters derived from QTAIM, NBO and EDA approaches. It was found that the complexes are connected through C-X···C and X-X···C halogen bonds. Figure II.9 shows the linear correlation between total electron energy density at the BCP located at the BP between carbon and halogen centres, $H_{X\cdots C}$, and steric energy resulting from EDA analysis where the decomposition scheme expressed by eq. II.20 was applied. The steric interaction energy is the sum of the Pauli repulsion term and the electrostatic interaction energy, i.e., $\Delta E_{Pauli} + \Delta E_{elstat}$. In other decomposition schemes, the first-order

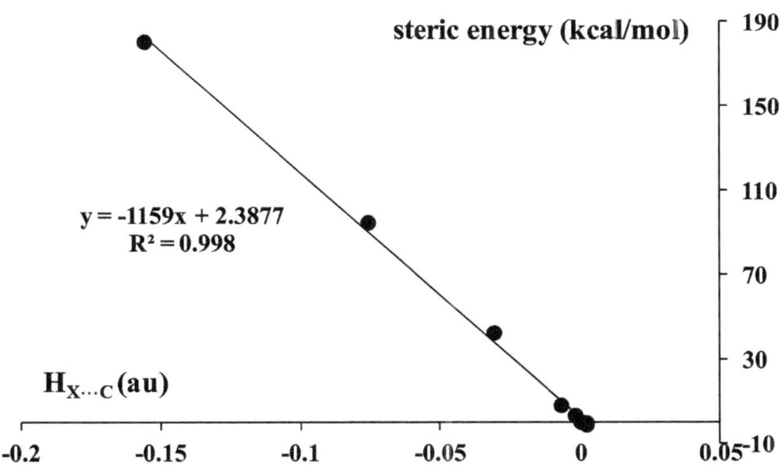

Fig. II.9 The dependence between the total electron energy density at the X···C BCP, $H_{C\cdots X}$, (au) and the steric energy (kcal/mol) for complexes of imidazol-2-ylidene linked with the XCCH, XCN and X_2 units (X = Cl, Br and I) through halogen bonds. Reprinted with permission from ref. 91. Copyright 2023 The Royal Society of Chemistry.

Heitler-London energy term, $E_{H\text{-}L}^{(1)}$, that corresponds to the steric energy is used since it is the sum of the electrostatic energy term and the repulsive exchange energy. It is worth to note that for positive steric energy or Heitler-London energy, the repulsion interaction outweighs the attractive electrostatic interaction. Consequently, the interactions related to electron charge shifts, i.e., related to covalent character, are more important to make the complex energetically stable. It also means that the positive steric energy indicates the covalent character of interaction, similar to the QTAIM parameter H_{BCP}, where its negative value indicates at least partial covalent character. The correlation in Fig. II.9 confirms these explanations, where the steric interaction energy is more positive and thus the total electron energy density is more negative.

These examples and discussions of various kinds of interactions lead to a clear picture of changes of properties of these interactions with the change of their strengths, from very weak to strong interactions, up to covalent bonds. These properties were usually described by geometrical, energetic and other parameters derived from QTAIM or NBO approaches, for example. Such descriptions concern mainly the hydrogen-bonded systems; however, one may apply these parameters to other types of interactions. Table II.14 summarises selected energetic, topological and geometric properties of interactions, which are related mainly to the strength of interactions. The first two characteristics, the absolute value of the energy of interaction, $|E_{int}|$, and the Z···B distance between the Lewis acid and Lewis base centres refer mainly to ranges typical for hydrogen-bonded systems,[18,98,99] but it seems that they work also for other

Table II.14 The change of parameters of various interactions if their strengths change.

Parameter	Weak	Medium	Strong	Very Strong
$\|E_{int}\|$	<5	5–15	15–40	>40
Z···B	>2.1	1.8–2.1	1.2–1.8	<1.2
ρ_{BCP}	<0.02	0.02–0.07	0.07–0.1	>0.1
$\nabla^2\rho_{BCP}$	>0	>0	>0	<0
H_{BCP}	>0	>0	<0	<0
$-V_{BCP}/G_{BCP}$	<1	<1	1–2	>2
$\Delta E_{Pauli}+\Delta E_{elstat}$	<0	<0	>0	>0
$\Delta E_{orb}/\Delta E_{elstat}$	<1	<1	~1	>1
character	mainly electrostatic	mainly electrostatic	partly covalent	covalent

interactions,[69,107] at least approximately. The same may be said for the electron density at the Z···B BCP, ρ_{BCP}. The remaining characteristics of interactions presented in Table II.14 are more universal since they do not concern ranges of values but the general "mathematical properties" of parameters considered. The negative value of the Laplacian of electron density at BCP, $\nabla^2\rho_{BCP}$, is observed for covalent bonds and for very strong interactions covalent in nature since it confirms the accumulation of electron density in the atom-atom region. For weaker interactions, this parameter should be characterised by a positive value. The total electron energy density at BCP, H_{BCP}, is negative for covalent and partly covalent interactions, and in other cases it is positive. The $-V_{BCP}/G_{BCP}$ ratio is related to the $\nabla^2\rho_{BCP}$ and H_{BCP} parameters by eqs. II.32 and II.33. Hence it is an additional parameter that covers characteristics of $\nabla^2\rho_{BCP}$ and H_{BCP} parameters. It is greater than 2 for covalent interactions, between 1 and 2 for partly covalent interactions and lower than 1 for medium and weak interactions between closed-shell systems. This parameter was introduced in the analysis of systems characterised by a broad range of distances between the fluorine and hydrogen centres.[153] Next, it was applied in other studies, for hydrogen bonds mainly.[18] However, the $\nabla^2\rho_{BCP}$ and H_{BCP} parameters are related to this ratio (eqs. II.32 and II.33) that concern all atom-atom contacts (or, similarly, centre-centre contacts for multicentre interactions). Hence this ratio is applicable for other interactions, and not only for the hydrogen-bonded systems.

The last two parameters presented in Table II.14, $\Delta E_{Pauli} + \Delta E_{elstat}$ and $\Delta E_{orb}/\Delta E_{elstat}$ concern the interaction energy terms derived from the EDA approaches. They were discussed before and only their approximate characteristics are presented in Table II.14. It is worth to note that the latter ratio was compared with other energetic, geometrical and topological parameters in an early study on the broader spectrum of the hydrogen-bonded systems.[154]

The localisation $\lambda(A)$ (eq. II.34) and delocalisation $\delta(A,B)$ (eq. II.35) indices are other useful QTAIM parameters that may provide additional characteristics of interactions.[155,156]

$$\lambda(A) = -\int_A (2\Gamma(r_1,r_2) - \rho(r_1)\rho(r_2))dr_1dr_2 \qquad (II.34)$$

$$\delta(A,B) = -2\int_{A,B} (2\Gamma(r_1,r_2) - \rho(r_1)\rho(r_2))dr_1dr_2 \qquad (II.35)$$

$\rho(r)$ and $\Gamma(r_1,r_2)$ are one- and two-electron densities, respectively, where integrations are performed through one and two atomic basins. The localisation index $\lambda(A)$

corresponds to the number of electrons localised in one atom while the delocalisation index δ(A,B) corresponds to the number of electrons delocalised between two A and B atoms. In such a way the δ(A,B) index may be understood as a measure of the degree of covalency or the number of shared electrons. The relationships between the delocalisation index and parameters related to the hydrogen bond strength were analysed;[157] however, the H-bonded systems of a non-homogeneous sample of complexes were used, thus correlations appear rather poor.[157]

Figure II.10 presents correlation between the delocalisation index, DI (δ(A,B)), and the Wiberg index[158,159] that also corresponds to the strength of interaction. The latter term, similar to DI, also corresponds to the bond order and may be understood as the number of shared electron pairs within an interatomic region. This relationship presented in Fig. II.10 concerns complexes of imidazol-2-ylidene with Lewis base units X_2, XCN and X^- (X designates the halogen centre).[91] The excellent second-order polynomial relationship between both indices is presented in the figure, indicating that for the homogeneous sample (of related species) both indices express almost the same properties related to the accumulation of electron charge density in the interatomic region.

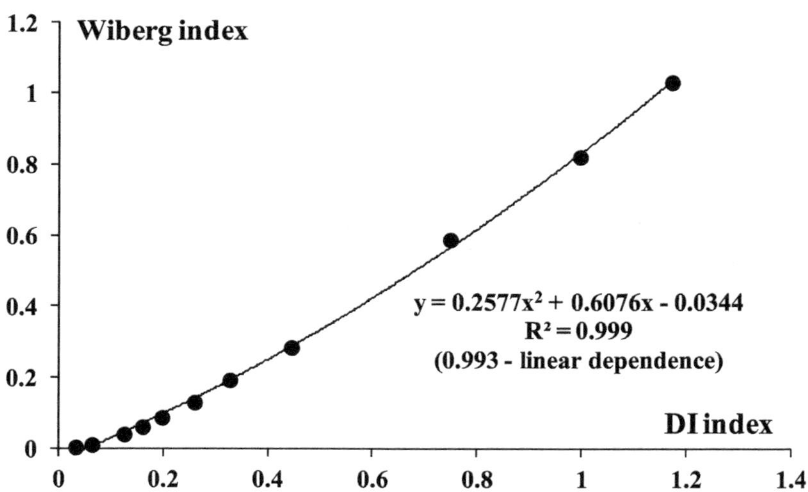

Fig. II.10 The dependence between the DI index and the Wiberg index for the X···C distance, for complexes of imidazol-2-ylidene linked with the XCCH, XCN and X_2 units (X = Cl, Br and I) through halogen bonds. Reprinted with permission from ref. 91. Copyright 2023 The Royal Society of Chemistry.

There are QTAIM parameters related to integrated atomic properties that come from the partitioning of electron density of a system into fragments attributed to atoms.[114] Atoms in a molecule or in another system like a cluster, complex or ion are bounded by the zero-flux surface in the gradient vector field of the electron density; this is defined by the following expression:

$$\nabla \rho(r) \cdot n(r) = 0 \qquad \text{(II.36)}$$

$\nabla \rho(r)$ is the gradient of the electron density and $n(r)$ is a unit vector which is normal to the surface at r. The sub-space corresponding to the atom is often marked as Ω. Thus, a number of electrons attributed to an atom is calculated by the integration over the atomic Ω space (eq. II.37).

$$N(\Omega) = \int_{\Omega} \rho(r) d\tau \qquad \text{(II.37)}$$

The atomic charge is the sum of negative charges of the electrons and positive charges of the nucleus, Z_{Ω}, hence the net atomic charge is expressed by eq. II.38.

$$q(\Omega) = [Z_{\Omega} - N(\Omega)] \qquad \text{(II.38)}$$

This charge, defined within the QTAIM approach, is observable because it may be determined experimentally in crystal structures by diffraction methods. It was noted that "An atomic population or charge, as defined within QTAIM, is the measurable expectation value of a Dirac observable and is now routinely determined in accurate X-ray diffraction experiments in crystals."[160] Other charges, like Mulliken or NBO, are defined within theoretical approaches and cannot be determined with the use of experimental techniques. However, charges defined in various ways may be useful to interpret numerous properties of structures, particularly the partitioning of the electron density space into orbitals (NBO approach) which is very useful to discuss and describe numerous chemical processes. However, one should be careful since net atomic charges that result from different population analyses may lead to opposite results and consequently to opposite conclusions.

It is worth mentioning that the QTAIM approach is often a subject of controversies and polemics. For example, the meaning of BP as an indicator of the local stabilising interaction is sometimes questionable.[122,123] Cioslowski and Mixon have found that

BPs in some systems do not show stabilising interactions, but correspond to repulsive local forces resulting from steric overflow.[161,162] Different types of BPs that do not indicate local stabilisations were described in detail in recent studies.[163,164] There are other reservations concerning QTAIM, but their detailed discussion is not included in the scope of this book.

II.9 The Electron Localisation Function

The electron localisation function (ELF) was introduced by Becke and Edgecombe within the framework of the HF method;[165] it is related to the Fermi hole curvature (eq. II.39).

$$\eta(r) = \frac{1}{1+\left(\frac{D_\sigma}{D_\sigma^0}\right)^2} \tag{II.39}$$

D_σ and D_σ^0 designate the curvature of the electron pair density of the same spins (i.e., the Fermi hole) for the system considered and for the homogeneous electron gas of the same density, respectively. The ELF function may be considered as expressing the excess kinetic energy density due to the Pauli repulsion.[166,167]

In regions of weak Pauli repulsion, the ELF function is close to unity and such a situation corresponds to the occurrence of a single electron or to an opposite spin pair of electrons. In the case of a high probability of finding same-spin electrons, the ELF function is close to zero. ELF is a scalar function, thus the analysis of its gradient field allows location of local maxima (attractors) with the corresponding basins. There are two types of basins: the C(X) core basins where X designates the atom symbol and that occur formally if the atomic number Z > 2, and the valence basins, designated as V(X_1, X_2...) where X_1, X_2... is the list of symbols of atoms. The number of atoms included in the valence basin is related to its synaptic order; monosynaptic, disynaptic and polysynaptic valence basins occur. It means that the synaptic order is the number of neighbouring core basins that share a common boundary with the valence basin. In the case of a proton located in the valence basin, the synaptic order increases by one since the proton is treated formally as a core.

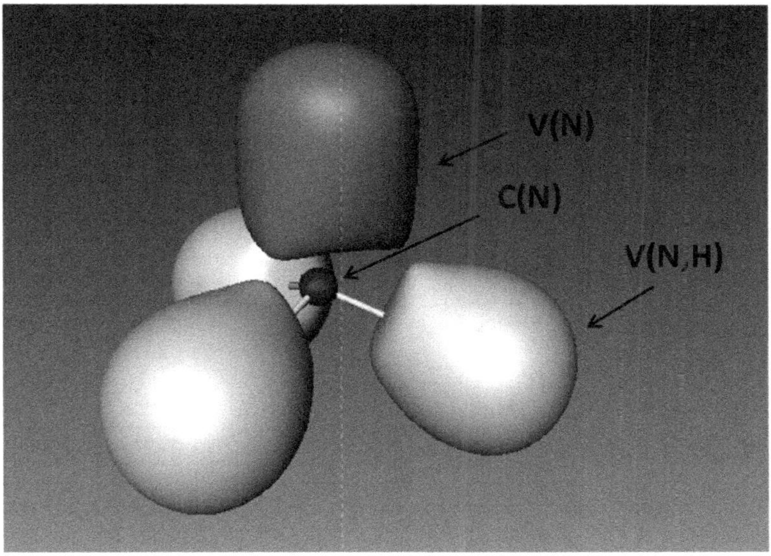

Fig. II.11 The ELF isosurface for $\eta(r) = 0.8$ that leads to the irreducible domains for NH_3 molecule. Reprinted with permission from ref. 168. Copyright 2014 Wiley-VCH.

For example, the partitioning of the molecular space of ammonia molecule, NH_3, leads to one core basin, C(N), one monosynaptic valence basin, V(N) and three protonated disynaptic valence basins, V(N,H) (see Fig. II.11).[168] The partitioning of the molecular space allows us to analyse the electronic structure of the considered chemical species and to deepen our understanding of the nature of intra- and intermolecular interactions. This analysis of the ELF function, $\eta(r)$,[168,169] is consistent with the Lewis valence concept.[170] The molecular space partitioning may be presented graphically (Fig. II.11). The ELF surfaces delimit space areas characterised by weak Pauli repulsion. For these areas, $\eta(r)$ is greater than the value fixed for the surface. In other words, these are areas characterised by electron charge concentration if $\eta(r)$ is large enough for the surface. These areas are often related to the concept of domains. The increase of $\eta(r)$ leads to greater partitioning of the molecular space where domains characterised by greater electron concentration occur, up to so-called irreducible domains that contain only one attractor. The domains which contain more than one attractor are marked as reducible.[171] Figure II.12 presents how the partitioning of the molecular space of the $FH\cdots NH_3$ complex changes with the increase of $\eta(r)$ for the surfaces surrounding

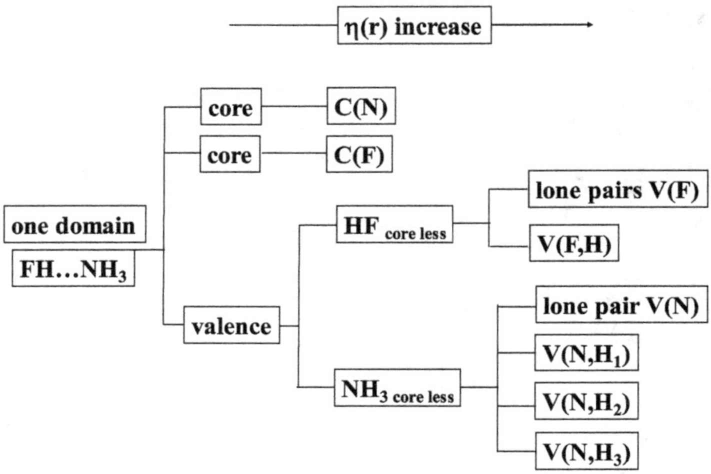

Fig. II.12 Localisation domain reduction tree diagrams of the FH...NH_3 complex. Modified with permission from ref. 172. Copyright 2000 Springer.

attractors.[172] $\eta(r)$ increase leads firstly to separation into irreducible core domains and one reducible valence domain that covers both HF and NH_3 units. Further increase of $\eta(r)$ leads to separation into irreducible domains. Figure II.12 is an example of the tree diagram that expresses the hierarchy of the bifurcation. In general, it is observed for tree diagrams that if the ELF function value for the surfaces is chosen close to zero, one domain covers all attractors. This is because the ELF function value equal to zero occurs for the infinitely great surface that covers the whole space.

Figures II.13 and II.14 show, for the $C_2H_2\cdots H^+\cdots C_2H_2$ complex, the graphical presentation and the tree diagram, respectively.[173] Figure II.13 presents the $\eta(r)$ localisation domains. One can see the trisynaptic V(H,C,C) valence domain that concerns the proton and the carbon centres of the acetylene molecule that is closer to the proton than the other acetylene moiety. It is reflected by the V(C,C) valence domains of both acetylene molecules. For that located closer to the proton the domain occupancy is equal to 4.08, while for the other a greater population of 4.97 is observed. This is because of electron charge transfer from the closer acetylene to the proton, i.e., to the V(H,C,C) domain. It means that this complex may be considered as the $C_2H_2\cdots H^+$ proton-donating species and the second C_2H_2 molecule as the proton acceptor (also discussed in Chapter IV).

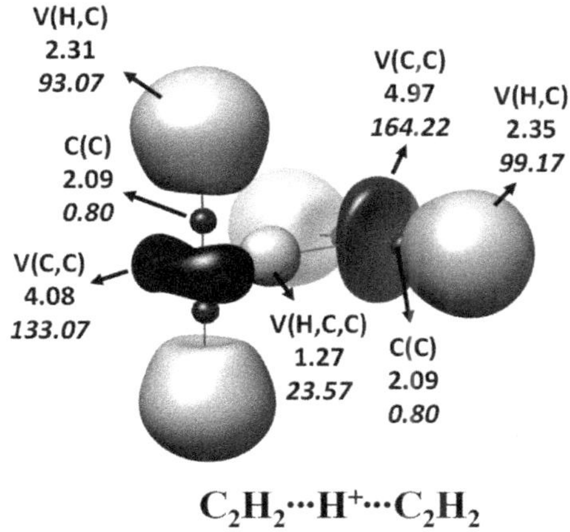

Fig. II.13 The $\eta(r) = 0.85$ localisation domains of the $C_2H_2\cdots H^+\cdots C_2H_2$ complex; the domain populations (top, in ē) and volumes (bottom, in au) are presented. Reprinted with permission from ref. 173. Copyright 2022 Wiley-VCH.

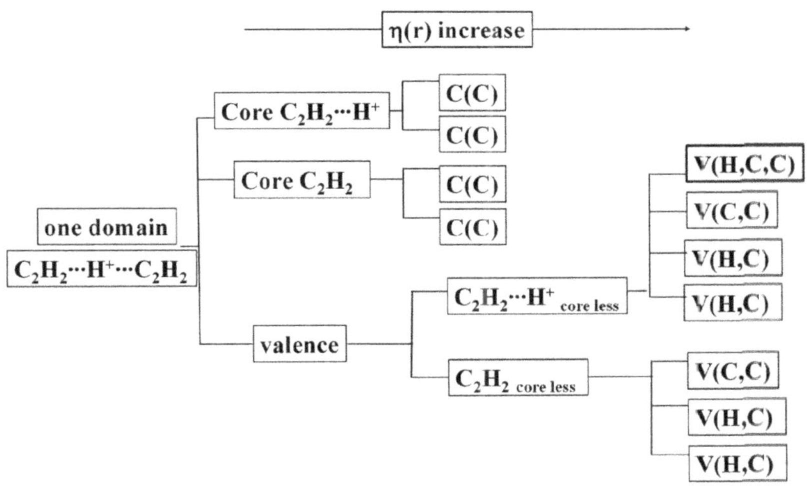

Fig. II.14 Localisation domain reduction tree diagram for $C_2H_2\cdots H^+\cdots C_2H_2$; V(C,C) valence domains may be divided into two valence basins. Reprinted with permission from ref. 173. Copyright 2022 Wiley-VCH.

Figure II.14 presents localisation domain reduction tree diagram for the $C_2H_2\cdots H^+\cdots C_2H_2$ complex. One can see here that the increase of $\eta(r)$ leads to the partitioning of one domain of the whole complex, into the core and valence fragments. The next splitting following further increase of $\eta(r)$ leads to four irreducible C(C) core domains (monosynaptic domains) containing core carbon attractors, valence disynaptic V(C,C) and V(H,C) domains, as well as one valence trisynaptic V(H,C,C) domain. It is worth mentioning here that the V(C,C) domains are reducible and that further increase of $\eta(r)$ results in their divisions.

Figure II.15 shows the graphical presentation of domains for the *trans* conformation of the dimer of thiosalicylic acid.[146] The valence basins are presented. The molecular graph of this dimer is shown in Fig. II.7a. The basins related to intermolecular O-H···O hydrogen bonds as well as to the intramolecular S···O chalcogen bonds are presented in Fig. II.15. This is a centrosymmetric dimer, thus two pairs of O-H···O and S···O interactions occur. This figure shows types of basins as well as their populations. One can see that the sums of populations of valence basins for S and O centres connected with hydrogen atoms are smaller than such sums for O centres playing the role of proton acceptor in the O-H···O hydrogen bonds. This is because the part of the electron charge of centres connected with hydrogen is moved to the V(S,H) and V(O,H) valence basins.

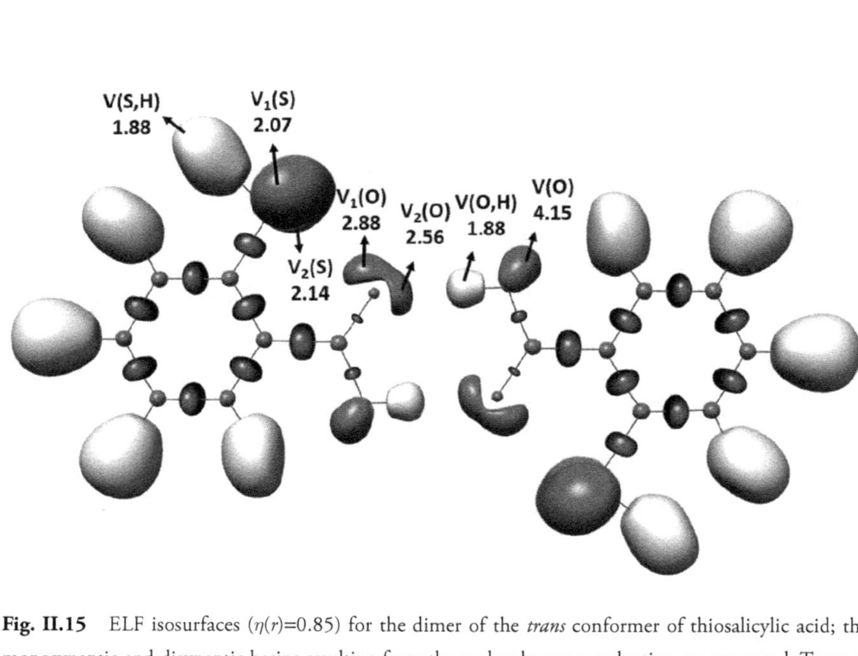

Fig. II.15 ELF isosurfaces ($\eta(r)$=0.85) for the dimer of the *trans* conformer of thiosalicylic acid; the monosynaptic and disynaptic basins resulting from the molecular space reduction are presented. Types of basins and their populations are shown. Reprinted with permission from ref. 146. Copyright 2021 American Chemical Society.

II.10 The Non-Covalent Interactions Approach

The Non-Covalent Interactions (NCI) approach that is useful to analyse intermolecular interactions, particularly σ-hole bonds and π-hole bonds, including hydrogen bonds, uses the reduced density gradient parameter, RDG. RDG is often marked by the letter s:[174,175]

$$s = \frac{1}{2(3\pi^2)^{1/3}} \frac{|\nabla \rho|}{\rho^{4/3}} \tag{II.40}$$

RDG has its roots in the DFT approach where it is used to describe the deviation of electron density of the system considered from homogeneous electron distribution.[176]

The name of the approach, "Non-Covalent Interactions", follows the term often applied in various studies to describe interactions that are not classified as covalent bonds. However, it is worth mentioning that there is no sharp border between covalent bonds and so-called non-covalent interactions. The covalency is related to the electron charge transfer from the Lewis base to the Lewis acid unit and it occurs for Lewis acid — Lewis base interactions, such as σ-hole and π-hole bonds. In this book the term non-covalent interactions is avoided; it is only used in this section for the NCI approach since such a name was proposed by its authors. In various studies this term is also used to name a broad class of interactions, including the Lewis acid — base interactions, which is not correct in the opinion of the author of this book.

RDG approaches zero for regions of covalent bonds as well as for regions of weaker interactions that correspond to electron density values of ~0.1-0.3 au and below ~0.05 au, respectively.[174] Hence, for isolated molecules or ions, the RDG values approach zero only for greater values of the electron density, ρ. For complexes linked by various weaker interactions, the corresponding RDG values close to zero occur for smaller values of ρ; however, RDG values close to zero also occur for greater ρ's that correspond to covalent bonds.

One can see that ρ vs. RDG plots may be characterised by spikes (RDG tends to zero) for bonds and a broad spectrum of stabilising interactions. However, such spikes may also imply repulsive interactions resulting from the steric crowding. Hence, to distinguish different types of interactions, particularly attractive from repulsive interactions, the λ_2 eigenvalue of the Hessian matrix of the electron density (eq. II.31) is taken into account to modify the plots.[174] For stabilising interactions, the BCP (3,–1) is observed with $\lambda_1 < \lambda_2 < 0 < \lambda_3$ while for the interaction resulting from steric crowding $\lambda_2 > 0$. Thus the sign(λ_2)ρ function was introduced instead of the electron density, ρ, to separate repulsive and attractive interactions.

Figure II.16 presents the case of the uracil — acetic acid complex which is linked by two hydrogen bonds: O-H···O and N-H···O.[177] Figure II.16A shows the ρ vs RDG plot where spikes are observed in three regions of about ~0.005 au, ~0.02 au and ~0.033 au that correspond to three types of interactions. The separation of these interactions is visible for the sign(λ_2)ρ vs RDG plot (Fig. II.16B) where the repulsive

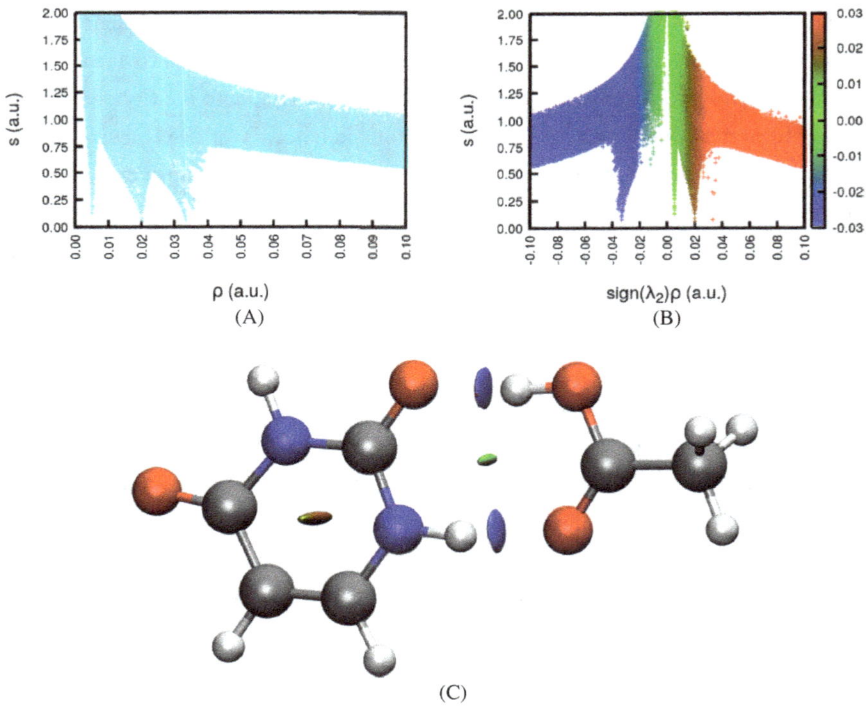

Fig. II.16 NCI analysis for the uracil — acetic acid complex. (A) ρ vs RDG plot; (B) sign(λ_2)ρ vs RDG plot; (C) molecular structure with RDG isosurfaces. Blue, green, and red colours indicate strong interactions, weak contacts, and steric clashes, respectively. Reprinted with permission from ref. 177. Copyright 2023 American Chemical Society.

interactions for positive sign(λ_2)ρ value at ~0.02 au occur that concern interactions in the uracil ring. The green spike at ~0.005 au of the latter plot indicates weak van der Waals interactions within the ring formed by the two hydrogen bonds mentioned above (positive and negative sign(λ_2)ρ values usually are attributed to very weak interactions). The spike at ~0.033 au of the sign(λ_2)ρ vs RDG plot corresponds to interactions moderate in strength, i.e., O-H···O and N-H···O hydrogen bonds. Figure II.16C presents the molecular structure of the uracil — acetic acid complex with RDG surfaces corresponding to interactions indicated at Figs. II.16A and II.16B by spikes.

Figure II.17 presents the centrosymmetric dimer of fluorophenylsilane, the sign(λ_2)ρ vs RDG plot as well as the molecular structure with RDG surfaces.[178] Surfaces ranging between two silicon centres and two CC bonds correspond to weak Si···π tetrel bonds. They are accompanied by spikes occurring at ~–0.005 au and ~+0.005 au. These two tetrel bonds are symmetry equivalent; the centre of inversion is situated between two units of the dimer. The π-electrons playing the role of donors in tetrel bonds come from one CC bond of the benzene rings. The red spike in sign(λ_2)ρ vs RDG plot corresponds to repulsive interactions in the benzene rings, which are also marked by red RDG surfaces located in the molecular structure.

One can see that the NCI approach is often applied to analyse interactions in different systems, and the areas of stabilising interactions are presented in corresponding

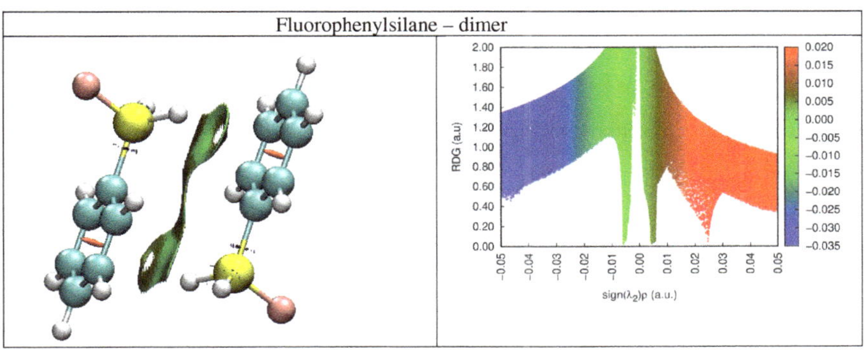

Fig. II.17 The RDG plot (right) and the corresponding molecular structure with RDG surfaces (left) for fluorophenylsilane dimer linked by Si···π interactions. Reprinted with permission from ref. 178. Copyright 2023 American Chemical Society.

molecular structures.[179–181] It is very important, however, that NCI was applied for the experimental electron density that comes from crystal structure measurements.[181,182]

It is worth mentioning that there are various modifications and even new methods related to NBO, QTAIM and NCI approaches. One can mention the electron localizability indicator[183–185] or the maximum probability domains concept.[186] The Density Overlap Regions Indicator is another approach to analyse different interactions, covering a broad range of interactions from strong covalent bonds to very weak interactions between closed-shell systems.[187]

II.11 The Car-Parinello Molecular Dynamics

The Car–Parrinello molecular dynamics (CPMD) method is useful to analyse and characterise various interactions.[5] This approach usually employs the following theoretical approaches and tools: MD, DFT, periodic boundary conditions and plane wave basis sets. In the CPMD method, the plane wave basis sets are often applied for valence electrons while the contribution and influence of core electrons are approximated by pseudopotentials. One can mention different applications of CPMD such as, for example, an analysis of structure and dynamics of liquid water, proton transfer in water clusters, or analyses of various chemical reactions.[188,189] It seems that among different inter- and intramolecular interactions, the analysis of hydrogen bonds with CPMD is the most often performed.[190–195]

What is the advantage of using the CPMD method in comparison with other approaches? The classical MD as a tool to analyse many-body systems uses predefined potentials, and it contains several limitations. Some may be removed. For example, many-body interaction energy terms may be incorporated that improves results of calculations concerning energetic parameters.[196,197] Numerous restrictions cannot be eliminated, however. These are, for example, interactions which need additional parametrisation, electronic structures of systems that change during simulations, and systems at high temperatures and/or high pressures. The CPMD approach overcomes many of such restrictions because interactions are calculated directly from the electronic structure of the system using DFT, while in the classical MD approach simulations rely on empirical potentials or in general on force fields that describe the interactions between atoms, and which do not change during simulations.

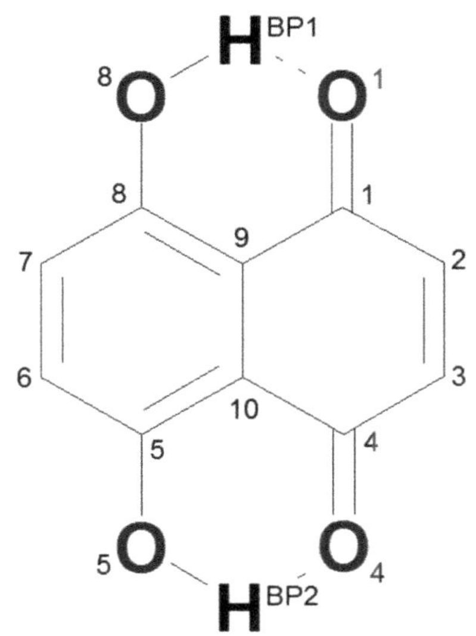

Scheme II.5 Naphthazarin with a numeration of atoms applied in descriptions included in this chapter. The dotted lines indicate intramolecular hydrogen bonds. BP1 and BP2 designate two H-bonded bridges. Reprinted with permission from ref. 198. Copyright 2021, Licensee MDPI, Basel, Switzerland.

Let us consider an example of the use of CPMD. The case considered concerns a hydrogen-bonded system.[198] Scheme II.5 presents the molecular structure of naphthazarin characterised by two intramolecular hydrogen-bonded bridges. CPMD calculations were performed here for the gas phase and for a crystal. As a starting point for these calculations, the experimental data of the naphthazarin crystal structure reported by Herbstein and co-workers were used.[199] Energy minimisation was performed with the use of the Hessian matrix of Schlegel,[200] with the Perdew, Burke and Ernzerhof functional[201] and with the use of the norm-conserving Troullier-Martins pseudopotentials.[202] A time step of 3 a.u. was fixed for CPMD calculations.

This allows us to investigate the processes of proton transfer for both intramolecular hydrogen bonds of naphtazarin. Figure II.18 shows time evolution of distances in hydrogen bonds of naphtazarin during the CPMD simulations, in the gas phase and in the crystal.[203] The time changes are expressed in picoseconds (1 picosecond, ps, = 10^{-12} second, s), the time step is usually expressed in parts of a femtosecond, i.e., in attoseconds (1 femtosecond, fs, = 10^{-15} second, while 1 attosecond, as, = 10^{-18} s). It is worth mentioning that the au unit of time is equivalent to about 24 as, thus the time

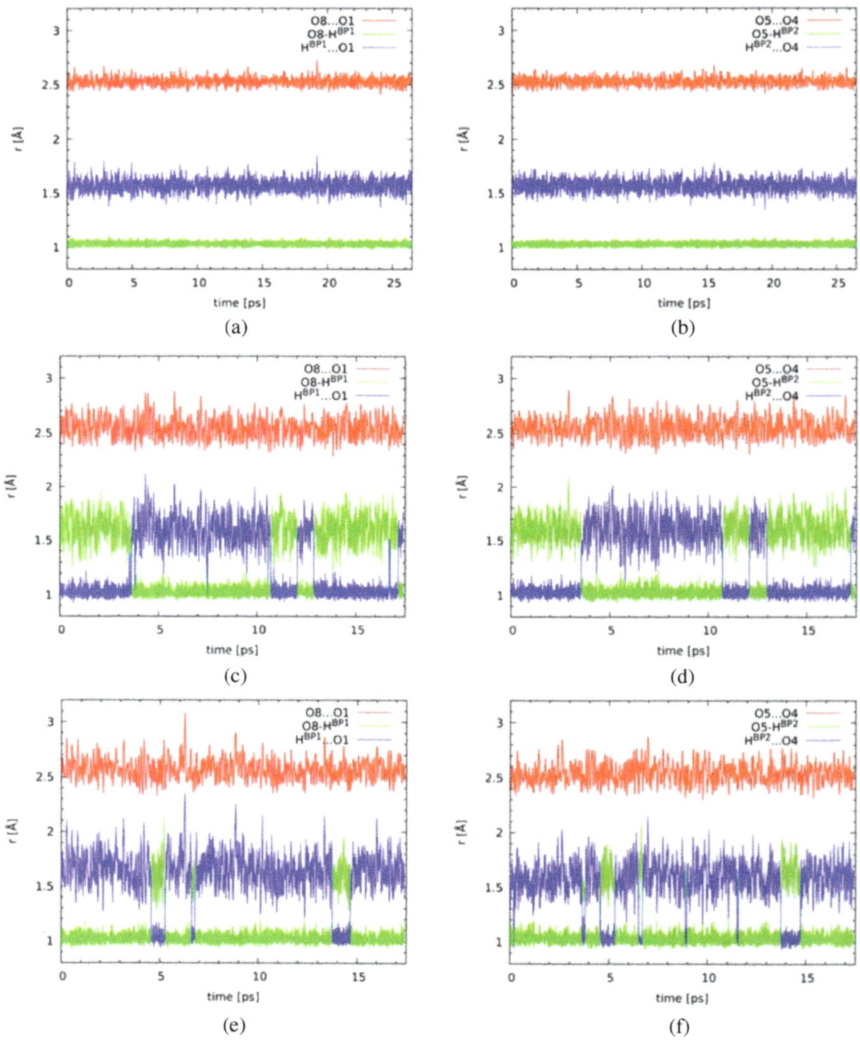

Fig. II.18 Time evolution of the metric parameters of the hydrogen bridges. Results of the CPMD simulation: (a) at 60 K, gas phase, bridge 1; (b) at 60 K, gas phase, bridge 2; (c) at 300 K, gas phase, bridge 1; (d) at 300 K, gas phase, bridge 2; (e) at 300 K, solid state, bridge 1; (f) at 300 K, solid state, bridge 2. Reprinted with permission from ref. 198. Copyright 2021, Licensee MDPI, Basel, Switzerland.

step applied amounts to about 72 as. The CPMD simulations on the naphtazarin molecular system performed at 60°C in the gas phase (Figs. II.18a and II.18b) show that protons of both hydrogen bond bridges are linked to O5 and O8 oxygen centres forming the O5-H and O8-H covalent bonds. In other words, the proton transfer is not observed with the change of time; during the time of 30 ps proton transfer for both bridges are not detected. The O···O and H···O distances oscillate at about 2.5–2.6 Å and 1.5–1.7 Å, respectively, while O-H bond length oscillates at 1.0–1.1 Å (see Figs. II.18a and II.18b). A very similar situation at 60°C is observed for these intramolecular hydrogen bonds in the crystal, where protons of hydrogen bonds are linked to O5 and O8 oxygen centres and proton transfer processes are not observed. It is not presented in Fig. II.18 since figures expressing the changes of geometrical parameters in time are very similar to those observed for the gas phase.

The situations change at the higher temperature of 300°C. First, the ranges of geometrical parameters are enlarged significantly. For example, the O···O distances change within ranges of 2.4–2.8 Å, but fluctuations outside these ranges are also observed. The same concerns H···O distances and O-H bond lengths, whose ranges are greater than those in the gas phase. Let us look at the left of Fig. II.18c. At the beginning the proton for O1-H-O8 bridge is close to O1 centre forming the O1-H bond since the corresponding distance oscillates at about 1 Å, and this proton is in contact with the O8 centre since the H···O distance oscillates at about 1.5 Å. After 4 ps, approximately, proton transfer is observed since the distances between the proton and O8 as well as O1 centres oscillate at about 1 Å and 1.5 Å, respectively. This state continues for 8 ps, approximately. The next proton transfers are observed later in time since the proton is close to O1 centre and next again to the O8 centre. Similar changes related to the proton transfer process are also observed for the O4-H-O5 bridge (Fig. II.18d).

The situation in the crystal (Figs. II.18e and II.18f) is in part like that observed in the gas phase (Figs. II18c and II.18d). There are the greater fluctuations here than those observed for the temperature of 60°C. However, the proton transfers, when they occur, are characterised by the shorter time of "changed states", less than 1 ps. Let us remember that for the gas phase at 300°C such "changed states" continue for 8 ps, approximately.

These results concerning the molecular structure of naphthazarin are in agreement with the corresponding early crystal structure which was chosen as a starting point of the CPMD simulations.[199] This structure has been determined by X-ray diffraction at 300 K and also by neutron diffraction at 60 K and 300 K. The neutron diffraction measurement at 300 K shows the disordered model with "one-half of the hydrogen atom" attached to each oxygen centre (Chapter III describes the crystal structures as well as methods of their determination).[199] The neutron diffraction results at 60 K show that the H-atoms are ordered. This means that the proton transfer processes that occur at higher temperature are detained at low temperatures, as was indicated by the CPMD simulations.

The same phenomenon of the order-disorder dependence on temperature of crystal structure measurement is observed for numerous other crystal structures. The structure of benzoic acid is an example; neutron diffraction measurements were performed at room temperature and at 130 K.[204] This structure consists of centrosymmetric dimers that are linked by two equivalent O-H···O hydrogen bonds through carboxylic groups (see Figs. II.2, II.4, II.7 and II.15 for examples of centrosymmetric carboxylic acids). The room temperature determination leads to the structure where the equalisation of C=O and C–O bonds in carboxylic groups is observed and where the "half-atoms" of hydrogen are located at oxygen centres of these groups. This clearly shows the disordered structure related to the double proton transfer phenomenon. On the other hand, the determination at 130 K leads to the ordered structure.

It was mentioned above that the CPMD approach is mainly applied to analyse hydrogen-bonded systems; the analysis of other Lewis acid — Lewis base interactions is rather rare. One can mention an example of the analysis of chalcogen bonds by the CPMD method that stabilise ligand conformation (acetazolamine) with species that mimic carbonic anhydrase.[205]

Summary

Various theoretical approaches are described in this chapter that are useful to analyse the Lewis acid — Lewis base interactions. The *ab initio* and DFT methods seem to be applied most often. They are often used to optimise molecular systems to obtain geometrical and energetic parameters. The binding and interaction energies inform of the strength of interactions, and they correlate with numerous other molecular parameters, such as geometrical ones. For example, the strength of interaction increases if the distance between interacting species decreases, up to the energetic minimum. The next shortening of this distance leads to weakening of the interaction up to the situation where the Pauli repulsion forces outweigh the attractive forces and the system becomes energetically unstable.

The deformation energy is the difference between the binding and interaction energies and it informs of "the cost" of linkage between interacting units. It means that the species that are in contact are usually deformed, and stronger interaction increases the possibility of greater deformation.

There are different schemes which allow us to decompose the energy of interaction. This consists of Pauli repulsion and of the attractive electrostatic, polarisation and dispersion terms. However, one should be careful because different decomposition schemes are characterised by terms possessing different physical meanings, and the names applied for terms are different in different schemes. For weak and moderate interactions, the electrostatic term is the most important among attractive terms. For strong and very strong interactions, the polarisation term related to the electron charge density shifts is more important as it is related to the covalent character of interactions. Very weak interactions, among them those which are not classified as the Lewis acid-base ones, are steered by dispersion forces.

QTAIM is another theoretical approach that provides additional parameters to characterise interactions. The characteristics of BCPs allow us to classify interactions as covalent bonds, partly covalent interactions and weaker interactions. Interactions between ions possess specific characteristics. The NBO approach provides information about orbital-orbital interactions; different kinds of interactions are characterised by specific orbital-orbital overlaps. For example, for the A-H···B hydrogen bond, the $n_B \rightarrow \sigma_{AH}^*$ orbital-orbital overlap is the most important, and is often treated as the signature of this interaction. Other methods are also often applied in various studies, such as NCI and CPMD. The theoretical tools that are described in this chapter were applied to analyse various molecular systems which are discussed in subsequent chapters.

Appendix

The following tools were used to construct figures presented in this chapter.

Figures II.5,6
AIMAll program — Reference: Keith, T.A. *AIMAll (Version 11.08.23)*; TK Gristmill Software. Overland Park, KS, USA, **2011**.

Scheme II.3, Figures II.1,2,3
Jmol program — Reference: Hanson, R.M. Jmol SMILES and Jmol SMARTS: specifications and applications. *J. Cheminform.* **2016**, *8*, 50.

Schemes II.2,4,5, Figures II.4,7,8,9,10,11,12,13,14,15,16,17,18
Reprinted with permission.

Scheme II.1
PowerPoint.

References

1. Levine, I.N. *Quantum Chemistry: Sixth Edition*, Pearson Educational Inc., New Jersey, **2009**.
2. Allinger, N.L. *Molecular Structure: Understanding Steric and Electronic Effects from Molecular Mechanics*, John Wiley & Sons, Hoboken, New Jersey, **2010**.
3. Leach, A.R. *Molecular Modelling: Principles and Applications Second Edition*, Pearson Education Limited, Harlow, **2001**.
4. Marx, D.; Hutter, J. *Ab Initio Molecular Dynamics*, Cambridge University Press, Cambridge, **2009**.
5. Car, R.; Parrinello, M. Unified approach for molecular dynamics and density-functional theory. *Phys. Rev. Lett.* **1985**, *55*, 2471–2474.
6. Allen, M.P.; Tildesley, D.J. *Computer Simulation of Liquids*, Oxford University Press, Oxford, New York, **1987**.
7. Hayashi, T.; Pertsin, A.J.; Grunze, M. Grand canonical Monte Carlo simulation of hydration forces between nonorienting and orienting structureless walls. *J. Chem. Phys.* **2002**, *117*, 6271–6280.
8. Tomasi, J.; Persico, M. Molecular interactions in solution: an overview of methods based on continuous distributions of the solvent. *Chem. Rev.* **1994**, *94*, 2027–2094.

9. Klamt, A.; Schüürmann, G. COSMO: a new approach to dielectric screening in solvents with explicit expressions for the screening energy and its gradient. *J. Chem. Soc. Perkin Trans. 2* **1993**, *2*, 799–805.
10. Klamt, A.; Moya, C.; Palomar, J. A comprehensive comparison of the IEFPCM and SS(V)PE continuum solvation methods with the COSMO approach. *J. Chem. Theory Comput.* **2015**, *11*, 4220–4225.
11. Sadlej, J. *Semi-Empirical Methods of Quantum Chemistry*, John Wiley & Sons, New York, Brisbane, Chichester, Toronto, **1985**.
12. Hoffmann, R. An extended Hückel theory. I. Hydrocarbons. *J. Chem. Phys.* **1963**, *39*, 1397–1412.
13. Piela, L. *Ideas of Quantum Chemistry*, Elsevier, Amsterdam, **2007**.
14. Lewars, E. *Computational Chemistry*, Kluwer Academic Publishers, Dordrecht, **2003**.
15. Cramer, C.J. *Essentials of Computational Chemistry: Second Edition*, John Wiley & Sons, Chichester, **2004**.
16. Jensen, F. *Introduction to Computational Chemistry: Second Edition*, John Wiley & Sons, Chichester, **2007**.
17. Hehre, W.J.; Radom, L.; Schleyer, P. v R.; Pople, J.A. *Ab Initio Molecular Orbital Theory*, John Wiley & Sons, New York, **1986**.
18. Grabowski, S.J. *Understanding Hydrogen Bonds: Theoretical and Experimental Views*, The Royal Society of Chemistry, Cambridge, **2021.**
19. Löwdin, P.-O. Correlation problem in many-electron quantum mechanics I. Review of different approaches and discussion of some current ideas. *Advan. Chem. Phys.* **1959**, *2*, 207–322.
20. Pople, J.A.; Head-Gordon, M.; Raghavachari, K. Quadratic configuration interaction. A general technique for determining electron correlation energies. *J. Chem. Phys.* **1987**, *87*, 5968–5975.
21. Schmitz, G.; Hättig, C. Accuracy of explicitly correlated local PNO-CCSD(T). *J. Chem. Theory Comput.* **2017**, *13*, 2623–2633.
22. Guo, Y.; Riplinger, C.; Becker, U.; Liakos, D. G.; Minenkov, Y.; Cavallo, L.; Neese, F. Communication: An improved linear scaling perturbative triples correction for the domain based local pair-natural orbital based singles and doubles coupled cluster method [DLPNOCCSD(T)]. *J. Chem. Phys.* **2018**, *148*, 011101.
23. Møller, C.; Plesset, M.S. Note on an approximation treatment for many-electron systems. *Phys. Rev.* **1934**, *46*, 618–622.
24. Roos, B.O.; Taylor, P.R.; Siegbahn, P.E.M. A complete active space SCF method (CASSCF) using a density matrix formulated super-CI approach. *Chem. Phys.* **1980**, *48*, 157–174.

25. Andersson, K.; Malmqvist, P.-A.; Roos, B.O. Second-order perturbation theory with a complete active space self-consistent field reference function. *J. Chem. Phys.* **1992**, *96*, 1218–1226.
26. Andersson, K.; Malmqvist, P.-A.; Roos, B.O.; Sadlej, A.J.; Wolinski, K. Second-order perturbation theory with a CASSCF reference function. *J. Phys. Chem.* **1990**, *94*, 5483–5488.
27. Angeli, C.; Cimiraglia, R.; Evangelisti, S.; Leininger.; Malrieu, J.P. Introduction of *n*-electron valence states for multireference perturbation theory. *J. Chem. Phys.* **2001**, *114*, 10252–10264.
28. Angeli, C.; Cimiraglia, R.; Malrieu, J.P. N-electron valence state perturbation theory: a fast implementation of the strongly contracted variant. *Chem. Phys. Lett.* **2001**, *350*, 297–305.
29. Angeli, C. ; Cimiraglia, R.; Malrieu, J.P. *n*-electron valence state perturbation theory: A spinless formulation and an efficient implementation of the strongly contracted and of the partially contracted variants. *J. Chem. Phys.* **2002**, *117*, 9138–9153.
30. Hohenberg, P.; Kohn, W. Inhomogeneous electron gas. *Phys. Rev.* **1964**, *136*, B864-B871.
31. Kohn, W.; Sham, L.J. Self-consistent equations including exchange and correlation effects. *Phys. Rev.* **1965**, *140*, A1133-A1138.
32. Perdew, J.P.; Ruzsinszky, A.; Constantin, L.A.; Sun, J.; Csonka, G.I. Some fundamental issues in ground-state density functional theory: a guide for the perplexed. *J. Chem. Theory Comput.* **2009,** *5*, 902–908.
33. Mardirossian, N.; Head-Gordon, M. Thirty years of density functional theory in computational chemistry: an overview and extensive assessment of 200 density functionals. *Mol. Phys.* **2017**, *115*, 2315–2372.
34. Zhao, Y.; Lynch, B.; Truhlar, D. Doubly hybrid meta DFT: New multi-coefficient correlation and density functional methods for thermochemistry and thermochemical kinetics. *J. Phys. Chem. A* **2004**, *108*, 4786–4791.
35. Grimme, S. Semiempirical hybrid density functional with perturbative second-order correlation. *J. Chem. Phys.* **2006**, *124*, 034108.
36. Santra, G.; Calinsky, R.; Martin, J.M.L. Benefits of range-separated hybrid and double-hybrid functionals for a large and diverse data set of reaction energies and barrier heights. *J. Phys. Chem. A* **2022**, *126*, 5492–5505.
37. Neugebauer, H.; Pinski, P.; Grimme, S.; Neese, F.; Bursch, M. Assessment of DLPNO-MP2 approximations in double-hybrid DFT. *J. Chem. Theory Comput.* **2023**, *19*, 7695–7703.

38. Grimme, S. Density functional theory with London dispersion corrections. *WIREs Comput. Mol. Sci.* **2011**, *1*, 211–228.
39. Slater, J.C. Atom shielding constants. *Phys. Rev.* **1930**, *36*, 57–64.
40. Boys, S.F. Electronic wave functions-I. A general method of calculation for the stationary states of any molecular system. *Proc. Roy. Soc.* **1950**, *200*, 542–554.
41. Gaussian 16, Revision C.01, Frisch, M. J.; Trucks, G. W.; Schlegel, H. B.; Scuseria, G. E.; Robb, M. A.; Cheeseman, J. R.; Scalmani, G.; Barone, V.; Petersson, G. A.; Nakatsuji, H.; Li, X.; Caricato, M.; Marenich, A. V.; Bloino, J.; Janesko, B. G.; Gomperts, R.; Mennucci, B.; Hratchian, H. P.; Ortiz, J. V.; Izmaylov, A. F.; Sonnenberg, J. L.; Williams-Young, D.; Ding, F.; Lipparini, F.; Egidi, F.; Goings, J.; Peng, B.; Petrone, A.; Henderson, T.; Ranasinghe, D.; Zakrzewski, V. G.; Gao, J.; Rega, N.; Zheng, G.; Liang, W.; Hada, M.; Ehara, M.; Toyota, K.; Fukuda, R.; Hasegawa, J.; Ishida, M.; Nakajima, T.; Honda, Y.; Kitao, O.; Nakai, H.; Vreven, T.; Throssell, K.; Montgomery, J. A., Jr.; Peralta, J. E.; Ogliaro, F.; Bearpark, M. J.; Heyd, J. J.; Brothers, E. N.; Kudin, K. N.; Staroverov, V. N.; Keith, T. A.; Kobayashi, R.; Normand, J.; Raghavachari, K.; Rendell, A. P.; Burant, J. C.; Iyengar, S. S.; Tomasi, J.; Cossi, M.; Millam, J. M.; Klene, M.; Adamo, C.; Cammi, R.; Ochterski, J. W.; Martin, R. L.; Morokuma, K.; Farkas, O.; Foresman, J. B.; Fox, D. J. Gaussian, Inc., Wallingford CT, **2016**.
42. Schmidt, M. W.; Baldridge, K. K.; Boatz, J. A.; Elbert, S. T.; Gordon, M. S.; Jensen, J. H.; Koseki, S.; Matsunaga, N.; Nguyen, K. A.; Su, S.; Windus, T. L.; Dupuis, M.; Montgomery, J. A. General atomic and molecular electronic structure system. *J. Comput. Chem.* **1993**, *14*, 1347–1363.
43. Neese, F.; Wennmohs, F.; Becker, U.; Riplinger, C. The ORCA quantum chemistry program package. *J. Chem. Phys.* **2020**, *152*, 224108.
44. te Velde, G.; Bickelhaupt, F.M.; Baerends, E.J.; Fonseca Guerra, C.; van Gisbergen, S.J.A.; Snijders J.G.; Ziegler, T. Chemistry with ADF. *J. Comput. Chem.* **2001**, *22*, 931–967.
45. Ditchfield, R; Hehre, W.J; Pople, J. A. Self-consistent molecular-orbital methods. IX. An extended Gaussian-type basis for molecular-orbital studies of organic molecules. *J. Chem. Phys.* **1971**, *54*, 724–728.
46. Dunning, T.H. Gaussian basis sets for use in correlated molecular calculations. I. The atoms boron through neon and hydrogen. *J. Chem. Phys.* **1989**, *90*, 1007–1023.
47. EMSL Basis Set Exchange: https://www.basissetexchange.org/
48. Sadlej, A.J. Medium-size polarized basis sets for high-level correlated calculations of molecular electric properties. *Collect. Czechoslov. Chem. Commun.* **1988**, *53*, 1995–2016.

49. Benkova, Z.; Sadlej, A.J.; Oakes, J.; Bell, S.E.J. Reduced-size polarized basis sets for calculations of molecular electric properties. I. The basis set generation. *J. Comput. Chem.* **2005**, *26*, 145–153.
50. Baranowska, A.; Sadlej, A.J. Polarized basis sets for accurate calculations of static and dynamic electric properties of molecules. *J. Comput. Chem.* **2010**, *31*, 552–560.
51. Roos, B.O.; Lindh, R.; Malmqvist, P-A.; Veryazov, V.; Widmark, P-O. New relativistic ANO basis sets for transition metal atoms. *J. Phys. Chem. A* **2005**, *109*, 6575–6579.
52. Ramachandran, K.I.; Deepa, G.; Namboori, K. *Computational Chemistry and Molecular Modeling*, Springer-Verlag, Berlin Heidelberg, **2008**.
53. ref. 1, p. 484.
54. Wiberg, K.B. Basis set effects on calculated geometries: 6-311++ G** vs. aug-cc-pVDZ. *J. Comput. Chem.* **2004**, *25*, 1342–1346.
55. Dunning Jr., T.H.; Hay, P.J. Gaussian Basis Sets for Molecular Calculations. Chapter in Schaefer III, H.F. (Ed.), *Modern Theoretical Chemistry*, Vol. 3, Plenum, New York, **1977**, pp. 1–28.
56. Feller, D. Application of systematic sequences of wave functions to the water dimer. *J. Chem. Phys.* **1992**, *96*, 6104–6114.
57. Helgaker, T.; Klopper, W.; Koch, H.; Noga, J. Basis-set convergence of correlated calculations on water. *J. Chem. Phys.* **1997**, *106*, 9639–9646.
58. Barbieri, P.L.; Fantin, P.A.; Jorge, F.E. Gaussian basis sets of triple and quadruple zeta valence quality for correlated wave functions. *Mol. Phys.* **2006**, *104*, 2945–2954.
59. Feller, D. Benchmarks of improved complete basis set extrapolation schemes designed for standard CCSD(T) atomization energies. *J. Chem. Phys.* **2013**, *138*, 074103.
60. Sheng, X.; Qian, S.; Hu, F. On the performance of an complete basis set extrapolation scheme tailored for the equilibrium distance applied to the helium dimer potential energy surface. *Comput. Theory Chem.* **2017**, *1105*, 14–17.
61. Varandas, A.J.C. CBS extrapolation in electronic structure pushed to the end: a revival of minimal and sub-minimal basis sets. *Phys. Chem. Chem. Phys.* **2018**, *20*, 22084–22098.
62. Pitoňák, M.; Hebelmann, A. Accurate intermolecular interaction energies from a combination of MP2 and TDDFT response theory. *J. Chem. Theory Comput.* **2010,** *6*, 168–178.
63. Jurečka, P.; Šponer, J.; Černy, J.; Hobza, P. Benchmark database of accurate (MP2 and CCSD(T) complete basis set limit) interaction energies of small model complexes, DNA base pairs, and amino acid pairs. *Phys. Chem. Chem. Phys.* **2006**, *8*, 1985–1993.
64. Halkier, A.; Helgaker, T.; Jørgensen, P.; Klopper, W.; Koch, H.; Olsen, J.; Wilson, A.K. Basis-set convergence in correlated calculations on Ne, N_2, and H_2O. *Chem. Phys. Lett.* **1998**, *286*, 243–252.

65. ref. 13, pp. 684–691.
66. Szalewicz, K.; Jeziorski, B. Comment on "On the importance of the fragment relaxation energy terms in the estimation of the basis set superposition error correction to the intermolecular interaction energy" [J. Chem. Phys. 104, 8821 (1996)]. *J Chem. Phys.* **1998**, 109, 1198–1200.
67. Boys, S.F.; Bernardi, F. The calculation of small molecular interactions by the differences of separate total energies. Some procedures with reduced errors. *Mol. Phys.* **1970**, *19*, 553–566.
68. Van Duijneveldt, F.B.; Van Duijneveldt-van de Rijdt, J.G.C.M.; Van Lenthe, J.H. State of the art in counterpoise theory. *Chem. Rev.* **1994**, *94*, 1873–1885.
69. Grabowski, S.J.; Sokalski, W.A. Are various σ-hole bonds steered by the same mechanisms? *ChemPhysChem* **2017**, *18*, 1569–1577.
70. Herman, K.M.; Aprà, E.; Xantheas, S.S. A critical comparison of CH···π versus π···π interactions in the benzene dimer: obtaining benchmarks at the CCSD(T) level and assessing the accuracy of lower scaling methods. *Phys. Chem. Chem. Phys.* **2023**, *25*, 4824–4838.
71. Peterson, K.A.; Woon, D.E.; Dunning Jr., T.H. Benchmark calculations with correlated molecular wave functions. IV. The classical barrier height of the H + H2 → H2 + H reaction. *J. Chem. Phys.* **1994**, *100*, 7410–7415.
72. Simon, S.; Duran, M.; Dannenberg, J.J. How does basis set superposition error change the potential surfaces for hydrogen-bonded dimers? *J. Chem. Phys.* **1996**, *105*, 11024–11031.
73. Simon, S.; Duran, M.; Dannenberg, J.J. Effect of basis set superposition error on the water dimer surface calculated at Hartree- Fock, Møller- Plesset, and density functional theory levels. *J. Phys. Chem. A* **1999**, *103*, 1640–1643.
74. Salvador, P.; Simon, S.; Duran, M. C–H··· O H-bonded complexes: how does basis set superposition error change their potential-energy surfaces? *J. Chem. Phys.* **2000**, *113*, 5666–5674.
75. Simon, S.; Bertran, J.; Sodupe, M. Effect of counterpoise correction on the geometries and vibrational frequencies of hydrogen bonded systems. *J. Phys. Chem. A*, **2001**, *105*, 4359–4364.
76. Foresman, J.B.; Frisch, Æ. *Exploring Chemistry with Electronic Structure Methods: Second Edition*, Gaussian Inc., Pittsburg, **1996**.
77. Alkorta, I.; Blanco, F.; Deyà, P.M.; Elguero, J.; Estarellas, C.; Frontera, A.; Quiñonero, D. Cooperativity in multiple unusual weak bonds. *Theor. Chem. Acta* **2010**, *126*, 1–14.
78. Grabowski, S.J. π-hole tetrel bonds — Lewis acid properties of metallylenes. *Crystals* **2022**, *12*, 112.

79. Kitaura, K.; Morokuma, K. A new energy decomposition scheme for molecular interactions within the Hartree-Fock approximation. *Int. J. Quantum Chem.* **1976**, *10*, 325–340.
80. Umeyama, H.; Morokuma,K.; Yamabe, S. Molecular orbital studies of electron donor-acceptor complexes. 4. Energy decomposition analysis of halo-complexes: ammonia-fluorine, -chlorine, -chlorine fluoride, methanamine-chlorine fluoride, formaldehyde-fluorine, hydrogen fluoride-chlorine fluoride, and fluorine-fluorine. *J. Am. Chem. Soc.* **1977**, *99*, 330–343.
81. Baerends, E.J.; Ziegler, T.; Atkins, A.J.; Autschbach, J.; Baseggio, O.; Bashford, D.; Bérces, A.; Bickelhaupt, F.M.; Bo, C.; Boerrigter, P.M. *et al.* ADF2019, SCM, Theoretical Chemistry; Vrije Universiteit: Amsterdam, **2019**; http://www.scm.com/
82. Ziegler, T.; Rauk, A. CO, CS, N_2, PF_3, and $CNCH_3$ as **σ** donors and **π** acceptors. A theoretical study by the Hartree-Fock-Slater transition-state method. *Inorg. Chem.* **1979**, *18*, 1755–1759.
83. von Hopffgarten, M.; Frenking, G. Energy decomposition analysis. *WIREs Comput. Mol. Sci.* **2012**, *2*, 43–62.
84. Politzer, P.; Murray, J.S.; Clark, T. Halogen bonding and other σ-hole interactions: a perspective. *Phys. Chem. Chem. Phys.* **2013**, *15*, 11178–11189.
85. Sokalski, W.A.; Roszak, S. Efficient techniques for the decomposition of intermolecular interaction energy at SCF level and beyond. *J. Mol. Struct. (Theochem)* **1991**, *234*, 387–400.
86. Sokalski, W.A.; Roszak, S.; Pecul, K. An efficient procedure for decomposition of the SCF interaction energy into components with reduced basis set dependence. *Chem. Phys. Lett.* **1988**, *153*, 153–159.
87. Politzer, P.; Murray, J.S.; Clark, T. Halogen bonding: an electrostatically-driven highly directional noncovalent interaction. *Phys. Chem. Chem. Phys.* **2010**, *12*, 7748–7757.
88. Jeziorski, B.; Moszyński, R.; Szalewicz, K. Perturbation theory approach to intermolecular potential energy surfaces of van der Waals complexes. *Chem. Rev.* **1994**, *94*, 1887–1930.
89. Parker, T.M.; Burns, L.A.; Parrish, R.M.; Ryno, A.; Sherrill, C.D. Levels of symmetry adapted theory (SAPT). I. Efficiency and performance for interaction energies. *J. Chem. Phys.* **2014,** *140*, 094106.
90. Aarabi, M.; Gholami, S.; Grabowski, S.J. S-H···O and O-H···O hydrogen bonds — comparison of dimers of thiocarboxylic and carboxylic acids. *ChemPhysChem* **2020**, *21*, 1653–1664.
91. Grabowski, S.J. Halogen bonds with carbenes acting as Lewis base units: complexes of imidazol-2-ylidene: theoretical analysis and experimental evidence. *Phys. Chem. Chem. Phys.* **2023**, *25*, 9636–9647.

92. Reed, E.; Curtiss, L.A.; Weinhold, F. Intermolecular interactions from a natural bond orbital, donor-acceptor viewpoint. *Chem. Rev.* **1988**, *88*, 899–926.
93. Weinhold, F.; Landis, C. *Valency and Bonding:: A Natural Bond Orbital Donor — Acceptor Perspective*, Cambridge University Press, Cambridge, **2005**.
94. Grabowski, S.J. unpublished results.
95. Glendening, E.D.; Badenhoop, J.K.; Reed, A.E.; Carpenter, J.E.; Bohmann, J.A.; Morales, C.M.; Landis, C.R.; Weinhold, F. *NBO 6.0*. Theoretical Chemistry Institute, University of Wisconsin, Madison, WI, **2013**; http://nbo6.chem.wisc.edu/
96. Pimentel, G.C.; McClellan, A.L. *The Hydrogen Bond*, W.H. Freeman and Company, San Francisco and London, **1960**, p. 6 and pp. 193–205.
97. Jeffrey, G.A.; Saenger, W. *Hydrogen Bonding in Biological Structures*, Springer, Berlin, **1991**.
98. Jeffrey, G.A. *An Introduction to Hydrogen Bonding*, Oxford University Press, New York, **1997**.
99. Scheiner, S. *Hydrogen Bonding: A Theoretical Perspective*, Oxford University Press, New York, **1997**.
100. Hobza, P.; Havlas, Z. Blue-shifting hydrogen bonds. *Chem. Rev.* **2000**, *100*, 4253–4264.
101. Alabugin, I.V.; Manoharan, M.; Peabody, S.; Weinhold, F. Electronic basis of improper hydrogen bonding: a subtle balance of hyperconjugation and rehybridization. *J. Am. Chem. Soc.* **2003**, *125*, 5973–5987.
102. Bent, H.A. An appraisal of valence-bond structures and hybridization in compounds of the first-row elements. *Chem. Rev.* **1961**, *61*, 275–311.
103. Weinhold, F.; Klein, R.A. What is a hydrogen bond? Mutually consistent theoretical and experimental criteria for characterizing H-bonding interactions. *Mol. Phys.* **2012**, *110*, 565–579.
104. The bold and underlined words of this definition cited here occur also in the original study of Weinhold and Klein (ref. 103).
105. Grabowski, S.J. Red- and blue-shifted hydrogen bonds: the bent rule from quantum theory of atoms in molecules perspective. *J. Phys. Chem. A* **2011**, *115*, 12789–12799.
106. Joseph, J.; Jemmis, E.D. Red-, blue-, or no-shift in hydrogen bonds: a unified explanation. *J. Am. Chem. Soc.* **2007**, *129*, 4620–4632.
107. Grabowski, S.J. Halogen bond and its counterparts: Bent's rule explains the formation of nonbonding interactions. *J. Phys. Chem. A* **2011**, *115*, 12340–12347.
108. Lipkowski, P.; Grabowski, S. J.; Leszczynski, J. Properties of the halogen-hydride interaction: an ab initio and "atoms in molecules" analysis. *J. Phys. Chem.* A **2006**, *110*, 10296–10302.

109. Shannon, R.D. Revised effective ionic radii and systematic studies of interatomic distances in halides and chalcogenides. *Acta Cryst.* **1976**, *A32*, 751–767.
110. Verhoeven, J.W. Glossary of terms used in photochemistry (IUPAC Recommendations 1996). *Pure and Appl. Chem.* **1996**, *68*, 2223–2286.
111. Mulliken, R.S. Electronic population analysis on LCAO–MO molecular wave functions. I. *J. Chem. Phys.* **1955**, *23*, 1833–1840.
112. Grabowski, S.J. Dihydrogen bond and X–H···σ interaction as sub-classes of hydrogen bond. *J. Phys.Org. Chem.* **2013**, *26*, 452–459.
113. Grabowski, S.J. A-H···σ hydrogen bonds: dihydrogen and cycloalkanes as proton acceptors. *ChemPhysChem* **2019**, *20*, 565–574.
114. Bader, R.F.W. *Atoms in Molecules: A Quantum Theory*, Oxford University Press, Oxford, **1990**.
115. Woodward, R.B.; Hoffmann, R. The conservation of orbital symmetry. *Angew. Chem. Int. Ed.* **1969**, *8*, 781–853.
116. Pearson, R.G. Symmetry rules for chemical reactions. *Acc. Chem. Res.* **1971**, *4*, 152–160.
117. Popelier, P. *Atoms in Molecules: An Introduction*, Prentice Hall, Pearson Education Limited, **2000**.
118. Matta, C.J.; Boyd, R.J. (Eds.). *Quantum Theory of Atoms in Molecules: Recent Progress in Theory and Application*, Wiley-VCH, **2007**.
119. Bader, R.F.W. Atoms in molecules. *Acc. Chem. Res.* **1985**, *18*, 9–15.
120. Bader, R.F.W. A quantum theory of molecules structure and its applications. *Chem. Rev.* **1991**, *91*, 893–928.
121. Grabowski, S.J. What is the covalency of hydrogen bonding? *Chem. Rev.* **2011**, *111*, 2597–2625.
122. Bader, R.F.W. A bond path: a universal indicator of bonded interactions. *J. Phys. Chem. A* **1998**, *102*, 7314–7323.
123. Bader, R.F.W. Bond paths are not chemical bonds. *J. Phys. Chem. A* **2009**, *113*, 10391–10396.
124. Bader, R.F.W. Pauli repulsions exist only in the eye of the beholder. *Chem. Eur. J.* **2006**, *12*, 2896–2901.
125. Desiraju, G.R.; Steiner, T. *The Weak Hydrogen Bond in Structural Chemistry and Biology*, Oxford University Press, New York, **1999**.
126. Nishio, M.; Hirota, M.; Umezawa, Y. *The CH/π Interaction: Evidence, Nature, and Consequences*, Wiley-VCH, **1998**.
127. Luaña, V.; Mori-Sánchez, P.; Costales, A.; Blanco, M.A.; Pendás, A.M. Non-nuclear maxima of the electron density on alkaline metals. *J. Chem. Phys.* **2003**, *119*, 6341–6350.

128. Platts, J.A.; Overgaard, J.; Jones, C.; Iversen, B.B.; Stasch, A. First experimental characterization of a non-nuclear attractor in a dimeric magnesium(I) compound. *J. Phys. Chem. A* **2011**, *115*, 194–200.
129. Grabowski, S.J. What Can Be Learnt from a Location of Bond Paths and from Electron Density Distribution?. In Chauvin, R.; Lepetit, C.; Silvi, B.; Alikhani, E. (Eds.), *Applications of Topological Methods in Molecular Chemistry*, Springer, **2016**, pp. 399–433.
130. Bader, R.F.W.; Essen, H. The characterization of atomic interactions. *J. Chem. Phys.* **1984**, *80*, 1943–1960.
131. Koch, U.; Popelier, P.L.A. Characterization of C-H-O hydrogen bonds on the basis of the charge density. *J. Phys. Chem.* **1995**, *99*, 9747–9754.
132. Pacios, L.F.; Gálvez, O.; Gómez, P.C. Variation of geometries and electron properties along proton transfer in strong hydrogen-bond complexes. *J. Chem. Phys.* **2005**, *122*, 214307.
133. Vener, M.V.; Manaev, A.V.; Egorova, A.N.; Tsirelson, V.G. QTAIM study of strong H-bonds with the O–H···A fragment (A = O, N) in three-dimensional periodical crystals. *J. Phys. Chem. A* **2007**, *111*, 1155–1162.
134. Gatti, C.; Cargnoni, F.; Bertini, L. Chemical information from the source function. *J. Comput. Chem.* **2003**, *24*, 422–436.
135. Mariam, Y.H.; Musin, R. Transition from moderate to strong hydrogen bonds: its identification and physical bases in the case of O–H···O intramolecular hydrogen bonds. *J. Phys. Chem. A* **2008**, *112*, 134–145.
136. Mó, O.; Yánez, M.; Elguero, J. Cooperative (nonpairwise) effects in water trimers: An *ab initio* molecular orbital study. *J. Chem. Phys.* **1992**, *97*, 6628–6638.
137. Mó, O.; Yánez, M.; Elguero, J. Cooperative effects in the cyclic trimer of methanol. An ab initio molecular orbital study. *J. Mol. Struct. (Theochem)* **1994**, *314*, 73–81.
138. Espinosa, E.; Molins, E.; Lecomte, C. Hydrogen bond strengths revealed by topological analyses of experimentally observed electron densities. *Chem. Phys. Lett.* **1998**, *285*, 170–173.
139. Galvez, O.; Gomez, P.C.; Pacios, L.F. Approximate kinetic energy density for intermolecular regions in hydrogen bond dimers. *Chem. Phys. Lett.* **2001**, *337*, 263–268.
140. Galvez, O.; Gomez, P.C.; Pacios, L.F. Variation with the intermolecular distance of properties dependent on the electron density in hydrogen bond dimers. *J. Chem. Phys.* **2001**, *115*, 11166–11184.
141. Galvez, O.; Gomez, P.C.; Pacios, L.F. Variation with the intermolecular distance of properties dependent on the electron density in cyclic dimers with two hydrogen bonds. *J. Chem. Phys.* **2003**, *118*, 4878–4895.

142. Pacios, L.F. Topological descriptors of the electron density and the electron localization function in hydrogen bond dimers at short intermonomer distances. *J. Phys. Chem. A* **2004**, *108*, 1177–1188.
143. Knop, O.; Rankin, K.N.; Boyd, R.J. Coming to grips with N–H···N bonds. 1. Distance relationships and electron density at the bond critical point. *J. Phys. Chem. A* **2001**, *105*, 6552–6566.
144. Knop, O.; Rankin, K.N.; Boyd, R.J. Coming to grips with N–H···N bonds. 2. Homocorrelations between parameters deriving from the electron density at the bond critical point. *J. Phys. Chem. A* **2003**, *107*, 272–284.
145. Grabowski, S.J. Intramolecular hydrogen bond energy and its decomposition — O-H···O interactions. *Crystals* **2021**, *11*, 5.
146. Gholami, S.; Aarabi, M.; Grabowski, S.J. Coexistence of intra- and intermolecular hydrogen bonds: salicylic acid and salicylamide and their thiol counterparts. *J. Phys. Chem. A* **2021**, *125*, 1526–1539.
147. Rozas, I.; Alkorta, I.; Elguero, J. Behavior of ylides containing N, O, and C atoms as hydrogen bond acceptors. *J. Am. Chem. Soc.* **2000**, *122*, 11154–11161.
148. Cremer, D.; Kraka, E. A description of the chemical bond in terms of local properties of electron density and energy. *Croat. Chem. Acta* **1984**, *57*, 1259–1281.
149. Jenkins, S.; Morrison, I. The chemical character of the intermolecular bonds of seven phases of ice as revealed by ab initio calculation of electron densities. *Chem. Phys. Lett.* **2000**, *317*, 97–102.
150. Arnold, W.D.; Oldfield, E. The chemical nature of hydrogen bonding in proteins via NMR: *J*-couplings, chemical shifts, and AIM theory. *J. Am. Chem. Soc.* **2000**, *122*, 12835–12841.
151. Gavezzotti, A. Comparing the strength of covalent bonds, intermolecular hydrogen bonds and other intermolecular interactions for organic molecules: X-ray diffraction data and quantum chemical calculations. *New J. Chem.* **2016**, *40*, 6848–6853.
152. Grabowski, S.J. Covalent character of hydrogen bonds enhanced by π-electron delocalization. *Croat. Chem. Acta* **2009**, *82*, 185–192.
153. Espinosa, E.; Alkorta, I.; Elguero, J.; Molins, E. From weak to strong interactions: A comprehensive analysis of the topological and energetic properties of the electron density distribution involving X–H···F–Y systems. *J. Chem. Phys.* **2002**, *117*, 5529–5542.
154. Grabowski, S.J.; Sokalski, W.A.; Dyguda, E.; Leszczynski, J. Quantitative classification of covalent and noncovalent H-bonds. *J. Phys. Chem. B* **2006**, *110*, 6444–6446.
155. Fradera, X.; Austen, M.A.; Bader, R.F.W. The Lewis model and beyond. *J. Phys. Chem. A* **1999**, *103*, 304–314.

156. Fradera, X.; Poater, J.; Simon, S.; Duran, M.; Solà, M. The calculation of electron localization and delocalization indices at the Hartree–Fock, density functional and post-Hartree–Fock levels of theory. *Theor. Chem. Acc.* **2002**, *107*, 362–371.
157. Poater, J.; Fradera, X.; Solà, M.; Duran, M.; Simon, S. On the electron-pair nature of the hydrogen bond in the framework of the atoms in molecules theory *Chem. Phys. Lett.* **2003**, *369*, 248–255.
158. Wiberg, K. Application of the Pople-Santry-Segal CNDO method to the cyclopropylcarbinyl and cyclobutyl cation and to bicyclobutane. *Tetrahedron* **1968**, *24*, 1083–1096.
159. Mayer, I. Bond order and valence indices: a personal account. *J. Comput. Chem.* **2007**, *28*, 204–221.
160. Bader, R.F.W.; Matta, C.F. Atomic charges are measurable quantum expectation values: a rebuttal of criticisms of QTAIM charges. *J. Phys. Chem. A* **2004**, *108*, 8385–8394.
161. Cioslowski, J.; Mixon, S.T. Topological properties of electron density in search of steric interactions in molecules: electronic structure calculations on ortho-substituted biphenyls. *J. Am. Chem. Soc.* **1992**, *114*, 4382–4387.
162. Cioslowski, J.; Mixon, S.T. Universality among topological properties of electron density associated with the hydrogen–hydrogen nonbonding interactions. *Can. J. Chem.* **1992**, *70*, 443–449.
163. Jabłoński, M. Bond paths between distant atoms do not necessarily indicate dominant interactions. *J. Comput. Chem.* **2018**, *39*, 2183–2195.
164. Jabłoński, M. On the uselessness of bond paths linking distant atoms and on the violation of the concept of privileged exchange channels. *ChemistryOpen* **2019**, *8*, 497–507.
165. Becke, A.D.; Edgecombe, K.E. A simple measure of electron localization in atomic and molecular systems. *J. Chem. Phys.* **1990**, *92*, 5397–5403.
166. Savin, A.; Becke, A.D.; Flad, J.; Nesper, R.; Preuss, H.; von Schnering, H.G. A new look at electron localization. *Angew. Chem. Int. Ed.* **1991**, *30*, 409–412.
167. Savin, A.; Silvi, B.; Colonna, F. Topological analysis of the electron localization function applied to delocalized bonds. *Can. J. Chem.* **1996**, *74*, 1088–1096.
168. Grin, Y.; Savin, A.; Silvi, B. The ELF Perspective of Chemical Bonding, in Frenking, G.; Shaik, S. (Eds.), *The Chemical Bond: Fundamental Aspects of Chemical Bonding*, Wiley-VCH, Weinheim, **2014**, pp. 354–382.
169. Silvi, B.; Savin, A. Classification of chemical bonds based on topological analysis of electron localization functions. *Nature* **1994**, *371*, 683–686.
170. Lewis, G.N. *Valence and the Structure of Atoms and Molecules*, American Chemical Society Monograph Series, New York, **1923**.

171. Silvi, B.; Fourré, I.; Alikhani, M.E. The topological analysis of the electron localization function. a key for a position space representation of chemical bonds. *Monastsh. Chem.* **2005**, *136*, 855–879.
172. Fuster, F.; Silvi, B. Does the topological approach characterize the hydrogen bond? *Theor. Chem. Acc.* **2000**, *104*, 13–21.
173. Gholami, S.; Aarabi, M.; Grabowski, S.J. Proton and lithium cations linked to π-electron and σ-electron systems: are such interactions beyond or within the definition of hydrogen/lithium bond? *ChemPhysChem* **2022**, *23*, e202200273.
174. Johnson, E.R.; Keinan, S.; Mori-Sánchez, P.; Contreras-García, J.; Cohen, A.J.; Yang, W. Revealing noncovalent interactions. *J. Am. Chem. Soc.* **2010**, *132*, 6498–6506.
175. Contreras-García, J. ; Johnson, E.R.; Keinan, S.; Chaudret, R.; Piquemal, J-P.; Beratan, D.N.; Yang, W. NCIPLOT: A program for plotting noncovalent interaction regions. *J. Chem. Theory Comput.* **2011**, *7*, 625–632.
176. Zupan, A.; Burke, K.; Ernzerhof, M.; Perdew, J.P. Distributions and averages of electron density parameters: Explaining the effects of gradient corrections. *J. Chem. Phys.* **1997**, *106*, 10184–10193.
177. Wieduwilt, E.K.; Boto, R.A.; Macetti, G.; Laplaza, R.;Contreras-García, J.; Genoni, A. Extracting quantitative information at quantum mechanical level from noncovalent interaction index analyses. *J. Chem. Theory Comput.* **2023**, *19*, 1063–1079.
178. Aarabi, M.; Gholami, S.; Grabowski, S.J. Double centrosymmetric Si···π tetrel bonds as new synthons– evidences from crystal structures and DFT calculations. *J. Phys. Chem. A* **2023**, *127*, 9995–10007.
179. Narth, C.; Maroun, Z.; Boto, R.A.; Chaudret, R.; Bonnet, M-L.; Piquemal, J-P.; Contreras-Garcia, J. A Complete NCI Perspective: From New Bonds to Reactivity, in Chauvin, R.; Lepetit, C.; Silvi, B.; Alikhani, E. (Eds.), *Applications of Topological Methods in Molecular Chemistry*, Springer, **2016**, pp. 491–527.
180. Boto, R.A.; Contreras-Garcia, J. Beyond QTAIM: NCI Indexes as a Tool to Reveal Intermolecular Bonds in Molecular Aggregates, in Novoa, J.J. (Ed.), *Intermolecular Interactions in Crystals: Fundamentals of Crystal Engineering*, The Royal Society of Chemistry, United Kingdom, **2018**, pp. 222–257.
181. Gatti, C.; Forni, A. Beyond Revealing the Intermolecular Bonds in Molecular Crystals Through Charge Density Methods, in Novoa, J.J. (Ed.), *Intermolecular Interactions in Crystals: Fundamentals of Crystal Engineering*, The Royal Society of Chemistry, United Kingdom, **2018**, pp. 617–672.
182. Saleh, G.; Gatti, C.; Presti, L.L.; Contreras-Garcia, J. Revealing non-covalent interactions in molecular crystals through their experimental electron densities. *Chem. Eur. J.* **2012**, *18*, 15523–15536.

183. Kohout, M. A measure of electron localizability. *Int. J. Quantum Chem.* **2004**, *97*, 651–658.
184. Kohout, M.; Pernal, K.; Wagner, F.R.; Grin, Y. Electron localizability indicator for correlated wavefunctions. I. Parallel-spin pairs. *Theor. Chem. Acc.* **2004**, *112*, 453–459.
185. Kohout, M. Bonding indicators from electron pair density functionals. *Faraday Discuss.* **2006**, *135*, 43–54.
186. Savin, A. Chemical bonding and interpretation of time-dependent electronic processes with maximum probability domains. *Acta Phys. –Chim. Sin.* **2018**, *34*, 528–536.
187. de Silva, P.; Corminboeuf, C. Simultaneous visualization of covalent and noncovalent interactions using regions of density overlap. *J. Chem. Theory Comput.* **2014**, *10*, 3745–3756.
188. Marx, D.; Hutter, J. *Ab Initio Molecular Dynamics*, Cambridge University Press, Cambridge, **2009** and references therein.
189. Rapaport, D. Many-Body Interactions, in *The Art of Molecular Dynamics Simulation*. Cambridge University Press, Cambridge, **2004**, pp. 326–343.
190. Jezierska, A.; Panek, J.J.; Koll, A.; Mavri, J. Car-Parrinello simulation of an O–H stretching envelope and potential of mean force of an intramolecular hydrogen bonded system: Application to a Mannich base in solid state and in vacuum. *J Chem. Phys.* **2007**, *126*, 205101.
191. Dopieralski, P.; Perrin, C.L.; Latajka, Z. On the intramolecular hydrogen bond in solution: Car–Parrinello and path integral molecular dynamics perspective. *J. Chem. Theory Comput.* **2011**, *7*, 3505–3513.
192. Pirc, G.; Stare, J.; Mavri, J. Car–Parrinello simulation of hydrogen bond dynamics in sodium hydrogen bissulfate. *J. Chem. Phys.* **2010**, *132*, 224506.
193. Stare, J.; Mavri, J.; Grdadolnik, J.; Zidar, J.; Maksić, Z.B.; Vianello, R. Hydrogen bond dynamics of histamine monocation in aqueous solution: Car–Parrinello molecular dynamics and vibrational spectroscopy study. *J. Phys. Chem. B* **2011**, *115*, 5999–6010.
194. Panek, J.J.; Jezierska, A. *N*-oxide derivatives: Car–Parrinello molecular dynamics and electron localization function study on intramolecular hydrogen bonds. *J. Phys. Chem. A* **2018**, *122*, 6605–6614.
195. Kizior, B.; Michalczyk, M.; Panek, J.J.; Zierkiewicz, W.; Jezierska, A. Unraveling the nature of hydrogen bonds of "proton sponges" based on Car-Parinello and metadynamics approaches. *Int. J. Mol. Sci.* **2023**, 24, 1542.
196. Clementi, E.; Kołos, W.; Lie, G.C.; Ranghino, G. Nonadditivity of interaction in water trimers. *Int. J. Quantum Chem.* **1980**, *18*, 377–398.
197. Ortega-Blake, I.; Novaro, O.; Leś, A.; Rybak, S. A molecular orbital study of the hydration of ions. The role of nonadditive effects in the hydration shells around Mg^{2+} and Ca^{2+}. *J. Chem. Phys.* **1982**, *76*, 5405–5413.

198. Jezierska, A.; Błaziak, K.; Klahm, S.; Lüchow, A.; Panek, J.J. Non-covalent forces in naphthazarin — cooperativity or competition in the light of theoretical approaches. *Int. J. Mol. Sci.* **2021**, *22*, 8033.
199. Herbstein, F.H.; Kapon, M.; Reisner, G.M.; Lehman, M.S. ; Kress, R.B.; Wilson, R.B.; Shiau, W.-I.; Duesler, E.N.; Paul, I.C.; Curtin, D.Y. Polymorphism of naphthazarin and its relation to solid-state proton transfer. neutron and X-ray diffraction studies on naphthazarin C. *Proc. R. Soc. Lond. A* **1985**, *399*, 295–319.
200. Schlegel, H.B. Estimating the Hessian for gradient-type geometry optimizations. *Theor. Chem. Acc.* **1984**, *66*, 333–340.
201. Perdew, J.P.; Burke, K.; Ernzerhof, M. Generalized gradient approximation made simple. *Phys. Rev. Lett.* **1996**, *77*, 3865; Erratum in **1997**, *78*, 1396.
202. Troullier, N.; Martins, J.L. Efficient pseudopotentials for plane-wave calculations. *Phys. Rev. B* **1991**, *43*, 1993–2006.
203. This figure is a combination of Fig. 8 and Fig. S5 from ref. 198.
204. Feld, R.; Lehmann, M.S. The crystal structure of benzoic acid: a redetermination with X-rays at room temperature; a summary of neutron-diffraction work at temperatures down to 5K. *Zeitsch. für Kristall.* **1981**, *157*, 215–231.
205. Wojtkowiak, K.; Michalczyk, M.; Zierkiewicz, W.; Jezierska, A.; Panek, J.J. Chalcogen bond as a factor stabilizing ligand conformation in the binding pocket of carbonic anhydrase IX receptor mimic. *Int. J. Mol. Sci.* **2022**, *23*, 13701.

Chapter III

Experimental Studies on Crystal Structures as a Source of Information About Interactions

Abstract: Crystal structures are examined here as a source of information about interactions. Hence, methods that allow us to solve crystal structures are presented, mainly X-ray and neutron diffraction methods. The accuracy of results of solved crystal structures is discussed. Additionally, it is described that X-ray diffraction may provide information on electron density of crystal structures. Hence Quantum Theory of Atoms in Molecules, QTAIM, may be applied for the experimental electron density. Another approach, the Hirshfeld surface analysis that is useful to calculate parameters describing intermolecular interactions, is also reviewed.

Keywords: X-ray diffraction, neutron diffraction, electron density, Quantum Theory of Atoms in Molecules, Hirshfeld surface analysis.

III.1 Determinations of Crystal Structures

X-ray and neutron diffraction are powerful methods that allow us to determine crystal structures, including geometry of species constituting crystals.[1–3] Furthermore, they also allow us to determine other features and parameters of crystals, among them symmetry. Hence, results of crystal structure determination are useful to analyse intermolecular interactions since they allow designation of arrangement of molecules and/or ions.

The Cambridge Structural Database (CSD)[4,5] contains information on crystal structures of organic and metal-organic compounds. Over 1.3 million crystal structures are stored in this base, mainly from X-ray diffraction results; other types of measurements, such as neutron diffraction or synchrotron radiation, are not so common. The number of neutron diffraction crystal structures is about 2,600 while about twenty thousand synchrotron radiation crystal structures are stored in CSD. CSD contains information of symmetry, conditions of diffraction measurements, and geometry of species constituting crystals since their atomic coordinates are stored within. It is worth mentioning that this base also contains programs that allow one to perform various statistical analyses. Hence one can see that CSD is a useful tool to analyse intra- and intermolecular interactions. It is very important to explain which results that are related to parameters describing interactions are provided by diffraction measurements. It is also important to describe how these parameters are extracted from diffraction measurements.

In a diffraction experiment, the intensity of the diffracted radiation is measured.[1–3] Hence it is important to analyse relations between intensities of diffracted beams and physical observables that are related to geometries and mutual arrangements of species constituting crystals. Hence the interrelation between the measured intensity, which is a function of reciprocal space, and the electron density, which is described in direct space and that is related to location of moieties constituting crystals, is discussed further here.

The electron density, at a point defined by the $\vec{r}$ vector, is designated as $\rho(\vec{r})$, and it is expressed by eq. III.1.

$$\vec{r} = x\vec{a} + y\vec{b} + z\vec{c} \tag{III.1}$$

$\vec{a}$, $\vec{b}$ and $\vec{c}$ are basis vectors while x, y and z are fractional coordinates of this point that are expressed in a, b and c units. They are named as lattice units or cell units that determine the unit cell, a simple fragment of a crystal lattice. There are standard conditions of the choice of the unit cell; it is described in monographs on crystallography and crystal chemistry.[1–3,6–8] The unit cell is the basic, simple block that reproduces the crystal lattice if it is translated in three non-parallel directions. These directions are the same as directions of the crystallographic (lattice) axes. One can assume for the crystal structure considered that contents of unit cells are approximately the same; there are additional effects (physical and those related to conditions of measurements) for some structures which may disturb this regularity, disorder, pseudosymmetry, etc. However, usually the translation of the unit cell together with its content in the directions of crystallographic axes reproduces the crystal structure.

Figure III.1a presents the unit cell with its content for the crystal structure of acetic acid.[9] One can see that there are four molecules in the unit cell, all possessing the same geometry since they are related by the symmetry operations (the space group of this structure is $P2_1/n$ that corresponds to the 2/m point group — both are characterized by the same number of symmetry-equivalent points, four, since the cell is primitive). One could expect from Fig. III.1a that not exactly four molecules belong to the unit cell since molecules are cut through its walls. However, parts of other molecules

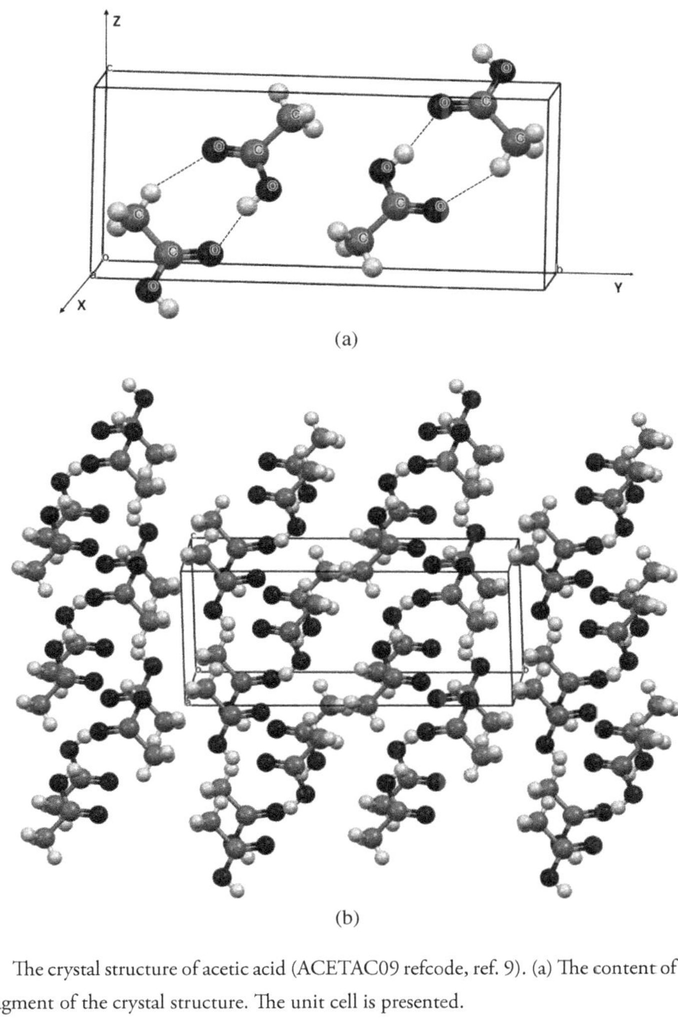

(a)

(b)

Fig. III.1 The crystal structure of acetic acid (ACETAC09 refcode, ref. 9). (a) The content of the unit cell. (b) The fragment of the crystal structure. The unit cell is presented.

contribute to the cell in question; Figure III.1b shows this situation. Hence the unit cell contains the number and kinds of atoms which exactly correspond to four acetic acid molecules.

One can see that the dimers of acetic acid are linked by two hydrogen bonds, O-H···O and C-H···O where H···O distances are equal to 1.84 Å and 2.62 Å, respectively, with the corresponding O-H···O and C-H···O angles amounting to 161.8° and 153.4°, respectively. The H···O distance for the C-H···O bridge is close to the sum of van der Waals radii (see Table I.1) that indicates weak interaction, weaker than for the O-H···O hydrogen bond where the H···O distance is shorter. The O-H···O system is closer to linearity than the C-H···O arrangement indicating also that the former interaction is stronger. It is interesting that in carboxylic acid crystal structures, the centrosymmetric dimer motifs are often observed (see, for example, Figures II.2, II.4, II.7, and II.15 in Chapter II).[10] This is not the case for acetic acid in the solid state that is analysed here.[9]

It is worth mentioning here that the crystal lattice is the mathematical (geometrical) concept while the crystal structure concerns the real chemical species, usually molecules or ions. The unit cell is the piece (brick) of the crystal lattice while its content is the piece of the crystal structure. Fig. III.1a shows a unit cell with lattice units *a*, *b*, and *c* (along X, Y and Z crystallographic axes, respectively). The angles between crystallographic axes are defined as follows: α is an angle between axes Y and Z, β between X and Z, and γ between X and Y. These angles are not necessarily equal to 90° as in the Cartesian coordinate system. In general, according to the commonly accepted conditions for the unit cell selection, this unit is a special kind of polyhedron, i.e., it is a parallelepiped.

One can see that the necessary characteristic of a crystal as a solid state is the long-range order that is referred to as translational symmetry — the unit cell together with its content reproduces the crystal structure if it is translated in three non-parallel directions. The sharp diffraction pattern is another characteristic of a crystal; particularly in early studies, this condition was specified as sufficient to classify a system as a crystal. There are two types of solid-state materials: crystals and amorphous solids (i.e., non-crystalline materials). For non-crystalline materials, translational symmetry is not observed. Not only translational symmetry may occur in crystal structures; however, it is outside of the scope of this book to describe them in detail (look at more specialized monographs on crystallography and crystal chemistry).[1–3,6–8]

The positions of atoms in the unit cell in X-ray diffraction may be determined from the maxima of the electron density; these maxima are determined from the diffracted intensities. The intensity of the diffracted radiation is expressed in the reciprocal space where points are represented by $\vec{H}$ reciprocal space vectors (eq. III.2).

$$\vec{H} = h\vec{a}^* + k\vec{b}^* + l\vec{c}^* \tag{III.2}$$

The reciprocal space (it is often called "Fourier space") basis vectors $\vec{a}^*$, $\vec{b}^*$ and $\vec{c}^*$ have dimensions of reciprocal length (for example, $Å^{-1}$) while h, k and l are integers which are the coordinates of the point in the reciprocal space. The h, k and l values also correspond to so-called Miller indices. The (hkl) point in the reciprocal space defines the family of lattice planes orthogonal to the $\vec{H}$ vector expressed by eq. III.2 and it is also known as the Miller index of the plane. The product of vectors $\vec{r}$ and $\vec{H}$ leads to the following expression:

$$\vec{r}\vec{H} = hx + ky + lz \tag{III.3}$$

This is a consequence of scalar products of the real space and the Fourier space vectors (eq. III.4).

$$\vec{a}\vec{a}^* = \vec{b}\vec{b}^* = \vec{c}\vec{c}^* = 1 \tag{III.4}$$

The remaining "mixed" scalar products of vectors of real and reciprocal spaces are equal to zero; e.g., products such as $\vec{a}\vec{b}^* = 0$.

The intensities of the diffracted beams are functions of the reciprocal space and they depend on the structure factor, $F(\vec{H})$, that is also a function of the reciprocal space.

$$I(\vec{H}) \equiv I(hkl) \sim F(\vec{H})F^*(\vec{H}) \tag{III.5}$$

The structure factor is the Fourier transform of the electron density, $\rho(\vec{r})$:

$$F(\vec{H}) = \int_V \rho(\vec{r})e^{2\pi i(\vec{H}\vec{r})}dV \tag{III.6}$$

The integration is performed here over the real space cell unit volume, V; i is the imaginary unit. The electron density, as the V volume, concerns the real space while the result of integration, the structure factor, concerns the reciprocal space. $F^*(\vec{H})$ is the conjugate function to the complex $F(\vec{H})$ function.

The electron density, ρ, is the inverse Fourier transformation of the structure factor:

$$\rho(\vec{r}) = \int_{V^*} F(\vec{H}) e^{-2\pi i(\vec{H}\vec{r})} dV^* \qquad \text{(III.7)}$$

The integration is performed over the volume of the reciprocal space cell unit, V^*. The volume V^* as well as the structure factor, $F(\vec{H})$, concern the Fourier space while the result of integration, the electron density $\rho(\vec{r})$, is a function of the real space. The structure factor and the electron density expressed by eqs. III.6 and III.7, respectively, are restricted to the lattice cells only. It seems to be justified because the translational symmetry of the unit cell with its content in real space reproduces the whole crystal structure.

The expressions presented above show that insights into the electron density function, $\rho(\vec{r})$, allow us to determine positions of atoms in the crystal structure since they correspond to the electron density maxima. The $\rho(\vec{r})$ function may be obtained from the structure factor, $F(\vec{H})$, that should be recovered from the diffraction measurement, i.e., from the intensities of diffracted beams measured experimentally. However, only the amplitude of the structure factor, $|F(\vec{H})|$ (eq. III.8), or the product of the structure factor and its conjugate function (eq. III.5), may be extracted from the diffraction experiment, not the structure factor which is the complex function.

$$I(\vec{H}) \sim |F(\vec{H})|^2 \qquad \text{(III.8)}$$

The structure factor is a complex function that may be expressed as:

$$F(\vec{H}) = |F(\vec{H})| e^{i\phi_{\vec{H}}} \equiv F(hkl) = |F(hkl)| e^{i\phi_{hkl}} \qquad \text{(III.9)}$$

The phase angle, ϕ_{hkl}, for each $\vec{H}$ vector cannot be determined in the diffraction experiment, thus the density function is not recovered from the diffracted beams. It is known in crystal chemistry as the "phase problem".

However, there are different ways to extract information of the electron density from the radiation intensity. One can mention the Patterson method where such extraction is performed via the Patterson function, $P(\vec{u})$. The use of $F(\vec{H})F^*(\vec{H})$ product is another way, known as "direct methods". They are most often applied to determine crystal structures. There are other approaches to solve the "phase problem", such as the Maximum Entropy Method,[11,12] but they are not applied as often as the direct methods.

It is assumed in the process of determination of the crystal structure that the electron density of the unit cell depends on electron densities of individual atoms. This leads to the expression for the structure factor that is the sum (eq. III.10) instead of the integration (eq. III.6).

$$F(\vec{H}) = \sum_{j=1}^{N} f_j e^{2\pi i(\vec{H}\vec{r}_j)} \quad \text{(III.10)}$$

The above approximation leads to simplification of calculations in comparison with the integral equation for the structure factor. In eq. III.10 the atomic scattering factor, f_j, occurs; the summation for N atoms in the unit cell is performed here.

The spherical approximation of the atomic electron densities, ρ_j's, for the refinement of crystal structure is usually assumed as it also relates to the spherical symmetry of the atomic scattering factors. These approximations related to scattering factors computed for isolated non-interacting atoms or ions are named as the independent atom model (IAM).[13] The simplicity of IAM makes it easy to perform calculations leading to the solution of crystal structures. However, this model loses information regarding the bonds and leads to incorrect determination of positions of hydrogen atoms.

The atomic scattering factor that occurs in eq. III.10 concerns the stationary atom; however, atoms vibrate in crystal structures. This is why the correction related to such thermal motions is usually introduced with the assumption that the considered atom in the crystal does not depend on vibrations of other atoms; additionally, chemical bonds between atoms are not considered in this correction. Hence this correction allows us to persist with the IAM model. The expression of the scattering factor where vibrations of the atom are considered, f_T, has the following form:

$$f_T = f\exp\left[-B\left(\frac{\sin\theta}{\lambda}\right)^2\right] = fe^{-T} \quad \text{(III.11)}$$

In this expression, B is known as the Debye-Waller factor, λ is the wavelength of the radiation and θ is the diffraction angle according to Bragg's law.

According to this law, a beam of radiation that passes through a family of (*hkl*) planes (parallel and equally spaced planes) is diffracted only if the following conditions are met:

$$n\lambda = 2d_{hkl}\sin\theta \quad \text{(III.12)}$$

n is an integer, λ is the wavelength of the radiation, d_{hkl} is the distance between the neighbouring planes of the (*hkl*) family, and θ is the diffraction angle between the radiation beam and the (*hkl*) plane.

In the isotropic model the T-term occurring in the exponential part of eq. III.11 (e^{-T}) is expressed by eq. III.13.

$$T_{iso} = (Bh^2a^{*2} + Bk^2b^{*2} + bl^2c^{*2} + 2Bhka^*b^*\cos\gamma^* + 2Bhla^*c^*\cos\beta^* + 2Bklb^*c^*\cos\alpha^*)/4 \quad \text{(III.13)}$$

The isotropic model assumes that atoms move in the same way in all directions; B is named as the isotropic temperature factor. In the anisotropic model the B-term of eq. III.13 is replaced by the B_{ij} tensor components.

$$T_{anis} = (B_{11}h^2a^{*2} + B_{22}k^2b^{*2} + B_{33}l^2c^{*2} + 2B_{12}hka^*b\cos\gamma^* + 2B_{13}hla^*c^*\cos\beta^* + 2B_{23}klb^*c^*\cos\alpha^*)/4 \quad \text{(III.14)}$$

The anisotropic temperature factors, B_{ij}, may be expressed as a symmetric matrix (eq. III.15).

$$\begin{pmatrix} B_{11} & B_{12} & B_{13} \\ B_{12} & B_{22} & B_{23} \\ B_{13} & B_{23} & B_{33} \end{pmatrix} \quad \text{(III.15)}$$

The isotropic and anisotropic temperature factors are named in recent studies as the atomic displacement parameters. The sphere is a geometrical representation of the isotropic thermal vibrations of an atom while the ellipsoid represents thermal vibrations in the anisotropic approximation. For more accurate fit to the thermal vibrations other approximations were proposed, for example to consider the anharmonicity.[14]

The anisotropic and isotropic approximations of the thermal vibrations of atoms (eqs. III.14 and III.13, respectively) are dictated by the number of parameters that are established as results of the crystal structure determination. The following parameters are usually determined: atomic coordinates, atomic displacement parameters, a scale factor, extinction parameters, occupancy factors and twin component fractions. An appropriate number of intensities of diffracted beams is needed for successful and accurate refinement of parameters, usually ten and recommended to be not less than six). In the X-ray diffraction approach, the isotropic approximation is applied for H-atoms to reduce the number of parameters to be determined. Hence for each H-atom where the isotropic approximation is applied, four parameters are determined: three coordinates and one thermal parameter. In the anisotropic approximation there are nine parameters, i.e., three coordinates and six thermal parameters, B_{ij} ($i \leq j = 1, 2, 3$), see eq. III.15.

Figure III.2 presents the fragment of the crystal structure of 1-(bromomethyl)-2-(dibromomethyl) naphthalene where short intermolecular contacts are presented, the C-H···Br hydrogen bonds and Br···Br dihalogen bonds.[15] One can see that thermal vibrations for non-hydrogen atoms of this structure are approximated by ellipsoids while the hydrogen atoms are marked by spheres of the same size that do not correspond to any thermal approximation (however, the isotropic approximation was applied for this structure for H-atoms in the refinement process). Displacement ellipsoids for non-H atoms are shown with 50% probability.

The solvation of the phase problem mentioned above leads to the approximate determination of positions of atoms from the electron density maps. The accuracy of these positions and of atomic displacement parameters is improved in cycles of the refinement process of the solution of the crystal structure. Roughly speaking, this process involves matching parameters from the model structure known from the electron density to the parameters from experimental diffraction data. One can say of several

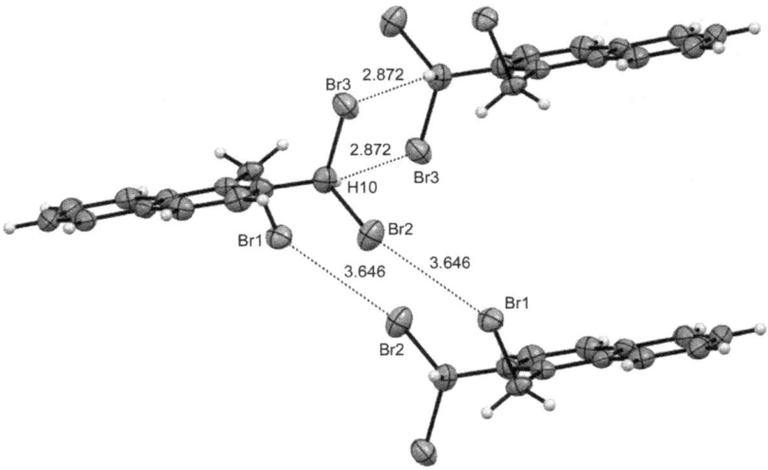

Fig. III.2 The fragment of the crystal structure of 1-(bromomethyl)-2-(dibromomethyl) naphthalene. Reprinted with permission from ref. 15. Copyright 2023 IUCr.

factors that inform of the accuracy of the solution of the crystal structure, i.e., the final structure after cycles of its refinement.

The R-value (R-factor) is expressed by the following equation:

$$R = \frac{\sum_{\vec{H}} \left| |F(\vec{H})_{obs}| - |F(\vec{H})_{calc}| \right|}{\sum_{\vec{H}} |F(\vec{H})_{obs}|} \qquad \text{(III.16)}$$

The R-factor is called also as residual index or agreement index; it compares the amplitudes of structure factors calculated from the model, $|F(\vec{H})_{calc}|$'s, with those obtained directly from experiment, $|F(\vec{H})_{obs}|$'s (eq. III.8). The R-factor decreases during the refinement process and for the final structure it is usually within the 0.08–0.05 range, or even lower.

The weighted R-value is usually presented in various studies concerning crystal structures. The $|F|$, F^2 or I function is inserted in the expression of the weighted R-factor. The expression on the R-index based on the F^2 function has the form given below:

$$wR = \left\{ \frac{\sum_{\vec{H}} [w(F(\vec{H})^2_{obs} - F(\vec{H})^2_{calc})]^2}{\sum_{\vec{H}} w[F(\vec{H})^2_{obs}]^2} \right\}^{1/2} \qquad \text{(III.17)}$$

The goodness-to-fit is another accuracy factor; it is close to unity for a correct refinement process. The data-to-parameter ratio is the ratio of the number of intensities measured to the number of parameters determined (atomic coordinates, atomic displacement parameters and others mentioned above). This ratio should be at least 6:1; however, most journals where crystal structures are published demand a higher ratio. The estimated standard deviations inform directly about the precision of determined parameters. The estimated standard deviation for the bond length is presented in parentheses and it is related to the last digit. For example, for a bond length of 1.395(3) Å, the estimated standard deviation is equal to 0.003 Å.

III.2 Neutron Diffraction Method

In the neutron diffraction approach, neutrons that may be produced by different sources (reactor or spallation sources, for example[3,16]) are diffracted by nuclei of crystal structures. Neutrons applied in scattering experiments are slowed down to energies of 1 eV or even lower. The selected advantages of neutron diffraction for the solution of the crystal structure are listed below. This list is based partly on the monograph on neutron diffraction by C. Wilson.[16]

- The positions of hydrogen atoms are determined more accurately by neutron diffraction than by X-ray diffraction. For neutron diffraction, the precision of the determination of H-atoms is similar to that of heavier atoms.
- In the neutron diffraction approach, neighbouring elements in the periodic table have different scattering cross-sections, while in the X-ray approach such neighbouring elements have a very small contrast of X-ray scattering. Thus, neutron diffraction allows us to distinguish neighbouring elements.
- Neutron scattering depends on the nucleus, thus different scattering is observed for isotopes of the same element. Hence, isotopic substitution may be applied to yield structural and dynamic characteristics.
- For neutron diffraction, structures may be analysed to very high resolution, but this is not the case for X-ray diffraction.

The precise positions of hydrogen atoms are very important in various analyses on systems linked by hydrogen bonds. That is why the application of results of the neutron diffraction approach seems to be beneficial. It was described that the neutron diffraction method shows the H-atom positions to be shifted by about 0.13–0.15 Å in comparison with such positions found by X-ray techniques.[16] For example, these shifts lead to the "X-ray bond lengths" being shorter than the neutron equivalents. There are a few reasons for such differences. The neutron diffraction approach provides information of positions of nuclei while X-ray techniques provide information about the maxima of electron density. For the A-H bond the H-atom electron density maximum is closer to the A-centre than the H-atom nucleus, and greater shifts of such maxima are observed for highly electronegative A-centres. This is also related to the spherical scattering factor approximation in the X-ray approach that results in the "X-ray hydrogen atom position" being closer to the A-centre than the hydrogen nucleus. It is worth mentioning here that the H-atom position determined with the spherical assumption is not the same as the hydrogen electron density maximum.

Figure III.3 presents the complex of urea with phosphoric acid from the crystal structure solved by the neutron diffraction method.[17] It is interesting that all atoms are refined here with the application of the anisotropic model (six atomic displacement parameters – eqs. III.14 and III.15). This complex is linked through a strong O-H···O hydrogen bond (O5-H4···O4). At 150 K, the O5···O4 distance is equal to 2.411(2) Å. The H-atom (H4) is closer to one of the oxygen centres (O5), forming an O-H bond with the length equal to 1.168(4) Å. Another H···O distance (H4···O4) is equal to 1.252(4) Å. The O-H···O angle is equal to 169.8(4)°. One can see that the longest axis of the ellipsoid that approximates the thermal vibrations of the H4 atom has a direction close to the O···O line that indicates the direction of greatest motion of the H-atom. The situation changes for the same crystal structure measured at 350 K. The H-atom is close to the mid-point of the distance of 2.420(5) Å since the H···O distances are equal to 1.210(7) Å and 1.217(8) Å. And the H-bond bridge is closer to linearity since the O-H···O angle is equal to 171.5(6)°. The greatest thermal motion of the H4 atom is in the same direction as for 150 K because the longest axis of the thermal ellipsoid is in the direction of the O···O distance, approximately. However, one can see that this vibration is much greater for the higher temperature (the longest axis of thermal ellipsoid of the H4 atom is longer than that for 150 K). In general, all vibrations are greater for the structure at 350 K than for that at 150 K. This is also reflected in the values of the estimated standard deviations, which are greater for the greater temperature.

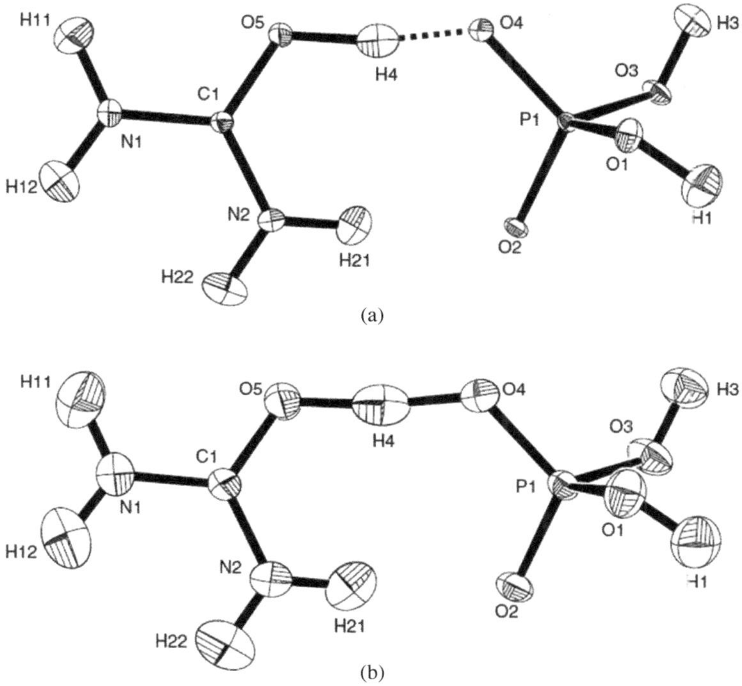

Fig. III.3 An ORTEP (ref. 18) plot showing urea-phosphoric acid at (a) 150 K and (b) 350 K, with anisotropic displacement ellipsoids plotted at 50% probability. The magnitude of the ellipsoid for atom H(4), and its elongation along the hydrogen bond direction, increases systematically with increasing temperature and is fairly pronounced at 350 K. Reprinted with permission from ref. 17. Copyright 2001 Oldenbourg Wissenschaftsverlag, München.

III.3 Intermolecular Distances in Crystal Structures, Coordinates of Atoms

Symmetry in crystal structures allows us to find out arrangements of molecules, ions or other species that constitute the crystal. In such a way it is possible not only to have information of geometry of species and intramolecular contacts in crystals, but also to know intermolecular atom-atom distances. All other geometrical information like angles for atomic triads, dihedral angles or best planes for groups of atoms are also accessible since coordinates of all atoms are usually available for solved crystal structure.

Let us look at an example of the crystal structure of $C_{24}H_{24}N_4Se_2^{2+}·2CF_3O_3S^-$.[19] A fragment of this structure is presented in Figure III.4. Different types of interactions are presented in this figure: hydrogen bonds, anion-π interactions and chalcogen bonds. The displacement ellipsoids drawn at the 50% probability level are presented for non-hydrogen atoms. Only H atoms involved in hydrogen bonds refined with isotropic approximation are shown. The atoms of reference $C_{24}H_{24}N_4Se_2^{2+}·2CF_3O_3S^-$ species are determined by the (*x*,*y*,*z*) coordinates, and another species is generated by the (−*x* + 1,

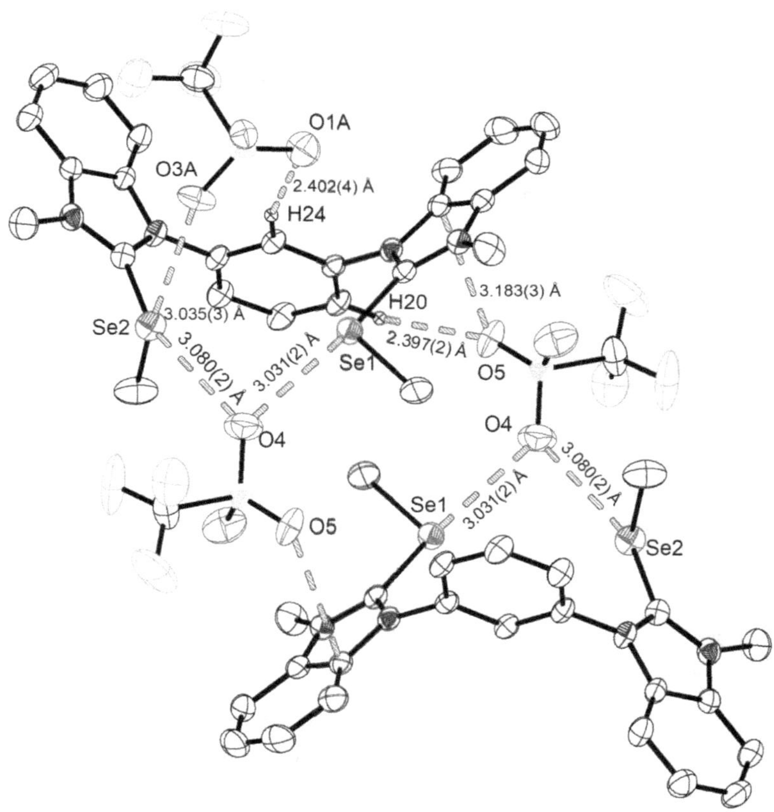

Fig. III.4 Structure of $C_{24}H_{24}N_4Se_2^{2+}·2CF_3O_3S^-$ with the chalcogen bonding, hydrogen bonding and anion–π interactions marked, and with displacement ellipsoids drawn at the 50% probability level. For the sake of clarity, only H atoms involved in hydrogen bonding are shown. Symmetry operation to generate the second species of $C_{24}H_{24}N_4Se_2^{2+}$ and trifluoromethanesulfonate is: −*x* + 1, −*y* + 1, −*x* + 1. Reprinted with permission from ref. 19. Copyright 2023 IUCr.

$-y+1$, $-x+1$) symmetry operation. It means that the second species is related to the reference by an inversion symmetry operation and translation along X, Y and Z crystallographic axes by a single *a*, *b* and *c* unit cell, respectively. Two C-H···O hydrogen bonds are observed with the H···O distances equal to 2.397(2) Å and 2.402(4) Å. The estimated standard deviations are given in parentheses. The anion-π interaction is marked in Fig. III.4 and the O5···C distance equal to 3.183(3) Å is presented where the C-centre belongs to the π-electron system. The latter distance is close to the corresponding sum of van der Waals radii, 3.22 Å. However, this contact concerns the interaction of the terminal oxygen centre (O5) of the trifluoromethanesulfonate anion with the ring composed of carbon and nitrogen atoms; the distance between the oxygen centre and the centre of the ring is slightly shorter than the O5···C distance. This interaction is rather weak, however. The chalcogen bonds, Se1···O4 and Se2···O4, are equal to 3.031(2) Å and 3.080(2) Å, respectively. These distances are much shorter than the corresponding sum of van der Waals radii that amounts to 3.42 Å. The difference between the selenium-oxygen distances and the van der Waals sum, close to ~0.4 Å, indicates that chalcogen bonds are rather strong. These two chalcogen bonds may be classified as bifurcated chalcogen bonds where there is the common O4 electron donor and two selenium electron acceptors. Describing more precisely the kinds of Lewis acid centres for these interactions, one may state that the σ-holes are located at selenium atoms in elongations of C-Se bonds or nearly so.

Let us consider another crystal structure of $C_{24}H_{24}N_4Se_2^{2+}{\cdot}2CF_3O_3S^{-}{\cdot}0.5C_2H_4Cl_2$, with chalcogen bonds to trifluoromethanesulfonate and 1,2-dichloroethane (Fig. III.5).[19] The displacement ellipsoids drawn at the 50% probability level for non-hydrogen atoms and H-atoms refined with isotropic radii are presented. The dichloroethane molecule and trifluoromethanesulfonate anion are generated by the ($x+1$, y, z) symmetry operation. It means that these species are related to those located in the reference unit cell only by the translation symmetry along the X-axis by a single unit cell. One can see here two chalcogen bonds, Se1A···Cl1A and Se2···O3 with distances equal to 3.535(24) Å and 3.133(4) Å, respectively, less than the corresponding sums of van der Waals radii that are equal to 3.65 Å and 3.42 Å, respectively The σ-holes of the selenium centres are located here at elongations of C-Se bonds, approximately. One can see that for the Se1A···Cl1A contact, the chlorine centre acts as the Lewis base centre in the direction roughly perpendicular to the C-Cl1A bond. One can expect

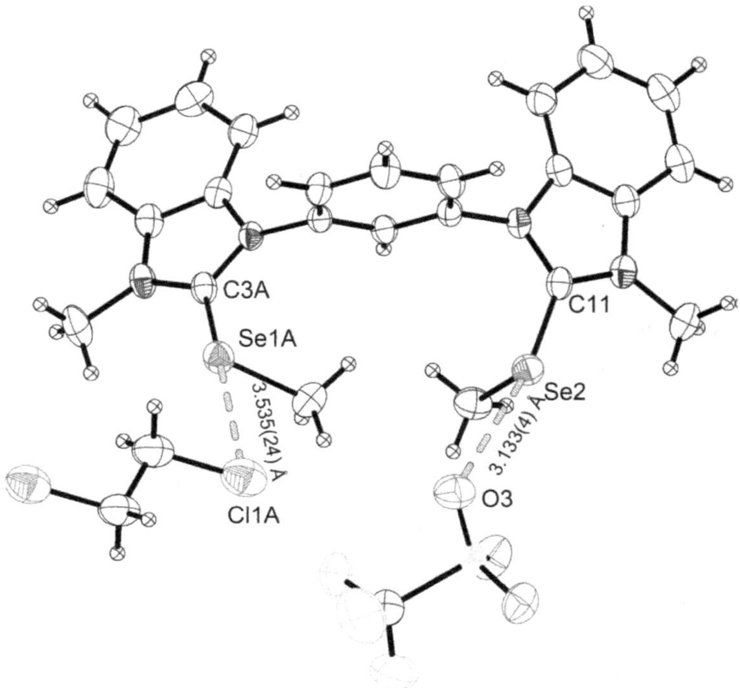

Fig. III.5 Structure of $C_{24}H_{24}N_4Se_2^{2+}\cdot 2CF_3O_3S^{-}\cdot 0.5C_2H_4Cl_2$, with chalcogen bonding to trifluoromethanesulfonate and 1,2-dichloroethane marked, and with displacement ellipsoids drawn at the 50% probability level. The dichloroethane molecule and trifluoromethanesulfonate anion are generated by ($x + 1, y, z$). Reprinted with permission from ref. 19. Copyright 2023 IUCr.

the occurrence of σ-hole in the elongation of the C-Cl1A bond of dichloroethane with weak Lewis acid properties. In other words, this chlorine centre possesses dual character; it may act as the Lewis acid and as the Lewis base.

The distances discussed above may be calculated from eq. III.18.

$$r_{AB} = [(x'_A - x'_B)^2 + (y'_A - y'_B)^2 + (z'_A - z'_B)^2]^{\frac{1}{2}} \tag{III.18}$$

This relationship is known as the Euclidean distance or 2-norm distance where A and B designate any pair of points considered possessing (x'_A, y'_A, z'_A) and (x'_B, y'_B, z'_B) coordinates; r_{AB} is the distance between A and B points. This may be any pair of atoms such as those in the chalcogen bond or hydrogen bond, for example, or in any intra- or intermolecular contact. However, to have the distances expressed in Å, the x', y', and z' coordinates must correspond to the Cartesian system with the same Å units for X′, Y′ and Z′ axes. The coordinates of atoms in any crystal structure are expressed in the crystallographic axes with a, b and c lattice units. Hence one needs to transform the coordinates expressed in lattice units, where the X, Y and Z axes are not necessarily perpendicular to one another, into Cartesian (x', y', z') coordinates where the units of axes are of the same length (say 1 Å). There are numerous matrices which allow us to perform such transformations; one of them is presented below:

$$\begin{pmatrix} x' \\ y' \\ z' \end{pmatrix} = \begin{pmatrix} a\sin\beta\sin\gamma^* & 0 & 0 \\ -a\sin\beta\cos\gamma^* & b\sin\alpha & 0 \\ a\cos\beta & b\cos\alpha & c \end{pmatrix} \begin{pmatrix} x \\ y \\ z \end{pmatrix}, \quad \gamma^* = 180^\circ - \gamma \qquad \text{(III.19)}$$

In general, the above equation shows the transformation of x, y, z coordinates of any point (atom) expressed in triclinic system into x', y' and z' Cartesian coordinates. This matrix is simplified significantly for monoclinic systems, where α and γ angles are equal to 90°, and even greater simplification occurs for the orthorhombic system (all crystallographic axes are perpendicular to one another), and also for the tetragonal ($a = b$) and cubic ($a = b = c$) systems.

The angle between two bonds may be calculated simply from the law of cosines. There are other geometrical results often discussed in various studies, torsional angles or the best planes passing through any number of atoms (more than three atoms since three points clearly define a single plane). It is worth mentioning that the majority of crystal structures of organic and organometallic compounds are stored in CSD[4,5] that possesses tools to perform calculations concerning arrangements of molecules or groups of atoms.

III.4 The Electron Density in Crystal Structures

The high-resolution X-ray diffraction measurements are often applied in various studies since they may provide information on the charge density. The experimental results concerning charge density are useful to analyse various characteristics of crystal structures, such as inter- and intramolecular interactions.[20–22] In the procedure to determine the experimental electron density of the crystal structure, the first stage is to solve this structure with the use of standard methods, which are usually direct methods and IAM. Next, the anisotropic refinement of non-hydrogen atoms and the isotropic refinement of hydrogen atoms (sometimes, quite rarely anisotropic) is applied. The approximation in the IAM model is that the electron density is a superposition of spherical atomic densities. Thus, the structure factor in eq. III.10 is the sum of products where the spherical atomic scattering factors occur. The structure determined with the IAM model is further applied in multipolar refinements; the Hansen-Coppens Multipolar Model[23] is often applied that allows us to consider atomic non-sphericity. Three terms of the atomic electron density, $\rho(r)_{atom}$, are considered in this model: the spherical core electron density, $\rho(r)_c$, the spherical valence electron density, $\rho(\kappa r)_v$, and the aspherical valence electron density, $\rho(\kappa' r)_d$.

$$\rho(r)_{atom} = P_c\rho(r)_c + P_v\kappa^3\rho(\kappa r)_v + \rho(\kappa' r)_d \tag{III.20}$$

$$\text{where} \quad \rho(\kappa' r)_d = \sum_{l=0}^{l_{max}} \kappa'^3 R_l(\kappa' r) \sum_{m=0}^{l} P_{lm\pm} d_{lm\pm}(\theta, \Phi)$$

The P_c and P_v are core and valence populations, respectively, R_l are normalized Slater-type radial functions, while $P_{lm\pm}$ are multipole populations and $d_{lm\pm}$ are real angular spherical harmonics. The indices l and m concern changes of angular and azimuthal numbers of spherical harmonics functions, while κ and κ' designate radial scaling parameters that allow enlargement or contraction of the spherical valence electron density and the aspherical valence electron density. The P_v, κ and κ' parameters are optimized in the least-squares refinement, in addition to the "traditional" nine parameters (coordinates and six atomic displacement parameters). The latter approach results in up to 36 parameters for each non-hydrogen atom; there is also an increase of parameters for hydrogen atoms in the case of multipolar refinements which are used to determine experimental electron density. Hence, for the determination of experimental crystal structure electron density, the number of refined parameters increases significantly.

The determination of crystal structure electron density is very useful. For example, it allows us to compare numerous parameters derived from this density with these parameters resulting from theoretical calculations. One can mention the Bader theory, Quantum Theory of Atoms in Molecules (QTAIM),[24] (described in Chapter II) since the analysis of characteristics of critical points that concern intra- and intermolecular interactions in crystals may be performed. Also, the electron density isolines concerning atoms and groups of atoms in crystal structures may be determined in appropriate planes.

For example, the crystal structure of 1-mesitylselanyl-8-(dimethylsilyl) naphthalene was determined by high-resolution X-ray diffraction measurements accompanied by multipole refinement.[25] The QTAIM topological calculations based on these experimental results were also performed, thus it was possible to analyse maps of electron density and its Laplacian. Figure III.6 presents the Laplacian of electron density for the fragment of the structure. This fragment covers the Si-H···Se arrangement that may be classified as the chalcogen-hydride bond. Fig. III.6 shows the concentration of the electron density at the H-centre (blue); on the other hand, there is depletion of the electron charge at the selenium and silicon centres (red). For the former Se-centre, two areas of the concentration of electron density (blue) are attributed to the lone electron pairs. One can see the bond path with the bond critical point (red point) linking the Se and H attractors.

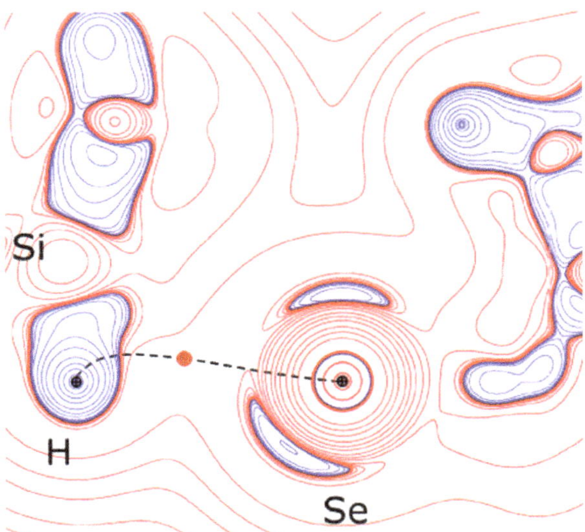

Fig. III.6 Laplacian of the experimental electron density of the Si-H···Se arrangement; positive values of Laplacian are in red, negative values in blue. Reprinted with permission from ref. 25. Copyright 2022 American Chemical Society.

The integrated QTAIM charges confirm the findings concerning the Si-H···Se bridge. The following experimental QTAIM charges are observed here (the values given in squared brackets correspond also to QTAIM charges but calculated at the M06-2X/def2-TZVP level). The negative charge of the H-centre of –0.8 [–0.7] e confirms it is the Lewis base site. This hydric hydrogen is situated between the considerably positively charged Si, 2.9 [2.8] e, and a positively charged Se centre, 0.3 [0.3] e. The bond path is located between areas of the electron charge concentration (lone electron pairs) of the selenium centre and goes through the Se σ-hole to the attractor of hydric hydrogen. This is why this interaction may be classified as a hydride bond on one hand and a chalcogen bond on the other hand.

Figure III.7 presents another example of the analysis of experimental electron density. The crystal of 2,3,5,6-tetrafluoro-1,4-diiodobenzene and 4-(dimethylamino) pyridine in 1:2 stoichiometry was analysed by high-resolution X-ray diffraction measurements at 100 K.[26] In this structure, the 2,3,5,6-tetrafluoro-1,4-diiodobenzene molecules form two halogen bonds by the iodine centres with the N-centres of two surrounding 4-(dimethylamino)pyridine molecules. These are very short I···N contacts equal to 2.6622(4) Å; the corresponding sum of van der Waals radii amounts to 3.53 Å (Table I.1 in Chapter 1), thus these contacts are much shorter than this sum.

Figure III.7 presents the Laplacian of the experimental electron density for the fragment of 2,3,5,6-tetrafluoro-1,4-diiodobenzene molecule and the nitrogen centre linked with the iodine. This N-centre belongs to the 4-(dimethylamino)pyridine molecule. The upper part of this figure corresponds to the experimental electron density and the bottom part to the single-point B3LYP/MIDIX calculations performed on the aggregate containing three molecules linked by two I···N halogen bonds (2,3,5,6-tetrafluoro-1,4-diiodobenzene molecule and two 4-(dimethylamino)pyridine molecules). The Laplacian of electron density resulting from these calculations is presented in the figure. The picture of experimental electron density shows negative Laplacian around the nitrogen centre that corresponds to the electron charge concentration (red; colours here are opposite to those presented in Figure III.6). However, the Laplacian of the electron density at the iodine centre is positive. There is depletion of the electron density at the valence shell area, and only its concentration is visible in the core region. The bottom part is similar to the upper part; however, here one can see two small valence

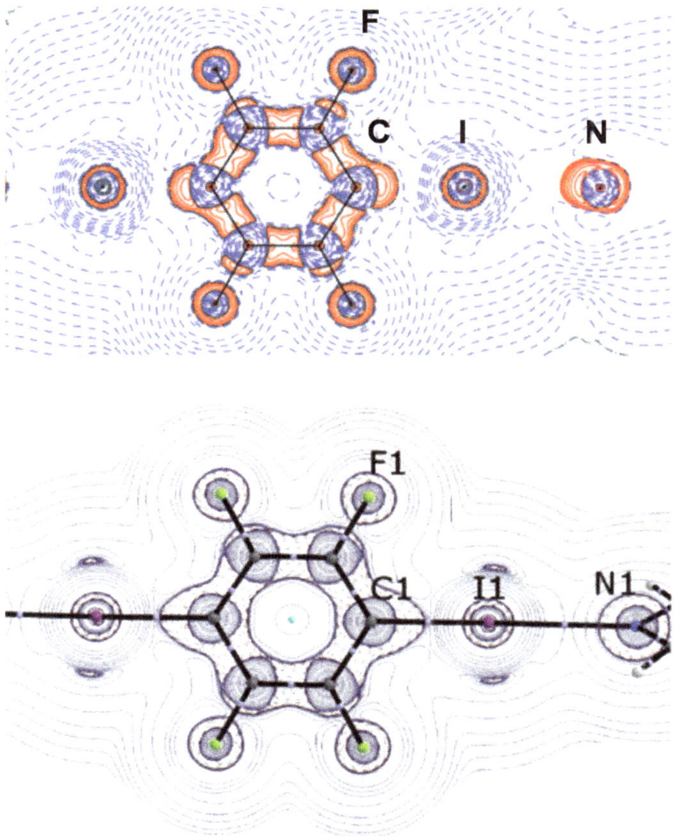

Fig. III.7 Laplacian of the electron density for 2,3,5,6-tetrafluoro-1,4-diiodobenzene. The nitrogen centre of 4-(dimethylamino)pyridine is visible on the right of both pictures. The upper part concerns the experimental electron density, the bottom part the theoretical electron density (B3LYP/MIDIX calculations). Positive values of Laplacian are in blue, negative values in red (opposite to colours of Figure III.6). Reprinted with permission from ref. 26. Copyright 2018 De Gruyter.

regions of the electron concentration for the iodine centre, similar to what was observed for the selenium centre in the chalcogen-hydride bond presented in Fig. III.6.

The Hansen-Coppens Multipolar Model[23] that was roughly described earlier requires time-consuming calculations as well as determination of numerous parameters (more than 30 for an atom, while in IAM only nine parameters in the anisotropic thermal vibrations model are needed). That is why other models that apply aspherical atom refinements are often used. One can mention the Hirshfeld atom refinement or transferable aspherical atom model.[27,28]

III.5 Crystal Structures as a Source of Information About Inter- and Intramolecular Links

The electron charge distribution influences properties of their sites; some of these sites may further act as electron donors and others as electron acceptors. In other words, the former sites are the Lewis base centres and the latter ones are the Lewis acid centres. This is connected, most often, with negative electrostatic potential (EP) and positive EP of these sites, respectively. Hence in numerous studies, the maps of EP are performed to know properties of different centres of molecules, ions or radicals. Undoubtedly, EP distribution is a result of the electron charge shifts. For example, the σ-hole and π-hole concept[29–31] explains how the electron charge shifts influence the creation of such holes that are characterised by deficiency of the electron charge and thus Lewis acid properties (see Chapter I).

The knowledge of the electron charge distribution and/or of the distribution of the EP allows us to predict the mutual arrangements of species constituting systems, complexes, clusters or even crystals. However, what is also important is that these distributions affect intermolecular connections. X-ray diffraction, synchrotron radiation or neutron diffraction results are a source of such connections that may be described by various geometrical parameters. The inter- and intramolecular atom-atom distances are often analysed since they partly describe the types and strength of interactions. Various interactions were discussed earlier in this chapter (Figs. III.1–III.5). The electron charge distributions for specific crystal structures were also discussed (Figs. III.6 and III.7).

Several examples of crystal structures with mutual arrangements of molecules and/or ions steered by various interactions are discussed further here. Figure III.8 presents the fragment of the crystal structure of (4,5-dichloro-1,3-dimesitylimidazol-2-ylidene) tris(trifluoromethyl) antimonide (WANDAQ refcode).[32] One can see here the (C)Cl···H(C) distance is 3.32 Å, which is greater than the corresponding sum of van der Waals radii of 2.85 Å (Table I.1) that excludes the existence of the C-H···Cl hydrogen bond. The other Cl···H contacts which could correspond to hydrogen bonds are also characterized by distances

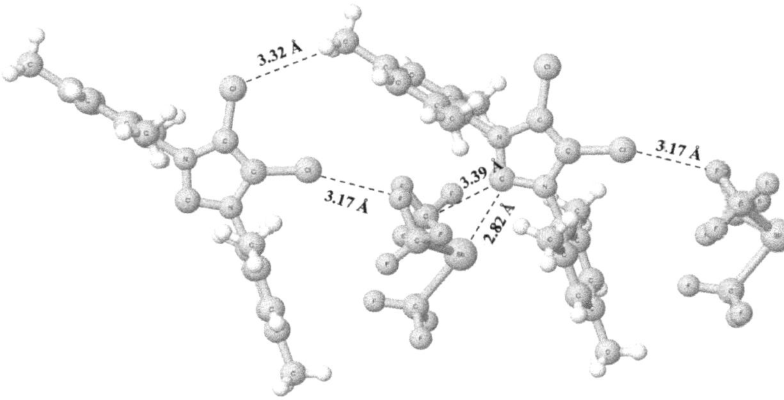

Fig. III.8 The fragment of the crystal structure of (4,5-dichloro-1,3-dimesitylimidazol-2-ylidene) tris(trifluoromethyl)antimonide (WANDAX refcode, ref. 32). Different kinds of interactions are designated for the atom-atom intermolecular contacts.

greater than this sum. These results may suggest that hydrogen bonds do not occur in this crystal structure. However, these may be weak hydrogen bonds since hydrogen bonds are mostly electrostatic in nature and it was stated that such interactions act far beyond the sum of van der Waals radii.[33] For the Cl···H contacts of this structure it is difficult to classify them as any kind of interactions without additional investigations. In other cases, if they are not classified as hydrogen bonds, they would be van der Waals interactions.

One can see in this structure the C-Cl···F contacts, with the Cl···F distances equal to 3.17 Å that is slightly lower than the sum of van der Waals radii, 3.22 Å. One can expect in this case the occurrence of weak halogen bonds. The chlorine centre acts as the Lewis acid site in the elongation of the C-Cl bond, approximately. The fluorine centre is the Lewis base site. The C-Cl-F angle is equal to 159° that in in line with the σ-hole concept.[29–31] The area characterised by the depletion of the electron charge in extension of the C-Cl bond occurs, thus the chlorine centre interacts with the fluorine centre possessing negative charge and negative EP at the whole fluorine surface probably. This interaction should be classified as halogen bond in the opinion of the author of this book and not as dihalogen bonds. In the case of dihalogen bonds, two halogen centres are also in contact, $X_1 \cdots X_2$, but both centres possess dual character, i.e., they may act as electron donor and electron acceptor simultaneously. Here the fluorine centre may act only as the electron donor, probably, while the chlorine centre possesses dual character. It may act as the Lewis acid in elongation of the C-Cl bond and as the Lewis base in the direction perpendicular to the C-Cl bond or nearly so.

Finally, one can see in this crystal structure the Sb···C and C···C interactions between the (4,5-dichloro-1,3-dimesitylimidazol-2-ylidene) anion and the tris(trifluoromethyl) antimonide cation which may be classified as pnicogen bond and tetrel bond, respectively. The C-Sb···C pnicogen bond is characterized by the Sb···C distance equal to 2.82 Å that is much shorter than the corresponding sum of van der Waals radii, 3.74 Å. The C-Sb-C angle is equal to 169°, close to linearity because the Sb-centre acts as the electron acceptor through the σ-hole in the extension of the C-Sb bond. The imidazole-2-ylidene ring possesses the active nucleophilic centre of carbon; this is the Lewis base centre for both singlet and triplet states characterized by a lone electron pair or two unpaired electrons, respectively.

The latter centre is involved in another interaction with another carbon centre that is the electron acceptor. Consequently, one can observe the C···C contact between electron donor and electron acceptor; the former belongs to the imidazole-2-ylidene ring and the other to the trifluoromethyl group of the tris(trifluoromethyl)antimonide cation. In the case of the latter cation three fluorine substituents withdraw electrons from the carbon centre that results in the areas of the depletion of electron charge in three F-C directions. Hence the positive EP very probably occurs in the elongation of F-C bonds. One of such positive EP areas, i.e., one of the σ-holes, is in contact with the nucleophilic carbon centre. The C···C distance is equal to 3.39 Å while the angle of this F- C···C tetrel bond amounts to 168°. Since the corresponding sum of van der Waals radii amounts to 3.4 Å, almost equal to the C···C distance, one may expect that this tetrel bond is rather weak and mostly electrostatic in nature.

Figure III.9 presents the fragment of the crystal structure of bis((μ2-ethynylsilyloxo)-dichloro-aluminum)[34] that was analysed later with the support of DFT, ωB97XD/aug-cc-pVTZ calculations.[35] One can see in this figure two types of tetrel bonds, O-Si···π and C-Si···Cl. In the former case, π electrons of the C≡C bond may be treated as the Lewis base site while the silicon centre is an acceptor of electrons through the σ-hole located approximately in the extension of the O-Si bond. The Si···C distances for the single O-Si···π interaction are equal to 3.583 Å and 3.662 Å. Two neighbouring bis((μ2-

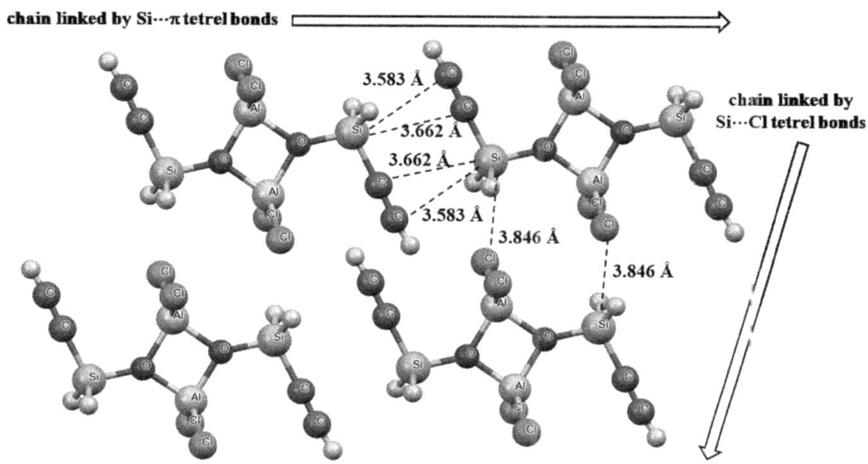

Fig. III.9 The fragment of the crystal structure of bis((μ2-ethynylsilyloxo)-dichloro-aluminum) (LECDUS refcode, ref. 34). The grey areas of circles correspond to H atoms. The Si···C and Si···Cl distances corresponding to the intermolecular contacts are given. The chains of molecules that are linked by the O-Si···π and C-Si···Cl tetrel bonds in the crystal structure are marked. Reprinted with permission from ref. 35. Copyright 2023 American Chemical Society.

ethynylsilyloxo)-dichloro-aluminum), or BESDA, molecules are linked by two equivalent O-Si···π tetrel bonds. This is the centrosymmetric dimer since the centre of inversion is located between these molecules. However, due to the symmetry of BESDA, both sites of this molecule are involved in double tetrel bonds. The chains of tetrel-bonded BESDA molecules are formed in the crystal structure (Fig. III.9). On the other hand, one can observe in another direction the dimer linked by two equivalent C-Si···Cl tetrel bonds, and consequently the chain bonded by pairs of such interactions. The silicon centre and the chlorine act here as the Lewis acid and the Lewis base sites, respectively. Here the area of electron charge depletion, i.e., the σ-hole, is formed in the elongation of the C-Si bond approximately. Hence this Si site acts as the electron acceptor.

A similar situation is often observed in the crystal structures of carboxylic acids. The latter structures were partly described in the former chapters (Chapter I, Figure I.5 and Chapter II, Figures II.4, II.7 and II.15). The crystal structure of benzoic acid is an example since centrosymmetric dimers occur.[36] However, the crystal structure of terephthalic acid[37] seems to be a better example of the structure that is very similar to the BESDA structure. Figure III.10 presents the fragment of the terphthalic acid structure (refcode TEPHTH15).

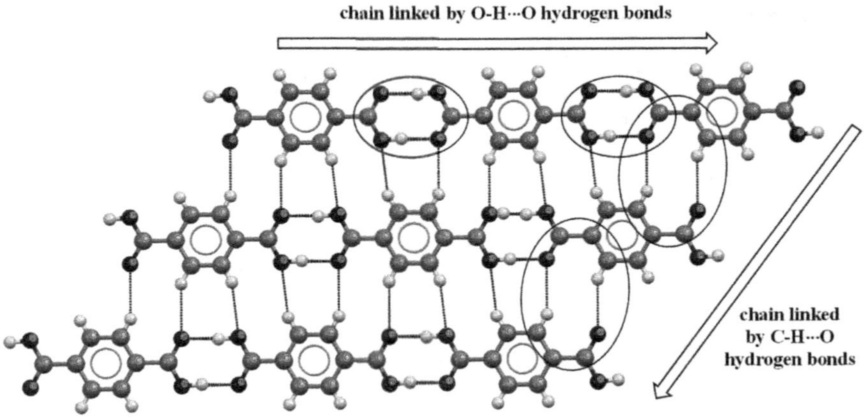

Fig. III.10 The fragment of the crystal structure of terephthalic acid (TEPHTH15 refcode, ref. 37). The chains of molecules linked by the O-H···O and C-H···O hydrogen bonds are indicated.

One can observe here the chains of molecules linked by the double and equivalent O-H···O hydrogen bonds between carboxylic groups. Other chains are linked by double and equivalent C-H···O hydrogen bonds. Thus, it is a very similar situation to that occurring for the BESDA structure. However, for the terephthalic acid structure the chains connected by hydrogen bonds occur while for the BESDA structure the chains are linked by tetrel bonds.

Figure III.11 presents the fragment of the crystal structure of trimethylselenonium selenocyanate (MOYREX refcode).[38] This is the structure where the selenium centre of the trimethylselenonium cation acts as the Lewis acid through σ-holes located in elongations of C-Se bonds. There is one contact of this centre with the nitrogen of the selenocyanate anion, and the Se···N distance is equal to 3.14 Å. The same selenium centre of the cation interacts with another Se-centre (Lewis base site) of the selenocyanate anion, and the Se···Se distance is equal to 3.76 Å. Both distances are lower than the corresponding sums of van der Waals radii, selenium plus nitrogen, 3.45 Å, and two times selenium, 3.80 Å (see Table I.1). The Se···Se distance is very close to the sum of van der Waals radii that suggests weak interaction.

Another situation is observed for the selenocyanate anion where the selenium centre acts as the Lewis base interacting with the cation; this is the Se···Se contact described above. The second interaction of the Se-centre of this anion, i.e., another Se···Se

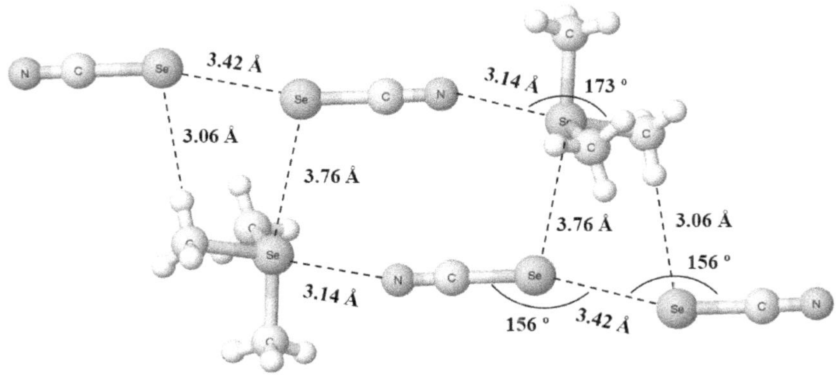

Fig. III.11 The fragment of the crystal structure of selenonium selenocyanate (ref. 38). Selected interactions are presented here: C-Se···N chalcogen bonds, Se···Se dichalcogen bonds and C-H···Se hydrogen bonds.

contact, is characterised by a distance equal to 3.42 Å. This is much lower than the corresponding sum of the van der Waals radii, 3.80 Å. Hence it should be an attractive interaction despite both units possessing a negative charge. One can expect here the σ-holes in extension of C-Se bonds. The C-Se···Se angle is equal to 156°. This is not so close to linearity. This interaction may probably be classified as dichalcogen interaction where the selenium centres possessing dual character (they may act as Lewis base and Lewis acid simultaneously) are in contact. One can see for this crystal structure (Fig. III.11) the H...Se contacts with distances equal to 3.06 Å. These interactions may be classified as very weak C-H...Se hydrogen bonds or as van der Waals interactions, since this distance is slightly greater than the sum of van der Waals radii, 3.0 Å.

Figure III.12 presents the fragment of the crystal structure of trimethyltelluronium cyanate (HUHBIV refcode).[39] In this structure the trimethyltelluronium cations are surrounded by cyanate anions. The tellurium centre is connected by covalent bonds with three methyl groups. Additionally, there are three contacts of this centre with the cyanate anions; this is the octahedron structure. One can expect that the Te-centre possesses three σ-holes in extensions of C-Te bonds. The trimethyltelluronium cation interacts through these holes with anions, forming contacts with nitrogen and oxygen centres. The Te···N/O distances are equal to 3.0–3.1 Å, approximately. The C-Te···N/O angles amount to 166–172°.

The examples of crystal structures presented in this section, as well as in previous chapters, show that not only hydrogen bonds are responsible for the arrangement of molecules and

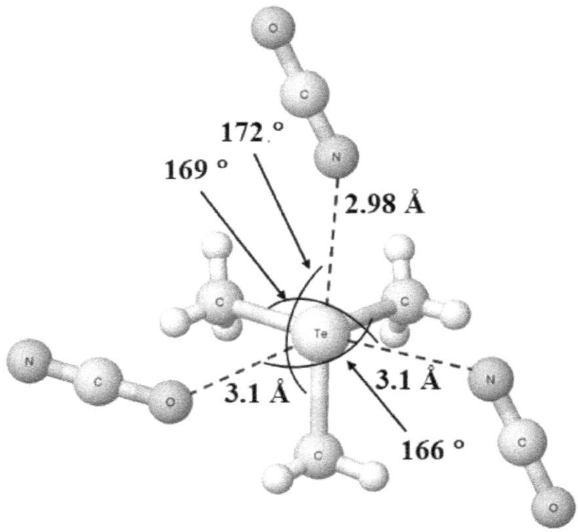

Fig. III.12 The fragment of the crystal structure of trimethyltelluronium cyanate (ref. 39). The coordination sphere of the tellurium centre is indicated.

ions. There may also be other interactions, mostly Lewis acid – Lewis base interactions that are often classified as σ-hole bonds or π-hole bonds. The latter interactions are directional, and they possess numerous characteristics typical for hydrogen bonds.

III.6 Hirshfeld Surfaces

There are various ways of dividing space to allocate appropriate areas to molecules or other chemical species that make up a system under consideration. One of such divisions comes from the QTAIM approach,[24] and it concerns the partitioning of the total electron density of a system. Hence the area belonging to the molecule because of such QTAIM division is related to its electron charge. It is worth mentioning that the QTAIM approach also offers further divisions into atomic fragments. One of the problems of this approach is the time-consuming process of this division and assigning parameters such as volumes and corresponding electron charges to atoms. The other problem concerns the borders of terminal atoms of the system. Their volumes are infinite until the electron density corresponding to the external borders of terminal atoms is determined. The electron density of 0.001 au is chosen most often. It was described (see Chapter I) that such a sphere of electron density agrees approximately with the van der Waals sphere of the species considered.[40,41] The QTAIM approach may be also applied for crystal structures, thus it is possible to consider the experimental crystal electron density and consequently its parts attributed to molecules or ions.[20–22] However, such analyses require high resolution X-ray diffraction measurements.

In the case of crystal structures another approach was proposed to divide the electron density space into molecular fragments.[42,43] This approach has its roots in the partitioning scheme proposed by Hirshfeld to divide molecular space into atomic fragments.[44] Thus, the weight function was defined for each atom designated by the subscript *a*.

$$w_a(r) = \rho_a^{at}(r) / \sum_{i \in molecule} \rho_i^{at}(r) \tag{III.21}$$

The $\rho_i^{at}(r)$ term marks the spherically averaged electron density of the *i*th atom. Consequently, the electron density of an atomic fragment is defined by the following equation:

$$\rho_a(r) = w_a(r)\rho^{mol}(r) \tag{III.22}$$

The $\rho^{mol}(r)$ term designates the molecular electron density. In analogy to the atomic weight function proposed by Hirshfeld (eq. III.21), the weight function for a molecule in a crystal may be defined:

$$W_A(r) = \sum_{i \in molecule} \rho_i^{at}(r) / \sum_{i \in crystal} \rho_i^{at}(r) = \rho_{promolecule}(r) / \rho_{procrystal}(r) \tag{III.23}$$

The numerator of the last equation is the sum of electron densities over atoms of molecule (pro-molecule) while the denominator is a sum over electron densities of atoms of a crystal (pro-crystal).

The $W_A(r)$ is a continuous function with $0 < W_A(r) < 1$. This function decays rapidly with the increase of distance from the molecule under consideration. The molecular surface with $W_A(r) = 0.5$ seems to be special since it covers the molecule and defines the volume where the pro-molecule electron density exceeds the electron density coming from all neighbouring species in a crystal.[43] It was found that the volume is approximately in agreement with molecular volumes determined by atomic van der Waals radii as well as with volume determined by the borders of the value of the electron density of 0.002 a.u.[43] However, there are exceptions if the terminal atoms are in close contacts with neighbouring molecules. In such cases, the van der Waals and/or electron density surface may protrude beyond the Hirshfeld surface.

It was pointed out that Hirshfeld surface analysis has an advantage over other divisions that lead to assignment of areas to molecular units. The Hirshfeld surface contains information about interactions of the molecule with neighbouring species while the other approaches do not.[43] It is worth mentioning, however, that other approaches that use the molecular geometry coming from crystal structures also contain such information. Nevertheless, this information is not so clear as in the case of Hirshfeld surfaces but one can see that changes of geometry of interacting units are detectable, at least for strong interactions.

There are various functions of distance and curvature that may be mapped on Hirshfeld surfaces (Table III.1). The shape index, *S*, and curvedness, *C*, partly decode information on intermolecular interactions; for example, they inform of the crystal packings, particularly on the stacking interactions. These indices are expressed by eqs. III.24 and III.25.

$$S = \frac{2}{\pi}\arctan\left(\frac{\kappa_1 + \kappa_2}{\kappa_1 - \kappa_2}\right) \tag{III.24}$$

$$C = \frac{2}{\pi}\ln\sqrt{(\kappa_1 + \kappa_2)/2} \tag{III.25}$$

κ_1 and κ_2 are functions introduced by Koenderink.[42]

Table III.1 Functions of distance and curvature that may be mapped on Hirshfeld surfaces. Modification of Table 1 from ref. 43.

Function	Symbol
Distance from a point on the surface to the nearest nucleus outside the surface	d_e
Distance from a point on the surface to the nearest nucleus inside the surface	d_i
Shape index, a measure of "which" shape, defined in terms of principal curvatures κ_1 and κ_2	S
Curvedness, a measure of "how much" shape, defined in terms of principal curvatures κ_1 and κ_2	C
Normalized contact distance, defined in terms of d_e, d_i and the van der Waals radii of the atoms	d_{norm}

There is another index, the normalized contact distance, d_{norm}, that informs of the shortest intermolecular contacts (eq. III.26) where d_e and d_i are the distance from a point on the surface to the nearest nucleus outside the surface and the distance from a point on the surface to the nearest nucleus inside the surface, respectively. The corresponding r_e^{vdW} and r_i^{vdW} van der Waals radii also occur in this equation.

$$d_{norm} = \frac{d_i - r_i^{vdW}}{r_i^{vdW}} + \frac{d_e - r_e^{vdW}}{r_e^{vdW}} \qquad \text{III.26}$$

The latter index is often mapped on Hirshfeld surfaces to indicate the atom-atom intermolecular contacts that are shorter than the corresponding sum of van der Waals radii. For example, Figure III.13 presents the fragment of crystal structure of 6-methyl-4-oxo-1,4-dihydropyridine-3-carboxylic acid.[45] The Hirshfeld surface for one of the molecules is presented with the mapped normalized contact distance, d_{norm}. The red

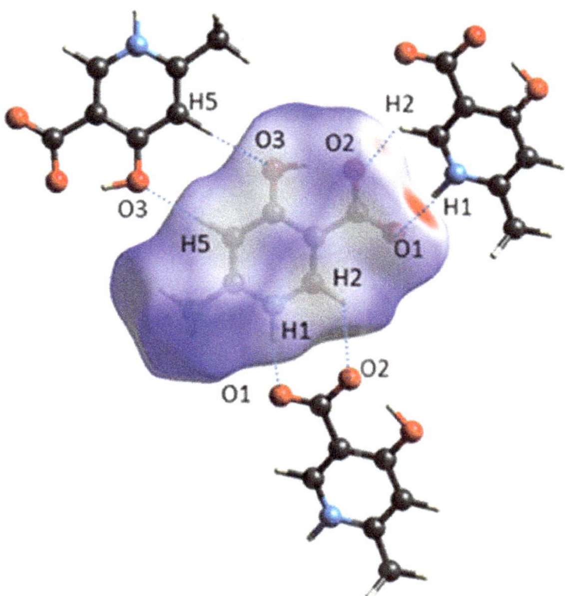

Fig. III.13 The fragment of the crystal structure of 6-methyl-4-oxo-1,4-dihydropyridine-3-carboxylic acid (ref. 45). The Hirshfeld surface for one of the molecules is presented with the mapped normalized contact distance, d_{norm}. The short intermolecular contacts are also indicated. Reprinted with permission from ref. 45. Copyright 2024 American Chemical Society.

areas on the Hirshfeld surface indicates contacts shorter than corresponding sums of van der Waals radii. These contacts with the neighbouring molecules are presented in the figure; one can see C-H···O and N-H···O hydrogen bonds.

Figure III.14 presents another example of the application of Hirshfeld surface analysis, the fragment of the crystal structure of bis(4-(2-hydroxy-4-oxopent-2-en-3-yl)-1,3,5-trimethyl-1H-pyrazol-2-ium) dibromide 1,2,4,5-tetrafluoro-3,6-di-iodobenzene dihydrate.[46] Figure III.14a shows the displacement ellipsoid plot where the bromine anion has close contacts with the 1,2,4,5-tetrafluoro-3,6-di-iodobenzene and water molecule (which is connected with the 4-(2-hydroxy-4-oxopent-2-en-3-yl)-1,3,5-trimethyl-1H-pyrazol-2-ium species) through the C-I···Br^- halogen bond and the O-H···Br^- hydrogen bond, respectively. The interactions between halogen centres and halogen anions are named sometimes in the literature as dihalogen bonds. However, it was stated here that only the interactions where both halogen centres being in contact possess dual character (as the Lewis acid and as the Lewis base centre simultaneously) may be classified as dihalogen bonds. Hence the C-I···Br^- interaction observed here is classified as halogen bond. Figure III.14b presents the same fragment as that of Fig. III.14a. However, in this case the Hirshfeld surface for the bromine anion is presented with the mapped normalized distance index. One can see here red areas corresponding to halogen and hydrogen bonds that indicate the I···Br and H···Br distances, respectively, which are shorter than corresponding sums of van der Waals radii.

It was pointed out that various parameters may be mapped on the Hirshfeld surfaces,[42,43] not only those presented in Table III.1. Figure III.15 presents the fragment of the cocrystal structure of 2,3,5,6-tetrafluoro-1,4-diiodobenzene and 4-(dimethylamino)pyridine.[26] The 2,3,5,6-tetrafluoro-1,4-diiodobenzene molecule is connected here with two 4-(dimethylamino)pyridine molecules by the C-I···N halogen bonds. The iodine centres act as the Lewis acid sites since the σ-holes are formed here in extensions of C-I bonds. The Hirshfeld surface for the central 2,3,5,6-tetrafluoro-1,4-diiodobenzene molecule is presented with the mapped EP.

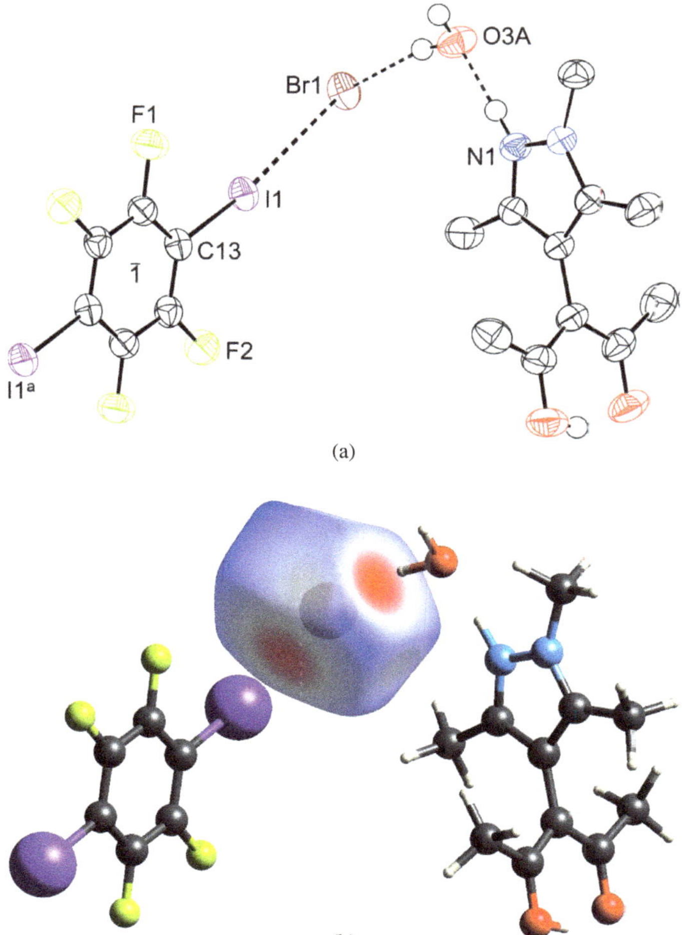

Fig. III.14 The fragment of the crystal structure of bis(4-(2-hydroxy-4-oxopent-2-en-3-yl)-1,3,5-trimethyl-1H-pyrazol-2-ium) dibromide 1,2,4,5-tetrafluoro-3,6-di-iodobenzene dihydrate (LAYTII refcode, ref. 46). (a) Displacement ellipsoid plot of 80% probability, with C-bonded H atoms omitted. (b) Hirshfeld surface about the Br^- anion with the normalized index, d_{norm}, mapped. Red areas at the surface indicate short intermolecular contacts, C-I$\cdots$$Br^-$ halogen bond and O-H$\cdots$$Br^-$ hydrogen bond. Reprinted with permission from ref. 46. Copyright 2022 IUCr.

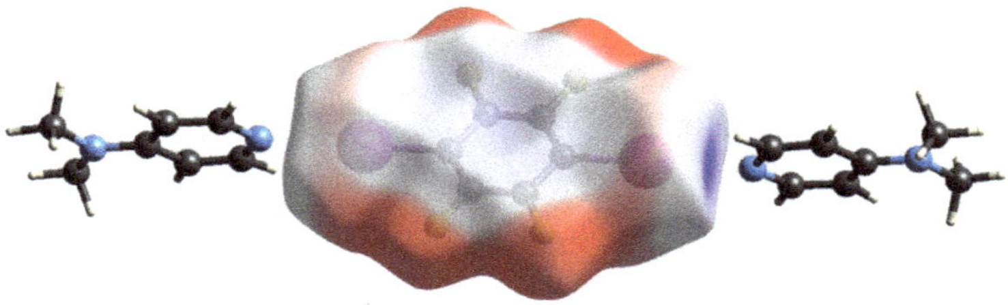

Fig. III.15 The fragment of the cocrystal of 2,3,5,6-tetrafluoro-1,4-diiodobenzene and 4-(dimethylamino) pyridine (ref. 26). The Hirshfeld surface of 2,3,5,6-tetrafluoro-1,4-diiodobenzene is presented with the map of the EP. Reprinted with permission from ref. 26. Copyright 2018 de Gruyter.

One can see the blue areas in elongations of the C-I bonds that indicate σ-holes with the corresponding positive EP values that are responsible for the Lewis acid properties of the iodine centres.

It is worth mentioning that the 2D fingerprint plot was introduced as a combination of d_i and d_e to provide useful information on intermolecular interactions.[47] These distance measures are often mapped on a 3D Hirshfeld molecular surface. In the fingerprint plot the (d_i, d_e) pairs are assigned to intervals of 0.01 Å (thus one obtains the 2D picture of pixels). Each pixel is coloured, usually from blue that corresponds to only a few points up to red (through green) for many points.

Figure III.16 presents the fingerprint plots for the S-benzyl 3-[1-(6-methylpyridin-2-yl)ethylidene]dithiocarbazate (BMPETC) crystal structure.[48] One can see the pseudo-symmetry of fingerprints according to the diagonal lines where $d_e = d_i$. This pseudo-symmetry is a consequence of the close packing of Hirshfeld surfaces.[47] This figure presents the scheme of BMPETC species as well as the overall fingerprint plot and plots delineated into the most important atom-atom contacts. Ideally, in the absence of errors, usually corresponding to disorder or the lack of full determination of packings (solvent molecules not determined, for example), the sum of percentage contributions should be equal to 100%. In the case of the BMPETC crystal structure, other interactions that are not presented in Fig. III.16 and that contribute less to all interactions also occur; these are C···S, S···N and C···C interactions.

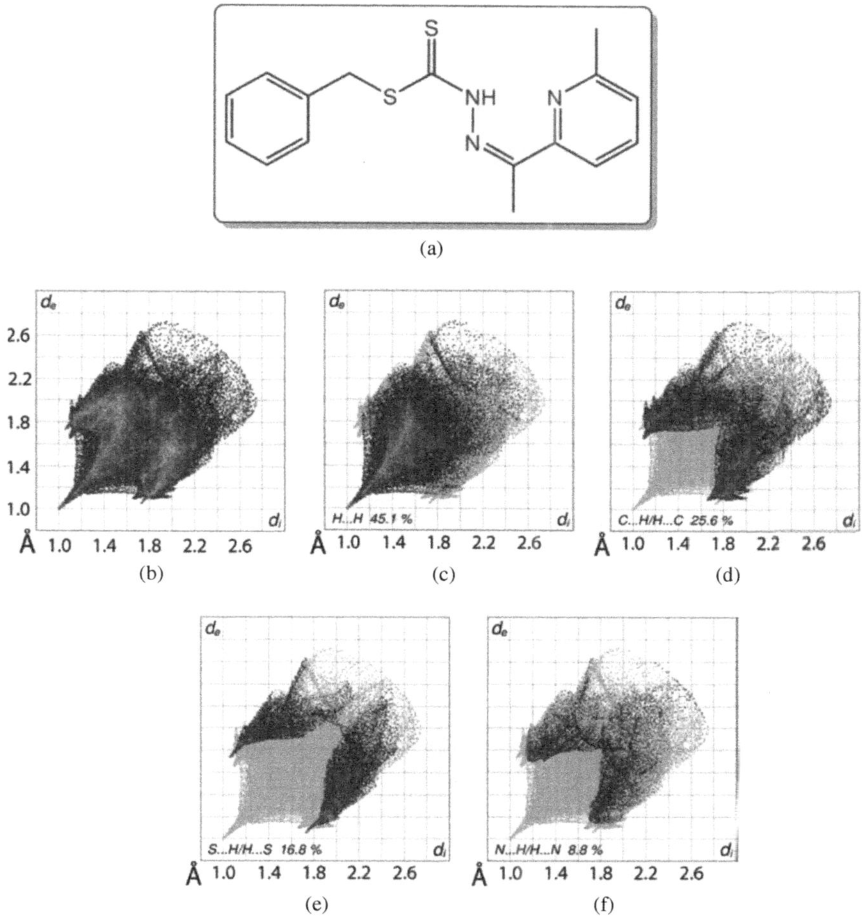

Fig. III.16 (a) Chemical diagram for S-benzyl 3-[1-(6-methylpyridin-2-yl)ethylidene]dithiocarbazate, BMPETC; (b) the full two-dimensional fingerprint plot for BMPETC and fingerprint plots delineated into (c) H···H, (d) C···H/H···C, (e) S···H/H···S and (f) N···H/H···N contacts. The outline of the full fingerprint plots is shown in grey. Reprinted with permission from ref. 48. Copyright 2019 IUCr.

Summary

There are various experimental techniques that provide insight into molecular geometry.[49] In general, the molecular geometry can be determined most often by two groups of methods: spectroscopic methods and diffraction methods. Infrared, nuclear magnetic resonance, microwave and Raman spectroscopy belong to the first group and they deliver information on molecular geometry from vibrational and rotational absorbance. X-ray, neutron and electron diffraction can give molecular structure to species in crystal structures on the basis of diffracted X-rays, neutrons or electrons.

X-ray and neutron diffractions methods are described in this chapter since they are often applied in investigations on molecular structures; particularly, the former method is very common. Over a million crystal structures are stored in the CSD,[4,5] constituting a source of information not only on molecular geometries but also on arrangement of molecules and ions in crystals. In other words, the CSD is a source of information on intra- and intermolecular interactions. This is why it is described in this chapter which information may be received from CSD and how to decode it from results stored therein. The information on positions of atoms in asymmetric units and on the symmetry of crystal structures allows us to obtain geometry of species constituting crystals and to obtain intermolecular distances.

The high-resolution X-ray diffraction measurements allow us to obtain information on electron density of crystal structures that leads to additional characteristics of interactions. It is also possible to apply additional approaches based on electron density such as QTAIM or the Hirshfeld structure analysis, for example.

Appendix

The following tools were applied to construct figures presented in this chapter.

Figures III.1,10

Graphical tools of the Cambridge Structural Database — References: (1) Groom, C.R.; Bruno, I.J.; Lightfoot, M.P.; Ward, S.C. The Cambridge structural database. *Acta Cryst.* **2016**, *B72*, 171–179; (2) Wong, R.; Allen, F.H.; Willett, P. The scientific impact of the Cambridge Structural Database: A citation-based study. *J. Appl. Cryst.* **2010**, *43*, 811–824.

Figures III.8,11,12

Jmol program — Reference: Hanson, R.M. Jmol SMILES and Jmol SMARTS: specifications and applications. *J. Cheminform.* **2016**, *8*, 50.

Figures III.2,3,4,5,6,7,9,13,14,15,16
Reprinted with permission.

References

1. Dunitz, J.D. *Analysis of the Structure of Organic Molecules*, Cornell University Press, Ithaca, **1979**.
2. Luger, P. *Modern X-Ray Analysis on Single Crystals: Second Fully Revised and Extended Edition*, Walter de Gruyter, Berlin, **2014**.
3. Giacovazzo, C. (Ed.). *Fundamentals of Crystallography: Second Edition*. Oxford University Press, New York, **2002**.
4. Groom, C.R.; Bruno, I.J.; Lightfoot, M.P.; Ward, S.C. The Cambridge structural database. *Acta Cryst.* **2016**, *B72*, 171–179.
5. Wong, R.; Allen, F.H.; Willett, P. The scientific impact of the Cambridge Structural Database: A citation-based study. *J. Appl. Cryst.* **2010**, *43*, 811–824.
6. Tilley, R. *Crystals and Crystal Structures*, John Wiley & Sons, Chichester, **2006**.
7. Schwarzenbach, D. *Crystallography*, John Wiley & Sons, Chichester, **1996**.
8. Li-ling, O. *Principles of X-ray Crystallography*, Oxford University Press, New York, **2010**.
9. Dawson, A.; Allan, D.R.; Parsons, S.; Ruf, M. Use of a CCD diffractometer in crystal structure determinations at high pressure. *J. Appl. Cryst.* **2004**, *37*, 410–416.
10. Leiserowitz, L. Molecular packing modes. Carboxylic acids. *Acta Cryst.* **1976**, *B32*, 775–802.
11. Bricogne, G. A multisolution method of phase determination by combined maximization of entropy and likelihood. III. Extension to powder diffraction data. *Acta Cryst.* **1991**, *A47*, 803–829.
12. Gilmore, C.J.; Shankland, K.; Bricogne, G. Applications of the maximum entropy method to powder diffraction and electron crystallography. *Proc. R. Soc. Lond. A* **1993**, *442*, 97–111.
13. The atomic scattering factors have been calculated from the spherical electron charge distribution for isolated atoms theoretically and tabulated: Cromer, D.T.; Waber, J.T. International tables for X-ray crystallography, **1974**, vol. IV, 71.
14. Bürgi, H.B.; Capelli, S.C.; Birkedal, H. Anharmonicity in anisotropic displacement parameters. *Acta Cryst.* **2000**, *A56*, 425–435.
15. Kuś, P.; Jones, P.G.; Kusz, J.; Książek, M. Dibromomethyl- and bromomethyl- or bromosubstituted benzenes and naphthalenes: C—Br···Br interactions. *Acta Cryst.* **2023**, *C79*, 83–93.
16. Wilson, C.C. *Single Crystal Neutron Diffraction From Molecular Materials*, World Scientific, Singapore, **2000**.

17. Wilson, C.C.; Shankland, K.; Shankland, N. Single-crystal neutron diffraction of urea-phosphoric acid: evidence for H-atom migration in a short hydrogen bond between 150 K and 350 K. *Z. Kristallogr.* **2001**, *216*, 303–306.
18. Johnson, C. K.: ORTEP, Thermal Ellipsoid Plotting Program, Report ORNL-3794, Oak Ridge National Laboratory, Tennessee, USA, 1971.
19. Steinke, T.; Engelage, E.; Huber, S.M. Chalcogen bonding in the solid-state structures of 1,3-bis(benzimidazoliumyl)benzene-based chalcogen-bonding donors. *Acta Cryst.* **2023**, *C79*, 26–35.
20. Koritsanszky, T.; Coppens, P. Chemical applications of X-ray charge-density analysis. *Chem. Rev.* **2001**, *101*, 1583–1627.
21. Stalke, D. (Ed.). *Electron Density and Chemical Bonding I: Experimental Charge Density Studies*, Springer-Verlag, Berlin, Heidelberg, **2012**.
22. Gatti, C.; Macchi, P. (Eds.). *Modern Charge-Density Analysis*, Springer, New York, **2012**.
23. Hansen, N.K.; Coppens, P. Testing aspherical atom refinements on small-molecule data sets. *Acta Cryst.* **1978**, *A34*, 909–921.
24. Bader, R.F.W. *Atoms in Molecules: A Quantum Theory*, Oxford University Press, Oxford, **1990**.
25. Keil, H.; Herbst-Irmer, R.; Rathej, S.; Girschik, C.; Müller, T.; Stalke, D. Si–H···Se chalcogen–hydride bond quantified by diffraction and topological analyses. *Inorg. Chem.* **2022**, *61*, 6319-6325.
26. Wang, R.; Hartnick, D.; Englert, U. Short is strong: experimental electron density in a very short N···I halogen bond. *Z. Kristallogr.* **2018**, *233*, 733–744.
27. Jha, K.K.; Kleemiss, F.; Chodkiewicz, M.L.; Dominiak, P.M. Aspherical atom refinements on X-ray data of diverse structures including disordered and covalent organic framework systems: a time–accuracy trade-off. *J. Appl. Cryst.* **2023**, *56*, 116–127.
28. Kulik, M.; Dominiak, P.M. Electron density is not spherical: the many applications of the transferable aspherical atom model. *Comput. Struct. Biotech. J.* **2022**, *20*, 6237–6243.
29. Politzer, P.; Murray, J.S.; Clark, T. Halogen bonding: an electrostatically-driven highly directional noncovalent interaction. *Phys. Chem. Chem. Phys.* **2010**, *12*, 7748–7757.
30. Politzer, P.; Murray, J.S.; Clark, T. Halogen bonding and other σ-hole interactions: a perspective. *Phys. Chem. Chem. Phys.* **2013**, *15*, 11178–11189.
31. Murray, J.S.; Politzer, P. Molecular electrostatic potentials and noncovalent interactions. *WIREs Comput. Mol. Sci.* **2017**, *7*, e1326.
32. Arduengo III, A.J.; Krafczyk, R.; Schmutzler, R.; Mahler, W.; Marshall, W.J. A Tris(trifluoromethyl)antimony adduct of a nucleophilic carbene: geometric distortions in carbene adducts. *Z. Anorg. Allg. Chem.* **1999**, *625*, 1813–1817.

33. Desiraju, G.R.; Steiner, T. *The Weak Hydrogen Bond in Structural Chemistry and Biology*, Oxford University Press, New York, **1999**.
34. Bissinger, P.; Paul, M.; Riede, J.; Schmid Baur, H. (Ethyl-, ethenyl- and ethynylsilyloxy) aluminum compounds generated by heterolytic cleavage of 1,3-disubstituted disiloxanes. *Chem. Ber.* **1993**, *126*, 2579–2584.
35. Aarabi, M.; Gholami, S.; Grabowski, S.J. Double centrosymmetric Si···π tetrel bonds as new synthons — evidence from crystal structures and DFT calculations. *J. Phys. Chem. A* **2023**, *127*, 9995–10007.
36. Wilson, C.C.; Shankland, N.; Florence, A.J. A single-crystal neutron diffraction study of the temperature dependence of hydrogen-atom disorder in benzoic acid dimers. *J. Chem. Soc., Faraday Trans.* **1996**, *92*, 5051–5057.
37. Karothu, D.P.; Weston, J.; Desta, I.T.; Naumov, P. Shape-memory and self-healing effects in mechanosalient molecular crystals. *J. Am. Chem. Soc.* **2016**, *138*, 13298–13306.
38. Klapotke, T.M.; Krumm, B.; Mayer, P.; Piotrowski, H.; Polborn, K.; Schwab, I. The first selenonium azides and a selenonium selenocyanate. *Z. Anorg. Allg. Chem.* **2002**, *628*, 1831–1834.
39. Klapotke, T.M.; Krumm, B.; Mayer, P.; Piotrowski, H.; Schwab, I.; Vogt, M. Synthesis and structures of triorganotelluronium pseudohalides. *Eur. J. Inorg. Chem.* **2002**, 2701–2709.
40. Bader, R.F.W.; Carrol, M.T.; Cheeseman, J.R.; Chang, C. Properties of atoms in molecules: atomic volumes. *J. Am. Chem. Soc.* **1987**, *109*, 7968–7979.
41. Gillespie, R.J.; Popelier, P.L.A. *Chemical Bonding and Molecular Geometry*, Oxford University Press, Oxford, **2001**, p. 137.
42. McKinnon, J.J.; Spackman, M.A.; Mitchell, A.S. Novel tools for visualizing and exploring intermolecular interactions in molecular crystals. *Acta Cryst.* **2004**, *B60*, 627–668.
43. Spackman, M.A,; Jayatilaka, D. Hirshfeld surface analysis. *CrystEngComm* **2009**, *11*, 19–32.
44. Hirshfeld, F.L. Bonded-atom fragments for describing molecular charge densities. *Theoret. Chim. Acta (Berl.)* **1977**, *44*, 129–138.
45. May, N.V.; Gál, G.T.; Holczbauer, T.; Bereczki, L.; Di Marco, V.B ; Bombicz, P. Gradual changes in the aromaticity in a series of hydroxypyridine-carboxylic acid derivatives and their effect on tautomerism and crystal packing. *Cryst. Growth Des.* **2024**, *24*, 1096–1109.
46. Van Terwingen, S.; Ebel, B.; Wang, R.; Englert, U. Weaving a 2D net of hydrogen and halogen bonds: cocrystal of a pyrazolium bromide with tetrafluorodiiodobenzene. *Acta Cryst.* **2022**, *C78*, 324–331.
47. Spackman, M.A.; McKinnon, J.J. Fingerprinting intermolecular interactions in molecular crystals. *CrystEngComm* **2002**, *4*, 378–392.

48. Tan, S.L.; Jotani, M.M.; Tiekink, E.R.T. Utilizing Hirshfeld surface calculations, non-covalent interaction (NCI) plots and the calculation of interaction energies in the analysis of molecular packing. *Acta Cryst.* **2019**, *E75*, 308–318.
49. Klebe, G. Experimental Methods of Structure Determination. In: Klebe, G. (Ed.), *Drug Design*, Springer, Berlin, Heidelberg, **2013**.

Chapter IV

Different Types of Hydrogen Bonds

Abstract: The hydrogen bond and its various types are examined here. Different early and contemporary definitions are considered. Special attention is paid to interactions that are hardly classified as hydrogen bonds according to the commonly accepted definitions. These are hydrogen bonds with multi-centre proton donors or σ-electrons as proton acceptors. The characteristics of intramolecular hydrogen bonds and the concept of resonance-assisted hydrogen bonds are discussed. The mechanism of the formation of the hydrogen bond and the phenomena of blue-shifting and red-shifting stretching modes of the proton-donating bonds are also presented.

Keywords: hydrogen bond, electron charge shift, blue-shifting hydrogen bond, intramolecular hydrogen bond, resonance-assisted hydrogen bond

IV.1 Definitions of the Hydrogen Bond

This chapter is devoted to the hydrogen bond interaction, its kinds and properties as well as mechanisms of its formation. Why devote a whole chapter to only one kind of interaction? The reason is that the hydrogen bond, in comparison to numerous other interactions, was analysed most often for over the last one hundred years in terms of the number of publications. One can say the same probably about the significance of this interaction; it is a very important link that plays a crucial role in numerous physical, chemical and biochemical processes.[1,2]

The hydrogen bond (or hydrogen bonding) is most often understood as an interaction characterised by an arrangement of three centres that is marked as A-H···B. The H-atom is positively charged while A and B centres are usually characterised by high values of electronegativity.[1,2] G.N. Lewis was probably the first person who introduced the term "hydrogen bond"; one can refer to the statement in his monograph that "Hydrogen when attached firmly to a pair of electrons, as in the hydrogen-hydrogen or hydrogen-carbon bond, shows no tendency whatsoever to become bivalent, or in other words, to form a hydrogen bond. On the other hand, when combined with an extremely negative element like nitrogen, oxygen or fluorine, towards which the electron pairs are very tightly drawn, the bonding pair may be drawn out of the first valence shell of hydrogen into a secondary shell, and it is apparently under some such circumstances that the hydrogen atom can form a loose attachment to another pair of electrons, thus forming the hydrogen bond."[3] Hence the A-H···B designation fits well to Lewis's statements since there is the A-H covalent, less or more polarised, bond and "a loose attachment", i.e., the H···B link. Additionally Lewis mentioned that the A centre is characterised by a high value of electronegativity ("an extremely negative element"). Thus, one can see that in these few statements Lewis had included the most important characteristics of typical (or — as often used in various studies — conventional) hydrogen bonds.

It is not the intention of the author of this book to present a history of ideas and definitions of the hydrogen bond, since it has been done in other studies. However, it is worth presenting the definition of Linus Pauling who stated that "under certain conditions an atom of hydrogen is attracted by rather strong forces to two atoms, instead of only one, so that it may be considered to be acting as a bond between them. This is called the hydrogen bond."[4] Thus the idea of a bridge formed between A and B centres and the H-atom linking these centres was introduced by Pauling. It is worth mentioning here that the term "hydrogen bridge" is used sometimes in various studies on interactions.[5]

One can refer to the definition of the hydrogen bond proposed by Jeffrey and Saenger where it is stated that "A hydrogen bond is therefore the attractive force that arises between the donor covalent pair X-H in which a hydrogen atom H is bound to a more electronegative atom X, and other noncovalently bound nearest neighbor electronegative acceptor atoms A, A′."[1] One can see that marks of centres in this definition differ from those applied in this book where the A-H···B designation is applied for the whole hydrogen bond bridge. This definition is similar to earlier definitions, particularly that proposed by Pimentel and McClellan[6] where it is described that "A H bond exists between a functional group A-H and an atom or a group of atoms B in the same or a different molecule when

(a) there is evidence of bond formation (association or chelation),
(b) there is evidence that this new bond linking A-H and B specifically involves the hydrogen atom already bonded to A."

The latter definition allows systems containing a negatively charged hydrogen centre that links two other centres to be classified as hydrogen bonds. These are, for example, the B-H-B bridges in boranes (B is the boron centre).[7]

Referring to the definition of Jeffrey and Saenger,[1] it is worth to note that two significant elements are included. The first is that the H-atom is less electronegative than the centre covalently bonded to it. This should result in a positive charge of the H-atom and this is often stated in various descriptions of the hydrogen bond. The second important element of this definition is that the H-centre is non-covalently bound with nearest electronegative atoms. Thus not only one proton acceptor centre may occur, but more. The latter case is known in various studies as so-called bifurcated hydrogen bonding. It seems that the most often analysed bifurcated hydrogen bond is an interaction with one proton-donating bond and two acceptors. Jeffrey classifies such an interaction as three-centred because the hydrogen atom is bonded to three centres: to one by a covalent bond and to two others by hydrogen bonds.[2] Jeffrey also proposes the term "chelated hydrogen bonds" for systems with one proton acceptor and two proton-donating bonds that belong to the same molecule.[2] Desiraju and Steiner[8] have proposed terms for the phenomenon of bifurcation: the bifurcated donor hydrogen bond and the bifurcated acceptor hydrogen bond for structures where the proton-donating bond interacts with two or more acceptors and where the proton acceptor centre interacts with two or more proton donors, respectively. Bifurcated hydrogen bonds are described in detail in the next section of this chapter.

The contemporary, most often cited and accepted definition of the hydrogen bond (named henceforth as the IUPAC definition)[9,10] includes the statements in former definitions of Pimentel and McClellan[6] and of Jeffrey and Saenger.[1] The IUPAC definition explains that "The hydrogen bond is an attractive interaction between a hydrogen atom from a molecule or a molecular fragment X-H in which X is more electronegative than H, and an atom or a group of atoms in the same or a different molecule, in which there is evidence of bond formation."[10] In this definition, similar to that of Jeffrey and Saenger,[1] the X-H designation is applied for the proton-donating bond. The term "evidence of bond formation" appears in the IUPAC definition and in the early definition of Pimentel and McClellan.[6] However, both definitions do not explain what this term means. The IUPAC definition is accompanied by two articles where additional criteria and footnotes are included and where it is stated that "The short definition is followed by a list of experimental and/or theoretical criteria, which can be used as evidence for the presence of the hydrogen bond."[9,10] This may simply mean that "the short definition" is not sufficient to characterise (define) the hydrogen bond and that numerous additional explanations are needed.

The two-site hydrogen bond definition, 2sHB, that was proposed recently seems to cover all interactions which possess typical characteristics commonly accepted as hydrogen bonds.[7,11] This definition states that "*the hydrogen bond is a local stabilizing interaction between the proton or the electron charge deficient region of hydrogen and the electron rich region attributed to one or more centres.*" It is interesting that only two sites are mentioned here, the electrophilic H-centre that acts as the part of the Lewis acid unit and the nucleophilic centre of the Lewis base. It is important that systems containing negatively charged H-centres, such as the boron systems like B-H-B, are excluded from being considered to have hydrogen-bonded interactions. This definition follows other definitions where interactions are classified on the basis of two centres being in contact. For example, Politzer and co-workers proposed the definition of the halogen bond where it is described that "A halogen bond is a highly directional, electrostatically-driven noncovalent interaction between a region of positive electrostatic potential on the outer side of the halogen X in a molecule R-X and a negative site B, such as a lone pair of a Lewis base or the π-electrons of an unsaturated system."[12] One can see that two sites are mentioned here, the positive electrostatic potential region of the halogen (X) centre and the negative site (B). However, the connection of the halogen with the R-centre is also mentioned. In other words, the R-X···B three-atomic system is mentioned, similar to several definitions of the hydrogen bond where the A-H···B arrangements are described.

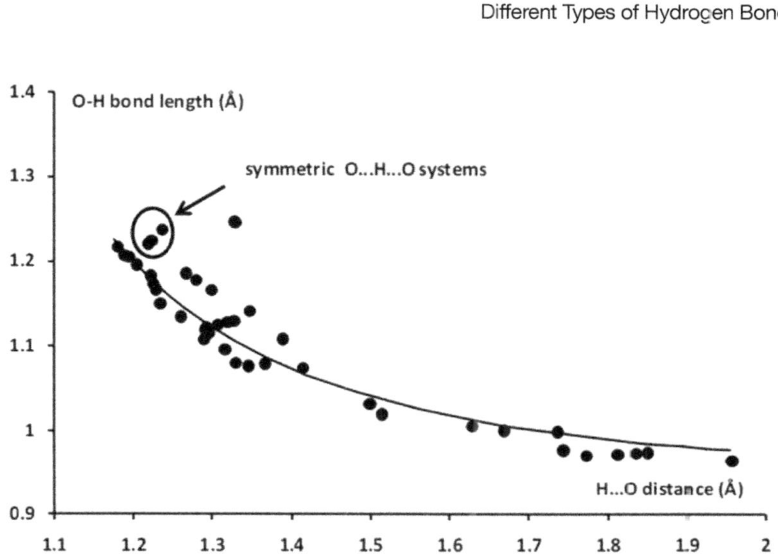

Fig. IV.1 The relationship between the H···O distance and O-H bond length (both in Å) for accurate neutron diffraction results of C=O···H-O-C systems. Three symmetric O···H· ·O systems with central position of the proton are shown. Solid line represents this relationship is obtained from the bond number conservation rule. Reprinted with permission from ref. 13. Copyright 2011 American Chemical Society.

Referring to the 2sHB definition, it is worth mentioning that there is no sharp border here between the hydrogen bonds and van der Waals interactions on one hand and between the hydrogen bonds and covalent bonds on the other hand. This is in line with the statement of Desiraju who claimed that the hydrogen bond is an interaction without borders.[5] Figure IV.1 presents the relationship between the H···O distance and the O-H bond length for the C=O···H-O-C systems taken from accurate neutron diffraction results of crystal structures.[13] The symmetric O···H···O systems with the central proton position are also shown. The solid line corresponds to the abovementioned relationship that is obtained from the bond number conservation rule.[14–17] This rule was applied for triatomic systems with the assumption that the sum of two bond numbers in the system is equal to unity (eq. IV.1).

$$10^{\frac{-\Delta r_1}{c}} + 10^{\frac{-\Delta r_2}{c}} = 1 \qquad \text{(IV.1)}$$

$\Delta r_1 = r_1 - r_S$ and $\Delta r_2 = r_2 - r_S$ correspond to r_1 and r_2 bond (contact) lengths within the considered system, r_S is usually the length of the single bond not involved in any disturbing interaction, and c is a constant. In this relation (eq. IV.1), two terms occur that are bond numbers of bonds (contacts) having r_1 and r_2 lengths. The bond number conservation rule may be applied for hydrogen-bonded systems, particularly for O-H···O type:

$$10^{\frac{-(r_{OH}-r_0)}{c}}+10^{\frac{-(r_{H...O}-r_0)}{c}}=1 \qquad \text{(IV.2)}$$

r_0 is the reference single O-H bond length not involved in interactions while $r_{\mathrm{H...O}}$ and r_{OH} are the H...O distance and O-H bond length in the O-H···O system. Such O-H···O systems are presented in Figure IV.1. The bond number conservation rule may be explained in the following way. The O-H bond not involved in any interaction is characterised by a bond number amounting to unity. If the O-H bond interacts with external centres, in the hydrogen bond, for example, it is elongated and its bond number decreases. However, such a decrease is compensated by the H···O contact and its bond number should be equal to the bond number loss of the O-H bond. Consequently, the sum of O-H and H···O bond numbers is equal to unity.

There is another interesting observation related to Figure IV.1; namely, there are symmetrical triatomic systems where two equivalent H···O contacts (or O-H bonds) are observed. It means that the H-atom of the hydrogen bridge is divalent or that we observe two H···O interactions (hydrogen bonds). This dilemma concerns not only the 2sHB definition, but also other definitions of the hydrogen bond. Another dilemma is related to weak hydrogen bonds, in terms of whether they should be classified as hydrogen bonds or as van der Waals interactions.

There is another matter related to the bond number conservation rule; namely, it is the weakening of the proton-donating bond that results from the hydrogen bond formation. This bond weakening relates to its lengthening and consequently to the decrease of the bond number. This results in so-called red-shifting hydrogen bonds.[1,2,11] There are also blue-shifting hydrogen bonds where the shortening of the proton-donating bond is observed as a result of the hydrogen bond formation.[11,18–21] In this way the bond number of the proton-donating bond exceeds unity and the bond number conservation rule is not fulfilled, especially since we have an additional value of the bond number from the interaction with the proton acceptor.

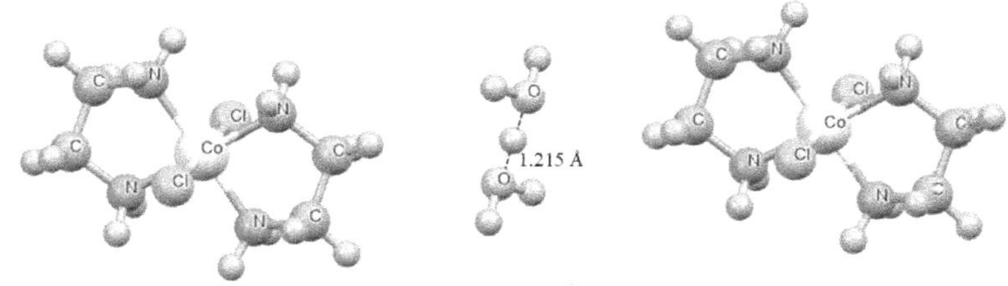

Fig. IV.2 A fragment of the crystal structure of *trans*-dichloro-bis(ethylenediamine) cobalt(iii) chloride hydrochloride dihydrate (ETDCOH11 refcode, ref. 22).

Figure IV.2 presents an example of a hydrogen-bonded system, the crystal structure of trans-dichloro-bis(ethylenediamine) cobalt(III) chloride hydrochloride dihydrate[22] which contains the $H_5O_2^+$ cation that is known as the Zundel cation.[23] The experimental and theoretical results on this ion show that its energetic minimum is characterised by the central position of the proton (in the mid-point of the O···O distance) or nearly so.[24,25] This is not exactly the proton since large electron charge shifts occur in the $H_5O_2^+$ cation and consequently its charge is lower than +1 au. According to the crystal symmetry ($P2_1/c$ space group), the proton is located at the inversion centre, thus the $H_5O_2^+$ cations are symmetrical. The $H_5O_2^+$ cation geometry is not disturbed by the intermolecular forces since the atom-atom distances between cation centres and centres of surrounding *trans*-dichloro-bis(ethylenediamine) cobalt(III) species are greater than the corresponding van der Waals radii. However, the most important point is that, for the $H_5O_2^+$ cation, both H···O contacts (bonds) are equivalent and the H···O distances are equal to 1.215 Å. If we assume that these are not covalent bonds but contacts rather, then there are two equivalent hydrogen bonds according to the 2sHB definition.

A slightly different situation is observed for the crystal structure of 2-nitro-indan-1,3-dione oxonium monohydrate.[26] The $H_5O_2^+$ cation is characterised by the asymmetrical position of the proton. The O-H···O bridge observed here, where the O-H bond length is 1.087 Å, is accompanied by a H···O contact of 1.390 Å. The latter link is classified as a hydrogen bond according to the 2sHB definition, and the former link is a covalent bond; the O-H-O angle is equal to 179.2°. One can see that the $H_5O_2^+$ cation may be considered as a complex of a hydronium cation, H_3O^+ with a water molecule, H_2O. Hence one of the O-H bonds of the H_3O^+ cation is involved in

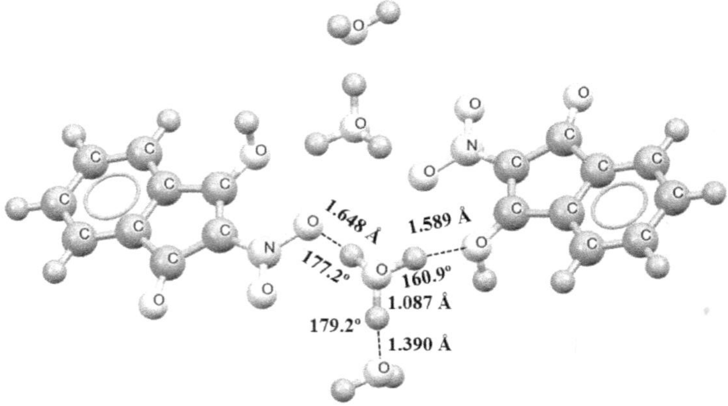

Fig. IV.3 A fragment of the crystal structure of 2-nitro-indan-1,3-dione oxonium monohydrate (NINDOD01 refcode, ref. 26).

hydrogen bond interaction with the H_2O molecule to form the $H_5O_2^+$ system, while two other O-H bonds of the hydronium cation are involved in weaker O-H···O hydrogen bonds since longer H···O distances are observed here (Fig. IV.3).

One can see that symmetrical and unsymmetrical hydrogen bonds in the $H_5O_2^+$ cation may be observed in crystal structures while in the gas phase this cation is characterised by the central, or nearly so, position of the proton. This situation is similar to that observed for the $(FHF)^-$ anion. In the gas phase this anion is linear with the H-centre located at the F···F mid-point,[27–29] thus it possesses $D_{\infty h}$ symmetry. Hence, similar to the $H_5O_2^+$ cation, two equivalent H···F hydrogen bonds may be indicated for the $(FHF)^-$ anion according to the 2sHB definition, or one may assume that the hydrogen centre is divalent here.

There are numerous other studies on this anion; one can mention the investigations related to the isotope effect.[30] The studies concerning crystal structures are very interesting since they show the $D_{\infty h}$ symmetry of the $(FHF)^-$ anion is disturbed.[2,31,32] This disturbance concerns not only the movement of the H-centre from the central position (the mid-point of the F···F distance) but also "the loss of linearity"; the F-H-F angle is often lower than 180°.[32] MP2/aug-cc-pVTZ calculations were also performed on the $(FHF)^-$ complexes with simple species such as HF, HCl, H_2O, C_2H_2, HCN, and H_2S.[32] The $(FHF)^-$ anion acts in these complexes as the Lewis base through the fluorine centres.

The latter examples show that the 2sHB definition that is restricted to two centres being in contact is a very useful tool to indicate if any interaction may be classified as hydrogen bonding. This is a simple and uncontroversial definition where all terms within are known and understandable.

IV.2 Bifurcated Hydrogen Bonds – a Case of Partial Interactions

Bifurcated hydrogen bonds were mentioned in the previous section of this chapter during the discussion on the Jeffrey and Saenger definition of the hydrogen bond.[1] Let us discuss a few examples of such bifurcated interactions that are taken from crystal structures.

Figure IV.4 presents a fragment of the crystal structure of 1-hydroxyethane-1, 1-diphosphonic acid monohydrate where the bifurcated proton donor hydrogen bond is observed.[33] This type of bifurcated bonding is mentioned in Jeffrey and Saenger's monograph[1] since one can observe here one O-H proton-donating bond coming from the water molecule that is involved in two intermolecular H···O contacts. In other words, the covalently bound H-centre interacts with two electronegative oxygen hydroxyl centres (Fig. IV.4). This results in the occurrence of two O-H···O bridges. For the shorter contact, amounting to 1.975 Å, the O-H-O angle is equal to 159.5º. For the longer contact that is equal to 2.446 Å, the O-H-O angle is far from linearity, at 111.5°.

Figure IV.5 shows a fragment of the mixed crystal structure of urea oxalic acid where one can observe the bifurcated proton acceptor hydrogen bond.[34] There is one oxygen proton acceptor (of the C=O carbonyl group) here and two proton-donating O-H bonds (of the COOH carboxylic groups) being in contact with this common acceptor. The acceptor comes from the urea molecule and the proton donor bonds from two oxalic acid molecules. In this structure, similar to Fig. IV.4, two O-H···O bridges are observed. The two H···O contacts are equivalent (1.542 Å) because of the crystal symmetry. Consequently, the same O-H-O angle (169.7°) occurs for both these contacts.

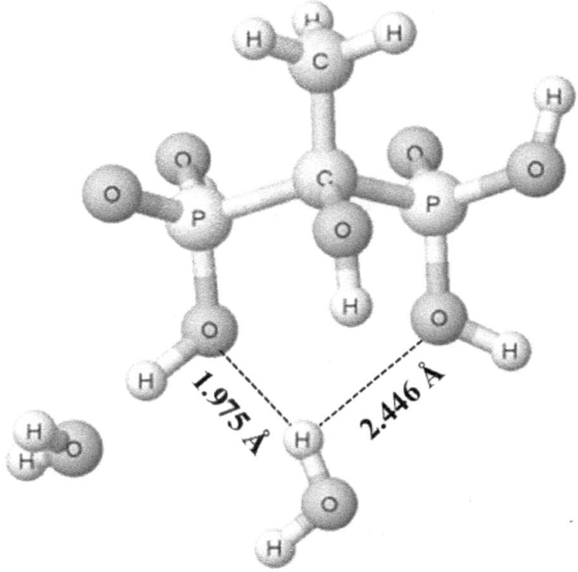

Fig. IV.4 A fragment of the crystal structure of 1-hydroxyethane-1,1-diphosphonic acid monohydrate (ETHDPH01 refcode, ref. 33). White circles correspond to hydrogen atoms. The H···O contacts are presented, and they correspond to the bifurcated donor hydrogen bond.

It was suggested that bifurcated hydrogen bonds may sometimes be treated as "partial hydrogen bonds." This idea was suggested on the basis of analyses of crystal structures of thioureidoalkylphosphonates[35] where bifurcated acceptor hydrogen bonds are observed since there are two N-H proton-donating bonds linked with the same oxygen proton acceptor, so-called chelated hydrogen bonds in the Jeffrey terminology.[2] For systems containing such chelated hydrogen bonds, the B3LYP/6-311++G(d) (B3LYP/6-311++G(d,p) for H-atoms participating in hydrogen bonds) calculations were performed.[35] Such calculations were also carried out for similar systems where "single", i.e., not bifurcated, (N-)H···O hydrogen bonds occur. It was found that the single hydrogen bonds are much stronger than those occurring in chelated systems.[35] It agrees with an earlier study where it was pointed out that the electron density at the bond critical point (BCP) may be treated as an additive parameter for some considered systems.[36] However, there are other studies where the additivity of electron densities at BCPs was reported. For example, the sum of electron densities at BCPs of all intermolecular contacts of a considered DNA base pair was discussed as a measure of the strength of links.[37]

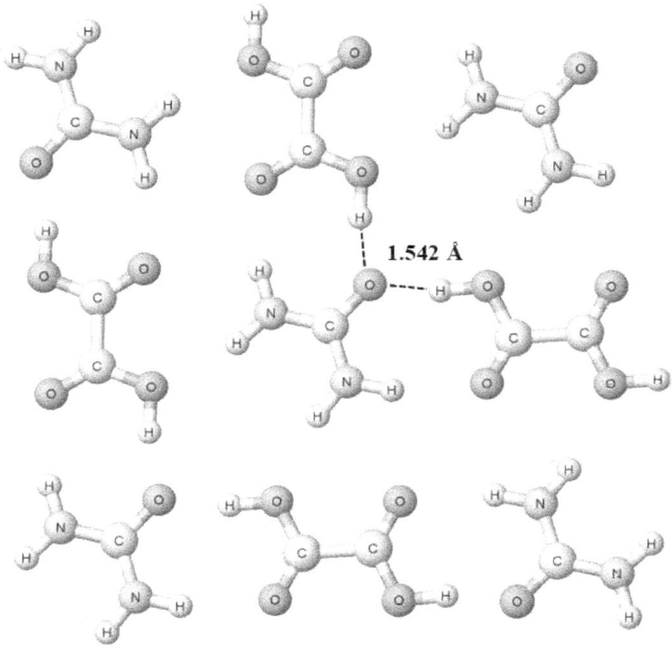

Fig. IV.5 A fragment of the crystal structure of urea oxalic acid (UROXAM01 refcode, ref. 34). White circles correspond to hydrogen atoms. The H···O contacts are presented, and they correspond to the bifurcated acceptor hydrogen bond.

Similarly, for the bifurcated acceptor hydrogen bonds in structures of thioureidoalkylphosphonates, the sum of electron densities at H···O BCPs has been considered as a measure of the hydrogen bond strength.[35] The relationship that was found by Espinosa and co-workers between the H···O distance and the electron density at the corresponding BCP[38] was also applied for bifurcated hydrogen bonds in thioureidoalkylphosphonates.[35] The dependence (eq. IV.3) was based on the experimental X-ray charge densities for hydrogen bridges in organic crystals.

$$\rho_{H\cdots O} = 9.63 \exp(-3.2\, r_{H\cdots O}) \tag{IV.3}$$

The coefficients of this equation are chosen in such a way to have the electron densities at the H···O BCP, $\rho_{H\cdots O}$, in atomic units, and the H···O distances, $r_{H\cdots O}$, in Å. For the bifurcated acceptor hydrogen bonds in thioureidoalkylphosphonates, there are two H···O contacts. If one inserts the sum of the $\rho_{H\cdots O}$ values for these contacts into eq.

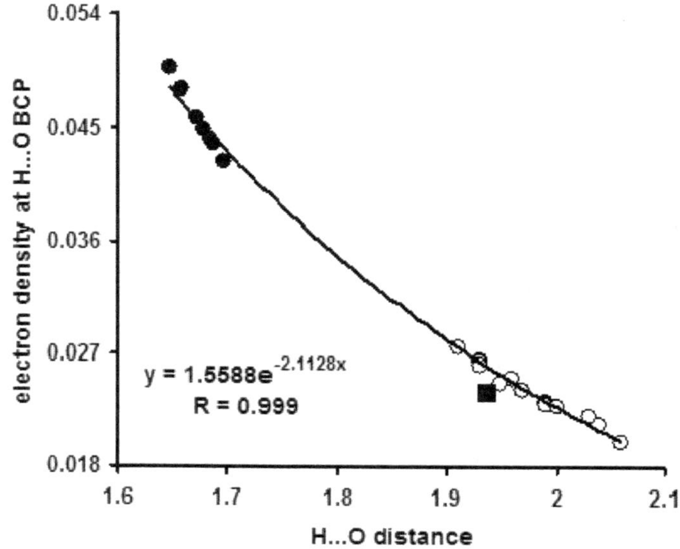

Fig. IV.6 The correlation between the H…O distance (in Å) and the electron density (in au) at the corresponding BCP. Full circles correspond to the modelled "single" hydrogen bonds and open circles to the "partial" interactions in bifurcated hydrogen bonds. The full square corresponds to the hydrogen bond in the water dimer. Reprinted with permission from ref. 35. Copyright 2005 American Chemical Society.

IV. 3, it is possible to obtain the single $r_{H\cdots O}$ distance which is equivalent to two H···O distances of the bifurcated interaction. Such model systems are presented in Figure IV.6 by full circles while the open circles correspond to "partial" hydrogen bonds in bifurcated systems. For all these systems (modelled and those in bifurcated hydrogen bonds) an exponential relationship, $\rho_{H\cdots O} = 1.56 \exp(-2.11\ r_{H\cdots O})$, was found; the correlation coefficient for this relationship is equal to 0.999. The full square presented in Fig. IV.6 corresponds to the H···O hydrogen bond in the trans-linear water dimer. For this system B3LYP/6-311++G(d,p) calculations were carried out, the same level as for the thioureidoalkylphosphonates.

The results presented here for thioureidoalkylphosphonates show that it seems appropriate to consider links that appear in bifurcated hydrogen bonds as partial interactions. Such "partial" contacts are usually weaker than corresponding links observed in complexes linked by "single" hydrogen bonds. Consequently, the proton — acceptor distances in partial interactions are longer than those in single hydrogen bonds.

IV.3 Cooperativity in Hydrogen Bonds

The term cooperativity is used for hydrogen-bonded systems most often, but is also applied for other types of interactions.[39] In the case of hydrogen bonds, it usually refers to the strengthening of the hydrogen bond in the complex if an additional hydrogen bond is formed between one of the complex units and the additional moiety.[40] The cooperativity in hydrogen-bonded systems is analysed in this chapter, while this effect related to other types of interactions is described in subsequent chapters. One of the first theoretical studies on the cooperativity phenomenon for links between water molecules was carried out by Del Bene and Pople, who justified that the water trimer chain configuration connected by a sequence of hydrogen bonds is characterised by stronger intermolecular interactions than two water molecules.[41]

The cooperativity effect is often a subject of theoretical calculations; however, it is also observed experimentally, for example in crystal structures where symmetry relations often force this phenomenon; particularly, the translational symmetry that is assigned to crystal structures by the definition.[42,43] This translation is often connected with the occurrence of chains of molecules linked by various interactions, particularly by hydrogen bonds. One can mention here the structures of carboxylic acids.[44] A fragment of the crystal structure of terephthalic acid was presented in Figure III.10 (Chapter III).[45] In this structure a chains of molecules are formed; such chains follow translational symmetry that is a necessary condition for a crystalline state. In these chains, double equivalent and centrosymmetric hydrogen bonds are formed between neighbouring molecules. This equivalency is also a result of the symmetry since the centre of inversion is located between linked carboxylic groups, in the centre of the eight-membered ring formed by these two groups. The O···O distances of the O-H···O hydrogen bond bridges are equal to 2.625(2) Å that indicates rather strong hydrogen bonds. These distances in the gas phase are usually longer. For example, for the formic acid dimers in the gas phase that are linked through the carboxylic groups, two equivalent O-H···O hydrogen bonds are observed with the O···O distances equal to 2.703(8) Å.[46]

The question arises as to what is the reason for the shorter O···O distances in the crystal structure than in the gas phase? Is it really the cooperativity effect in the case of the terephthalic acid chain structures? It was pointed out that the lengths of hydrogen bonds in crystal structures are shorter than those in the gas phase.[47] This effect is not only related to cooperativity but also to packing effects.[47] If cooperativity in terephthalic acid structure through benzene rings occurs, it would be a much more complex phenomenon. However, it seems that the cooperativity effect occurs in the hydrogen bond if the additional interaction which enhances (or, say, better influences) "primary interaction" concerns direct links with the proton donor or proton acceptor. Thus the possible cooperativity effects in the case of terephthalic acid need additional investigations. There are other effects that occur in the case of hydrogen bonds in carboylic acids, such as mesomeric effect of carboxylic groups, dynamic or orientational disorder in crystal structures, intermolecular resonance assistance of hydrogen bonds, etc. All these effects may enhance the strength of hydrogen bonds in carboxylic acid strucrures.[47] This is why examples of less complex structures, where cooperativity is possible, are discussed further here.

It is worth mentioning negative and positive cooperativity that were discussed by Scheiner.[48] Positive cooperativity occurs if the A-H···B hydrogen bond is involved in the additional A′-H···A interaction. Hence the A′-H···A-H···B sequence is formed. The strength of the A-H···B hydrogen bond is connected with the electron charge transfer from the B Lewis base unit to the Lewis acid unit. The additional A′-H···A hydrogen bond forces, in a similar way, electron charge transfer from the A-centre to the A′-H unit. This leads to the depletion of the electron charge in the A-H unit and further transfer from the B-centre. Consequently, the B → A-H electron charge transfer is greater than in the isolated A-H···B hydrogen bond bridge, and the H···B interaction is stronger than in the isolated bridge (Scheme IV.1). It is worth mentioning that the electron charge transfers described here are slightly more complicated. For example, for the A-H···B bridge the electron charge transfer from the B-unit to A-H is also accompanied by electron charge redistribution within the A-H unit –electron charge shift from the H-centre to the A-centre. Geometrical changes are also observed. The H···B distance is shorter in the sequence than in the isolated A-H···B bridge. It is important that the strength of the H···B local interaction depends on the sequence length. The greater number of units in the sequence, the stronger the interaction. Thus, this local interaction is stronger in the A″-H···A′-H···A-H···B sequence than in the A′-H···A-H···B system.[47,49]

Let us discuss an example of clusters of benzene with hydrogen fluoride molecules that were analysed theoretically by MP2/6-311++G(d,p) calculations.[50] Full optimisations

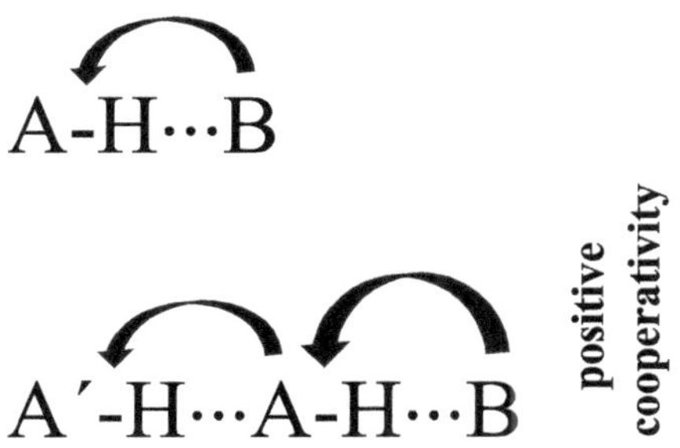

Scheme IV.1 Positive cooperativity. Arrows indicate the electron charge transfers.

were performed for the complex containing benzene and one hydrogen fluoride molecule, $C_6H_6\cdots HF$, and next for clusters containing more HF molecules, $C_6H_6\cdots HF\cdots HF\cdots$, up to four HF molecules. Such systems correspond to configurations characterised by energetic minima. In the optimised structures the HF molecule closest to the benzene molecule is directed to one of the carbon atoms, not to the centre of the benzene ring. Hence one can classify these interactions as F-H···C rather and not as F-H···π hydrogen bonds. For the $C_6H_6\cdots HF$ complex the shortest H···C distance is equal to 2.50 Å, and it decreases with the increase of the number of HF molecules in a cluster since the strength of this local interaction increases. For the $C_6H_6\cdots(HF)_4$ cluster (see Fig. IV.7), the shortest H···C distance is equal to 2.18 Å. The interaction energy for the $C_6H_6\cdots HF$ complex (BSSE included) is equal to –3.2 kcal/mol and it is "more negative" if the number of HF molecules increases in the cluster; for the $C_6H_6\cdots(HF)_4$ cluster it is equal to –8.2 kcal/mol. Thus, one can see the significant positive cooperative effect. The interaction energies in clusters were calculated such that the benzene molecule is one monomer and all HF molecules in the cluster are treated as the second monomer.

Similar MP2/6-311++G(d,p) calculations were performed for configurations with the geometric constraint that they possess the C_{6v} symmetry. In this way such optimised configurations do not correspond to energetic minima. However, the HF molecules are directed to the centre of the benzene ring. Positive cooperativity is also observed here; the strength of the interaction increases with the increase of the number of HF molecules and the H···C distances for contacts between benzene and the HF molecule nearest to benzene decrease. For the $C_6H_6\cdots HF$ complex the

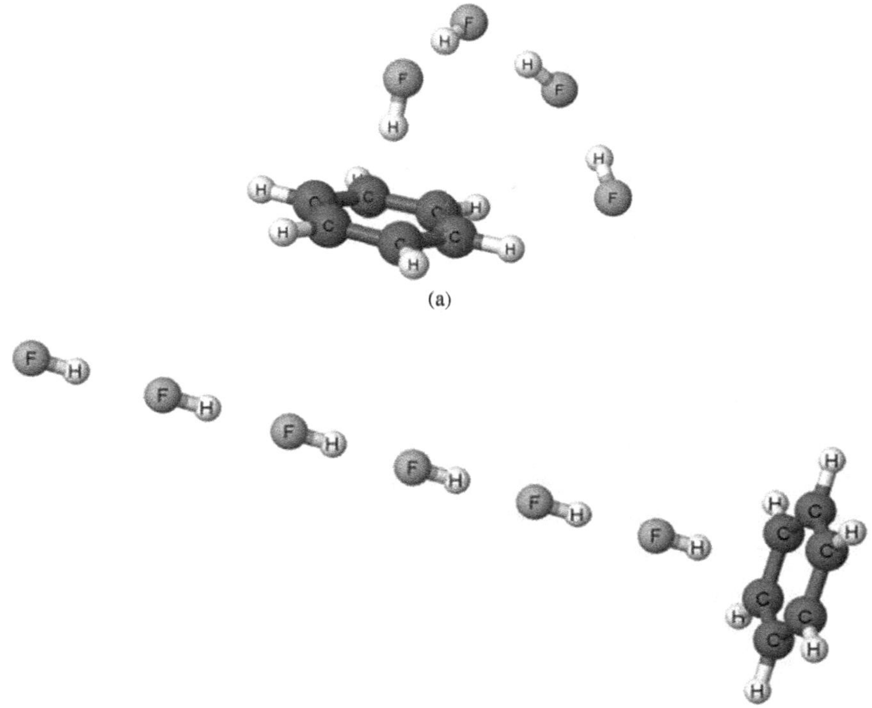

Fig. IV.7 The clusters of benzene and hydrogen fluoride molecules. (a) $C_6H_6 \cdots (HF)_4$ fully optimised cluster. (b) $C_6H_6 \cdots (HF)_6$ cluster possessing the C_{6v} symmetry.

interaction energy is equal to –3.2 kcal/mol (it is very close to the energy of the optimised complex since the difference between energies is equal to 0.06 kcal/mol only), while the H···C distance is equal to 2.7 Å. For the $C_6H_6 \cdots (HF)_6$ cluster (Fig. IV.7), the interaction energy is equal to –5.5 kcal/mol and the H···C distance is equal to 2.6 Å. Hence the shortening of the H···C distance for these C_{6v} systems is not so significant as for the optimised species. For the $C_6H_6 \cdots (HF)_n$ C_{6v} clusters, the interaction energies only slightly differ between themselves for larger number, n, of HF molecules. For n = 4, 5 and 6 the binding energy is equal to –5.3, –5.5 and –5.5 kcal/mol, respectively. This may be called as a "saturation effect" of the positive cooperativity that is common for a greater number of components in any cluster.

Negative cooperativity[48] occurs for the bifurcated donor and bifurcated acceptor hydrogen bonds (Scheme IV.2). The "negative cooperativity" is related here to the weakening of the H···B local interaction. In the case of the bifurcated acceptor hydrogen bond the electron charge transfer from the B-centre is directed to two units, A-H and A-H′ (Scheme IV.2). Hence both A-H···B and A′-H···B hydrogen bonds are weaker than any of them occuring separately. In the case of bifurcated proton acceptor hydrogen bond, this acceptor shares the delivered electron charge between two proton donors. The same concerns the bifurcated proton donor hydrogen bond (Scheme IV.2). The proton donor shares the received electron charge between two proton acceptors. In both cases of bifuracted hydrogen bonds negative cooperativity occurs, so one can indicate partial hydrogen bonds. It is worth mentioning that in the bifurcated hydrogen bonds, H···B distances are usually longer than in the corresponding single H-bonded systems.[48] It was described in various studies that the H···B distance roughly corresponds to the strength of interaction;[51]

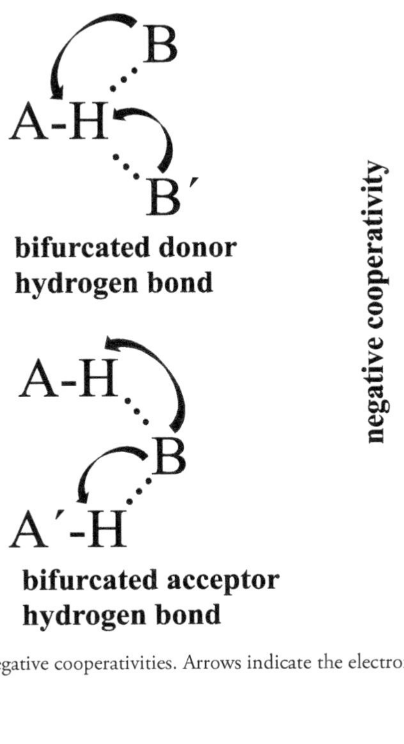

Scheme IV.2 Negative cooperativities. Arrows indicate the electron charge transfers.

however, in the case of a homogenous sample of interactions,[52] such interactions are characterised by the same-type pair of centres being in contact.

Let us consider another example of positive cooperativity. Figure IV.8 presents a fragment of the neutron diffraction crystal strucure of (15-crown-5) tetrachloro-gold(III) oxonium dihydrate (JOTQOY03 refcode).[53] One can see here the water dimer that interacts additionally with the hydronium cation. Thus the hydrogen bond in the dimer is enhanced by the electron-withdrawing properties of the H_3O^+ cation; this is the positive cooperativity effect. For the water dimer in this crystal structure, the H···O distance is equal to 1.579 Å and the O···O distance is equal to 2.574 Å. This is the neutron diffraction structure, thus the distances containing the H-centres are characterised by the same high precision as distances between non-hydrogen atoms.[42,54,55] The O···O distance for the water dimer in the gas phase is equal to ~2.98 Å,[56] thus the H...O distance is greater than 1.9 Å. Both distances in the gas phase are significantly greater than those in the presented crystal structure, thus it is not only a simple influence of packing forces. The H···O contact (1.318 Å) with the hydronium cation is responsible for the significant positive cooperativity.

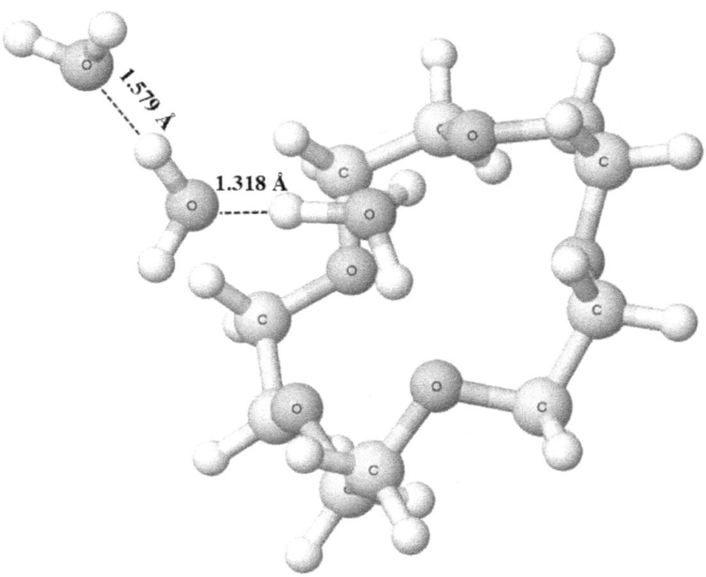

Fig. IV.8 A fragment of the crystal strucure of (15-crown-5) tetrachloro-gold(III) oxonium dihydrate (JOTQOY03 refcode, ref. 53).

There are various indices of cooperativity that were proposed in numerous studies. For example, Parra applied the following measure of cooperativity in an analysis of chains of Z,Z conformer molecules of N-formylformamide (MP2/6-31+G(d) calculations were performed).[57]

$$\text{cooperativity} = [\Delta Q_n - (n-1)\Delta Q_2]/(n-2) \qquad \text{(IV.4)}$$

Q is the property considered (interaction energy, frequency, intermolecular contact, bond length, etc.) and n designates the number of units in the cluster; for example, in the case of intermolecular distance, ΔQ_n is this distance in the cluster of size n and ΔQ_2 is that of the dimer (in the cluster, this may be the distance between terminal units). For the Z,Z conformer of N-formylformamide, the cooperativity index increases with the increase of the cluster size.[57] In the case of dissociation energy calculated with the zero-point vibrational energy correction, the cooperativity index (eq. IV.4) is equal to 3.03 and 5.01 for clusters containing 3 and 10 molecules, respectively. The cooperativity increase, measured by the index, is smaller if this increase concerns greater n values. The difference in the cooperativity index between clusters containing 3 and 2 molecules is equal to 0.67 kcal/mol, while between clusters containing 10 and 9 units it is equal to 0.11 kcal/mol, for example. This is the so-called "saturation" described earlier. It is worth mentioning that eq. IV.4 may be applied to chains containing the same type of units. Thus, it is not applicable for the clusters of benzene and hydrogen fluoride molecules described earlier (see Fig. IV.7).

Another measure of cooperativity, Coop-Energy, is often applied in analyses of triad chains.[39] Such chains may be marked as A···B···C.

$$\text{Coop-Energy} = E_i(\text{ABC}) - E_i(\text{AB}) - E_i(\text{BC}) - E_i(\text{AC}) \qquad \text{(IV.5)}$$

E_i(ABC) is the interaction energy of the triad, E_i(AB) and E_i(BC) are the interaction energies of the isolated dyads in their corresponding energy minima configurations and E_i(AC) is the interaction energy of the A and C units in the geometry they have in the triad. It is worth mentioning that the AC dyad possessing the triad geometry does not correspond to the energetic minimum. Eq. IV.5 may be applied not only for hydrogen-bonded systems but for triads linked by other types of interactions. A negative value of the Coop-Energy measure indicates favourable cooperativity.

IV.4 Classification of Hydrogen Bonds

There is a group of properties of the A-H···B hydrogen bond that is mentioned most often in various studies. These are: the high values of electronegativity of A and B centres, the positive charge of the H-centre, the elongation of the A-H bond as a result of complexation (the formation of the hydrogen bond), the H···B distance being lower than the corresponding sum of van der Waals radii, the increase of positive charge of the H-centre observed as a result of the complex formation, the local H···B interaction is stabilising, etc.[1,2,6–8,11] However, all these properties are questioned for various sub-classes of hydrogen bonds, thus it is really difficult to establish attributes of the hydrogen bond. The only property that is mentioned in the majority of definitions of the hydrogen bond is the positive charge of the H-centre.[7]

A property that is contested is the high electronegativity of A and B centres. Centres such as fluorine, oxygen, nitrogen or chlorine were considered in early studies as those that form hydrogen bonds.[58] However, sometimes, in these early studies C-H···B interactions were mentioned as those that may be classified as hydrogen bonds. For example, Suttor considered such systems and she analysed examples of crystal structures where they occur.[59] The study of Suttor was contested and arguments against classification of these interactions as hydrogen bonds were related to the long H···B contacts, close to the sum of van der Waals radii, and the low electronegativity of the carbon centre.[60] However, in the study of Taylor and Kennard it was justified, by applying statistical methods,[61] that the C-H···B interactions may be classified as hydrogen bonds.

Since then the number of studies on such interactions and others that were not previously classified as hydrogen bonds has increased enormously; these "new kinds of hydrogen bonds" were often named as unconventional hydrogen bonds in opposition to the conventional A-H···B hydrogen bonds possessing properties mentioned at the beginning of this section. For example, among these unconventional hydrogen bonds, one can mention those having multi-centre proton acceptors that were discussed in numerous studies. There are three great sub-classes of hydrogen bonds: A-H···B, A-H···π and A-H···σ.[7,11] The first group is characterised by the single-atom proton acceptor, while the latter two groups possess π-electrons and σ-electrons, respectively, as proton acceptors. Special attention may be paid to the C-H···π hydrogen bonds

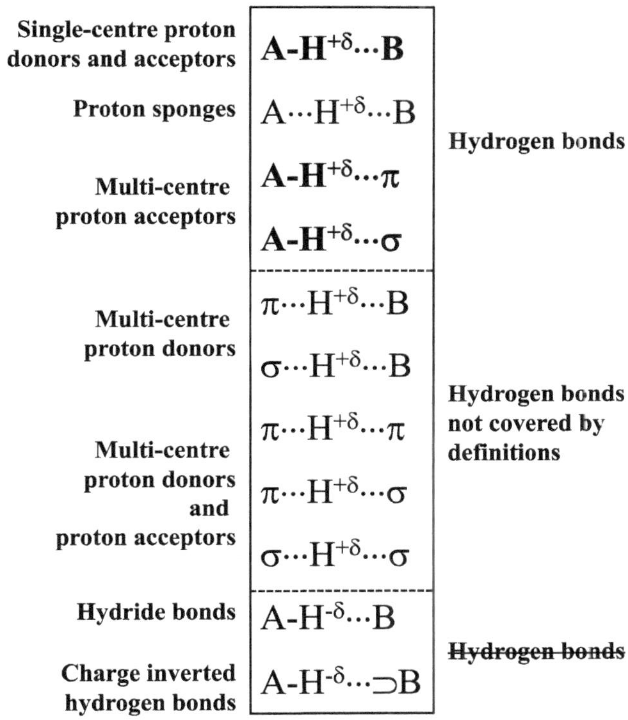

Scheme IV.3 The classification of hydrogen bonds. Reprinted with permission from ref. 7. Copyright 2024 The Royal Society of Chemistry.

(the group belonging to the A-H$\cdots\pi$ sub-class) that play an important role in arrangements of molecules and ions in crystal structures.[62]

Scheme IV.3 presents the classification of hydrogen bonds that was proposed in a recent study.[7] In the first group of interactions there are the three sub-classes mentioned above, A-H$^{+\delta}\cdots$B, A-H$^{+\delta}\cdots\pi$ and A-H$^{+\delta}\cdots\sigma$, and the systems where the proton is inserted between A and B centres, A$\cdots$H$^{+\delta}\cdots$B, but it is difficult to classify H$^{+\delta}\cdots$B and H$^{+\delta}\cdots$A interactions as covalent or hydrogen bonds since both are equivalent or nearly so. The positive charge of the H-centre is marked here to distinguish these interactions from those where we observe a negative charge on the hydrogen centre. The interactions of this group are classified as hydrogen bonds by the majority of commonly accepted definitions.[7]

It is worth mentioning here that the definition of Weinhold and Klein takes into account only the A-H$^{+\delta}\cdots$B systems;[63] this definition was shortly discussed in Chapter II (see section II.7). Hence the A-H$^{+\delta}\cdots\pi$ and A-H$^{+\delta}\cdots\sigma$ interactions are not considered. But they are by the IUPAC and 2sHB definitions. One can refer here to Table II.12 (Chapter II) where it is explained that for different kinds of interactions, the most important orbital-orbital overlaps may be specified. Thus for the three kinds of hydrogen bonds, there are the following orbital-orbital interactions: $n_B \rightarrow \sigma_{AH}{}^*$, $\pi \rightarrow \sigma_{AH}{}^*$ and $\sigma_{H2} \rightarrow \sigma_{AH}{}^*$, respectively. One can see that the last concerns complexes where molecular hydrogen is the proton acceptor. However, other σ-bonds may also play such a role; for example, cyclopropane and more complex cycloalkanes.[64] Thus $\sigma \rightarrow \sigma_{AH}{}^*$ is the more general form of the overlap that occurs for A-H$^{+\delta}\cdots\sigma$ hydrogen bonds.

It is worth mentioning here the special class of A-H$^{+\delta}\cdots$B hydrogen bonds where the negatively charged H-centre is the proton acceptor. Hence two hydrogen atoms characterised by opposite charges are in contact for such interactions. Named dihydrogen bonds,[65–67] they are designated as A-H$^{+\delta}\cdots{}^{-\delta}$H-B and $\sigma_{BH} \rightarrow \sigma_{AH}{}^*$ is the most important overlap.[68] Scheme IV.4 presents various "unconventional hydrogen bonds", and the dihydrogen bond is one of them. The dihydrogen bonds are characterised by a broad spectrum of strength, from very weak interactions to very strong ones possessing characteristics of covalent bonds.[69] MP2/6-311++G(d,p) calculations were performed on a large number of dihydrogen-bonded systems; for example, for HCCH$^{+\delta}\cdots{}^{-\delta}$HBeF and FCCH$^{+\delta}\cdots{}^{-\delta}$HBeF complexes the interaction energies are equal to –0.8 kcal/mol and these systems are stable owing to the correlation energy that is the most important attractive contribution to the total interaction energy.[69] The H···H distances are equal to 2.3 Å, approximately, close to the corresponding sum of van der Waals radii. On the other hand, for the $Cl_2OH^{+\delta}\cdots{}^{-\delta}$HBeH and $Cl_2OH^{+\delta}\cdots{}^{-\delta}$HBeF complexes, the interaction energies are equal to –29.0 kcal/mol and –19.0 kcal/mol while the corresponding H···H distances amount to 1.06 Å and 1.13 Å, respectively. These distances are typical for covalent bonds, not for intermolecular contacts. The Laplacian of the electron density for BCP at the H···H bond path is negative for both complexes. The delocalisation interaction energy term is the most important attractive (stabilising) contribution to the total interaction energy. This term in the decomposition scheme[70,71] applied in the study on DHBs[69] corresponds roughly to the sum of polarisation and charge transfer terms in the Morokuma scheme

Specific cases
Hydrogen bonds ?

$C\text{-}H^{+\delta}\cdots B$	
$A\text{-}H^{+\delta}\cdots C$	
$C\text{-}H^{+\delta}\cdots C$	**Single-centre proton donors and acceptors**
$A\text{-}H^{+\delta}\cdots Ng$	
$A\text{-}H^{+\delta}\cdots {}^{-\delta}H\text{-}B$	
$C\text{-}H^{+\delta}\cdots {}^{-\delta}H\text{-}C$	
$C\text{-}H^{+\delta}\cdots \pi$	**Multi-centre proton acceptors**
$C\text{-}H^{+\delta}\cdots \sigma$	
$(X_1HX_2)^-$	**Proton sponges?**
$(Ng_1HNg_2)^+$	

Scheme IV.4 Examples of so-called "unconventional hydrogen bonds". Reprinted with permission from ref. 7. Copyright 2024 The Royal Society of Chemistry.

(Chapter 2, eq. II.19).[72,73] It means that the electron charge shifts related to the covalency of interactions are very important here. The electron charge shifts from the beryllium Lewis base unit to the Cl_2OH Lewis acid are equal to 0.35 e and 0.23 e, respectively.

Let us discuss characteristics of other types of the hydrogen bond interaction. Table IV.1 presents parameters of a variety of hydrogen bonds that were analysed by MP2/aug-cc-pVTZ calculations.[74] One can see that there is a group of very strong hydrogen bonds where the absolute value of the MP2 interaction energy exceeds 40 kcal/mol. These are interactions that link the units in $(FHF)^-$, $H_2O\cdots H_3O^+$ ($H_5O_2^+$) and $HCN\cdots H_3O^+$ (linked by the $O\text{-}H^{+\delta}\cdots N$ or $O\cdots H^{+\delta}\cdots N$ interaction) complexes. The Laplacian of the electron density at BCP of the inter-units link for these complexes,

$\nabla^2\rho_{BCP}$, is negative, which indicates the covalent character of interaction according to the Quantum Theory of Atoms in Molecules (QTAIM) approach. The distances between units constituting complexes for this group do not exceed 1.35 Å. This group may be assigned to the $A\cdots H^{+\delta}\cdots B$ systems (see Scheme IV.3) since the proton is not clearly connected with any one of the electronegative centres. Proton sponges described in numerous studies are also assigned to this group.[75,76] To be accurate, the name "proton sponge" was attributed to compounds which contain non-bonded centres that are close to each other and that possess lone electron pairs, which implies strong interaction and consequently high basicity. That leads to favorable protonation of the proton sponge and the formation of the hydrogen bridge with the proton location between the lone electron pairs. If the proton is located at the mid-point, or nearly so, between the basic centres, the corresponding system may be classified as $A\cdots H^{+\delta}\cdots B$. However, often the proton is strongly connected with one of the centres, say A, thus the $A\text{-}H^{+\delta}\cdots B$ hydrogen bond is formed. 1,8-bis(dimethylamino)naphthalene is a classic example of the proton sponge; the N-H···N hydrogen bridge is observed in its protonated form.

Let us return to the results of Table IV.1, particularly to the systems where the hydrogen bond contacts are characterised by negative $\nabla^2\rho_{BCP}$ values. Such systems, as well as numerous protonated proton sponges, are often classified as short-strong hydrogen bonds (SSHBs).[77,78] It has been noted that the O-H···O SSHBs possess O···O distances ranging from 2.45 Å to 2.65 Å where the absolute value of the interaction energy exceeds 5 kcal/mol.[78] This is for hydrogen bonds in liquids, however. The absolute interaction energies for systems analysed in the gas phase exceed 12 kcal/mol according to one of the studies.[79] For the SSHBs in crystal structures, the O···O distances are lower than 2.45 Å, and often closer to 2.4 Å.[13] For the $H_5O_2^+$ cation the O···O distance amounts to 2.40–2.41 Å.[77] The calculated distance for this cation presented in Table IV.1 is equal to 2.39 Å, and the H···O distance is equal to 1.193 Å. The electron density at the H···O BCP, ρ_{BCP}, amounts to 0.1699 au that is typical in the case of very strong interactions; such values correspond to covalent bonds. The MP2 interaction energy for this cation is equal to −49.9 kcal/mol.

There are SSHBs that may be classified as low-barrier hydrogen bonds (LBHBs).[80,81] These are observed if the barrier height in the double potential well for the proton transfer is very low, below the zero-point energy. Various studies show that the proton

Table IV.1 Selected characteristics of different types of A-H···B hydrogen bonds. The proton acceptor distances, $R_{H...B}$, (in Å) and the QTAIM parameters (in au), ρ_{BCP}, $\nabla^2\rho_{BCP}$, H_{BCP}, and ellipticity ε are given. These are results taken from ref. 74. In the case of centrosymmetric dimers, $(HCOOH)_2$ and $(HCONH_2)_2$, the parameters presented in the table concern single hydrogen bonds (this concerns also interaction energy).

Complex	$R_{H...B}$	ρ_{BCP}	$\nabla^2\rho_{BCP}$	H_{BCP}	ε	ΔE^{MP2}
$(FHF)^-$	1.138	0.1787	-0.3229	-0.1988	1.5x10^{-9}	-61.13
$H_2O\cdots H_3O^+$	1.193	0.1699	-0.3313	-0.1734	0.0325	-49.86
$HCN\cdots H_3O^+$	1.347	0.1212	-0.0418	-0.0756	0.0052	-41.84
$OH^-\cdots H_2O$	1.376	0.1086	0.0965	-0.0518	0.0367	-32.54
$NH_3\cdots NH_4^+$	1.575	0.0267	0.0837	-0.0182	0.0451	-29.11
$(HCOOH)_2$	1.726	0.0408	0.1272	-0.0022	0.0197	-6.80
$(HCONH_2)_2$	1.903	0.0278	0.0955	0.0016	0.0464	-6.06
$H_2O\cdots H_2O$	1.950	0.0236	0.0898	0.0025	0.0401	-4.48
$C_5H_5^-\cdots HF$	2.007	0.0275	0.0562	-0.0015	0.2236	-16.96
$C_2H_2\cdots HF$	2.187	0.0167	0.0504	0.0028	0.4316	-3.18
$C_6H_6\cdots C_2H_2$	2.351	0.0069	0.0229	0.0011	26.8379	-2.15
$(HCCH)_2$	2.692	0.0068	0.0182	0.001	0.3130	-1.05
$H_3CH\cdots OCH_2$	2.706	0.0046	0.0161	0.0006	0.1054	-0.38

for the $H_5O_2^+$ cation is situated in the mid-point of the O···O distance or it is very close to this mid-point, thus the double-well potential for the proton is observed with the barrier height below the zero-point energy level.[77] This is why the O-H···O hydrogen bond in the $H_5O_2^+$ cation presented in Table IV.1 is classified as SSHB and LBHB. Similarly, the hydrogen bond that links units in the $HCN\cdots H_3O^+$ complex may be considered as SSHB and LBHB. However, in the $(FHF)^-$ anion the proton is located in the mid-point of the F···F distance with single well potential. Only in the crystal structures in the $(FHF)^-$ anion the proton is moved from the central position and the whole system is not linear sometimes.[32]

Another group of hydrogen-bonded complexes is characterised by the absolute values of energies of interaction in the range between 4 kcal/mol and 33 kcal/mol. These are the $A\text{-}H^{+\delta}\cdots B$ systems where there is the single-centre proton acceptor and the single-centre proton donor. For all hydrogen bonds of this group, the $\nabla^2\rho_{BCP}$ value is positive, but the total electron energy density at BCP, H_{BCP}, is negative that means these interactions are at least partly covalent in nature.[13,82–85] These are the charge-assisted hydrogen bonds in the $OH^{-}\cdots H_2O$ and $NH_3\cdots NH_4^+$ complexes, negatively and positively charge-assisted, respectively, i.e., CAHB(–) and CAHB(+). Similarly, charge-assisted hydrogen bonds are observed for the complexes of the first group, where the Laplacian of the electron density at the H···B contact BCP, $\nabla^2\rho_{BCP}$, is negative. For the second group of hydrogen-bonded complexes, such as for the $(HCOOH)_2$ dimer, hydrogen bonds partly covalent in nature are observed since H_{BCP} is negative for intermolecular contacts. These centrosymmetric dimers are connected through carboxylic groups, by two equivalent O-H···O hydrogen bonds; the centre of inversion is located between these COOH groups. A similar situation is observed for the structure of terephthalic acid discussed earlier,[45] where links through carboxylic groups occur (see Fig. III.10 in Chapter III), and in the structure of 4-iodobenzoic acid (see Fig. I.5 in Chapter I).[86]

There is one complex connected through the C-H···O hydrogen bond, the $H_3CH\cdots OCH_2$ species, where the interaction energy of –0.38 kcal/mol exhibits its low stabilisation. This low stability is mainly a result of the dispersion forces since the other attractive forces, electrostatic and polarisation, are negligible.[74] The H···O intermolecular distance of 2.71 Å in this complex (Table IV.1) is slightly greater than the corresponding sum of van der Waals radii, 2.65 Å (Table I.1).

The next group of complexes that are collected in Table IV.1 concerns hydrogen-bonded systems where there is a multi-centre proton acceptor. These are $C_2H_2\cdots HF$, $(HCCH)_2$ and $C_6H_6\cdots C_2H_2$ complexes with interaction energies amounting to –3.2 kcal/mol, –1.1 kcal/mol and –2.2 kcal/mol, respectively. The first two complexes are characterised by T-shaped configurations, posessing C_{2v} symmetry. The complex of benzene possesses C_{6v} symmetry since the acetylene proton-donating molecule is perpendicular to the plane of benzene and directed to the ring centre. A similar situation occurs in the mixed crystal structure of benzene and acetylene where the benzene-acetylene chains are obseved.[87] This situation is presented in Figure IV.9.

There is one complex that seems to have a multi-centre proton acceptor that is characterised by a large absolute value of the energy of interaction, 17.0 kcal/mol: the

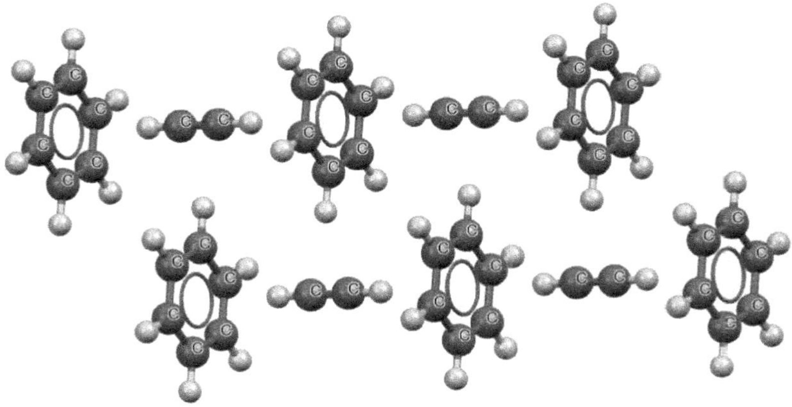

Fig. IV.9 A fragment of the mixed acetylene-benzene crystal strucure (BEVJUK refcode, ref. 87).

$C_5H_5^-\cdots HF$ complex. This is supported by the negative H_{BCP} value that suggests covalent character of this hydrogen bond, or at least partly covalent character. What is the reason for such strong interaction for this complex? One can mention two reasons, the strong Lewis acid properties of the hydrogen fluoride, and the charge-assisted hydrogen bond, CAHB(–).

Table IV.1 contains ellipticities of intermolecular contacts corresponding to hydrogen bonds, H···B contacts in the case of the one-centre proton acceptors. The ellipticity is expressed by the following equation:

$$\varepsilon = \lambda_1/\lambda_2 - 1 \qquad \text{(IV.6)}$$

where λ_1, λ_2 and λ_3 are the eigenvalues of the Hessian of the electron density (Chapter II, eq. II.30). The ellipticity, ε, is always positive with zero as the minimum value since $\lambda_1 < \lambda_2 < 0$. The ε value at BCP is low for cylindrical bonds (like single bonds) and increases if the bond order increases, up to double bonds. For the triple bond, like for acetylene and related compounds, there is again the cylindrical charge distribution and ε is equal or close to zero.

One can see, for complexes presented in Table IV.1, that the ε value decreases with the increase of the strength of interaction. Large values of ellipticity are observed for hydrogen bonds where π-electrons are proton acceptors. It was pointed out that in the case of multicenter proton acceptors, this value may likely result from the conical distribution of the electron density within the H···π contact.[74] For this group of hydrogen bonds, the ε value ranges from 0.2 to 26.9. The smallest value of 0.22 for this group occurs for the strongest interaction in the $C_5H_5^-$···HF complex.

Much lower ε values are observed for the A-$H^{+\delta}$···B hydrogen bonds with a single-centre proton acceptor. For the extremely weak interaction in the H_3CH···OCH_2 complex the ellipticity is equal to 0.11, and for the very strong hydrogen bond in the $(FHF)^-$ anion it is very close to zero, 1.5×10^{-9} (Table IV.1). Therefore, one can say that the latter results indicate unique properties of hydrogen bonds with a multi-centre proton acceptor, i.e., the great ε values.[74] However, the interrelation between the ellipticity and the multi-centre of the proton acceptor is not so clear. It is important that the ε value concerns the BCP located at the bond path linking two centres, usually attractors related to atoms. Of course, this is the situation that occurs for the 3c-4e A-$H^{+\delta}$···B hydrogen bonds where the ε values are related to the single BCP at the H···B bond path. For example, the hydrogen bonds that are observed in the dimer of the *trans* conformer of thiosalicylic acid are classified as 3c-4e systems (see Fig. II.7a in Chapter II).[88] A similar situation is observed for the A-H···π systems presented in Fig. IV.10. For the C_6H_6···C_2H_2 and $C_5H_5^-$···HF complexes, the bond paths between two centres are observed.[89] For the former complex, the H-attractor of acetylene is connected by six bond paths with C-attractors of benzene. The C_{6v} symmetry is also preserved for the electron charge distribution. One can have doubts if the connection between benzene and acetylene in this complex may be classified as C-H···π hydrogen bond or as bifurcated proton donor hydrogen bond. As six equivalent H···C links with corresponding bond paths are observed, one may classify them as partial C-H···C hydrogen bonds.

One should expect the occurrence of the F-H···π hydrogen bond in the case of the $C_5H_5^-$···HF complex. However, the H-attractor of the hydrogen fluoride is connected by the bond path here with the C-attractor of the cyclopentadienyl ring. This results in the most negative charge of this carbon centre in comparison with other carbon atoms of the Lewis base unit. This is why the F-H···C hydrogen bond materialises here.

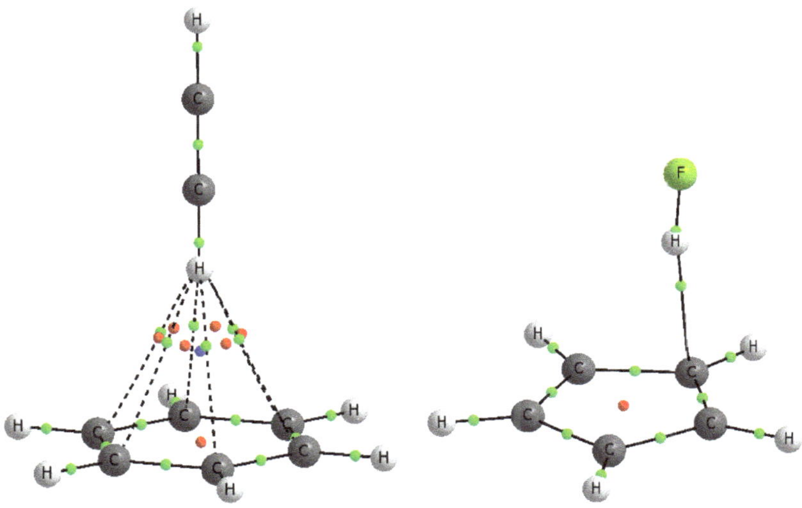

Fig. IV.10 Molecular graphs of two complexes: (a) $C_6H_6\cdots C_2H_2$ and (b) $C_5H_5^-\cdots HF$. Solid and broken lines correspond to bond paths, big circles to attractors and small circles to other critical points: green — bond critical points, BCPs, red — ring critical points, RCPs, blue — cage critical points, CCPs.

In the case of the $C_2H_2\cdots HF$ complex, the bond path links the H-attractor of the Lewis acid unit with the mid-point of the CC acetylene bond (see Fig. IV.11).[89] This depends on the level of calculations; it is either the non-nuclear attractor (MP2/aug-cc-pVTZ level) or the BCP (MP2/aug-cc-pVQZ level). Thus, one can see that in the case of A-H···π hydrogen bonds, the QTAIM approach shows the centre — centre links. These are atom — atom links or atom — critical point links. For atom — atom contacts, there may be single links between Lewis acid and Lewis base units or several links that may be classified as bifurcated proton donor hydrogen bonds. In the case of atom — critical point links, BCP or non-nuclear attractor may be the latter site. Thus, the term multi-centre proton acceptor may be contested, at least from the QTAIM point of view.

It is not the scope of this book to analyse in detail properties of the sub-classes of hydrogen bonds, i.e., the A-H···π systems. However, one can see that investigations with additional approaches are needed to solve the problem of multi-centre proton

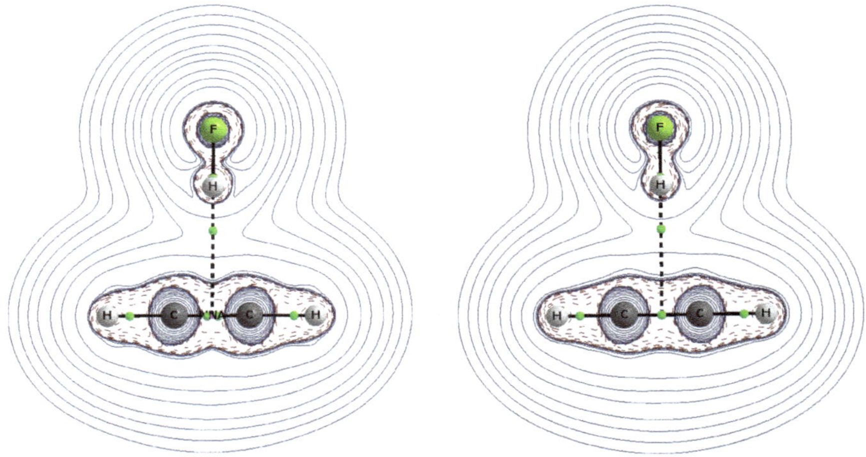

Fig. IV.11 The molecular graphs of the C_2H_2···HF complex, MP2/aug-cc-pVTZ (left) and MP2/aug-cc-pVQZ (right) levels. Solid and broken lines correspond to bond paths, big circles to attractors and small green circles to BCPs. The isolines of Laplacian of electron density are also presented; there is non-nuclear attractor (small red circle) between carbon atoms (left) for the MP2/aug-cc-pVTZ level. Reprinted with permission from ref. 89. Copyright 2016 Springer.

acceptors. Previous research partly describes this issue.[74,89] For example, for the $C_5H_5^-$··· HF complex, there is the A-$H^{+\delta}$···B hydrogen bond probably since the QTAIM approach indicates a single F-H···C link. The electron density at BCP, ρ_{BCP}, of the bond path corresponding to this link is equal to 0.028 au while the interaction energy for this complex amounts to −17.0 kcal/mol (Table IV.1). ρ_{BCP} is known as a measure of the strength of interaction, thus its low value for such high absolute value of the interaction energy seems to be strange. For the water dimer where the energy of interaction amounts to −4.5 kcal/mol, the ρ_{BCP} value is only slightly smaller, 0.024 au. In the case of the formic acid dimer where the interaction energy per hydrogen bond (this is the centrosymmetric dimer linked by two equivalent hydrogen bonds) is equal to −6.8 kcal/mol, a higher value of ρ_{BCP} is observed, 0.041 au, for the corresponding BCP at the H···O bond path. It means that in the case of the $C_5H_5^-$···HF complex, not only F-H···C hydrogen bond (with H···C distance equal to 2.0 Å) stabilises the complex but also other longer H···C links with distances equal to 2.4–2.9 Å. The carbon centre being closest to hydrogen fluoride possesses the most negative charge and the $n_C \rightarrow \sigma_{FH}^*$ overlap is observed here for the corresponding F-H···C hydrogen bond,[90] similarly as for other 3c-4e A-H···B hydrogen bonds. In the case of the C_2H_2··· HF complex, the most important $\pi_{CC} \rightarrow \sigma_{FH}^*$ overlap for A-H···π is observed.

Summarising, one can see that Scheme IV.3 presents three main groups of interactions classified as hydrogen bonds. The first group contains the "conventional" A-$H^{+\delta}$···B hydrogen bonds as well as the systems where the proton is inserted between two highly electronegative centres, A···$H^{+\delta}$···B, and those interactions where the π-electrons play the role of the proton acceptors, A-H···π interactions. These systems from the first group were described in this section so far. We next look at the A-H···σ interactions.

IV.5 A-H···σ Hydrogen Bonds

A similar situation to the A-H···π hydrogen bonds occurs often for the A-H···σ systems.[64] The bond path for the latter links the H-centre of the Lewis acid unit with the BCP of the σ-bond of the Lewis base unit (see Fig. IV.12). There are experimental and theoretical studies on complexes of molecular hydrogen acting as the Lewis base species where units are linked by the A–H···H_2 hydrogen bond. For example, high-resolution infrared spectra were analysed for the F-H···H_2 complex in the gas phase[91] and for its isotope analogues, F-H···D_2 and F-H···DH solvated in helium nanodroplets.[92–94] The T-shaped structure of these complexes was confirmed. A similar T-shaped structure of Cl-H···H_2 resulting from theoretical calculations is shown in Fig. IV.12a.[64] The figure also presents complexes linked by other A-H···σ hydrogen bonds where the C-C σ-bonds of cycloalkanes are the proton acceptors (Figs. IV.12b and IV.12c).

The infrared vibrational predissociation spectra were analysed for the OCH^+···H_2 complex and it was found that it is connected through C-H···σ interaction.[95] However, this interaction was not classified as hydrogen bonding in the original study. Similarly, in the complexes of hydrogen fluoride with molecular hydrogen mentioned earlier, their F-H···σ(H_2) links also were not considered as hydrogen bonds. In the case of the OCH^+···H_2 complex, its experimental spectra indicate that the H-H and C-H stretching vibrations occur at frequencies of 4060 cm^{-1} and 2840 cm^{-1}, respectively. Corresponding frequencies for isolated H_2 and OCH^+ species are equal to 4161 cm^{-1} and 3089 cm^{-1}, respectively. Thus, the decrease of the C-H stretching frequency

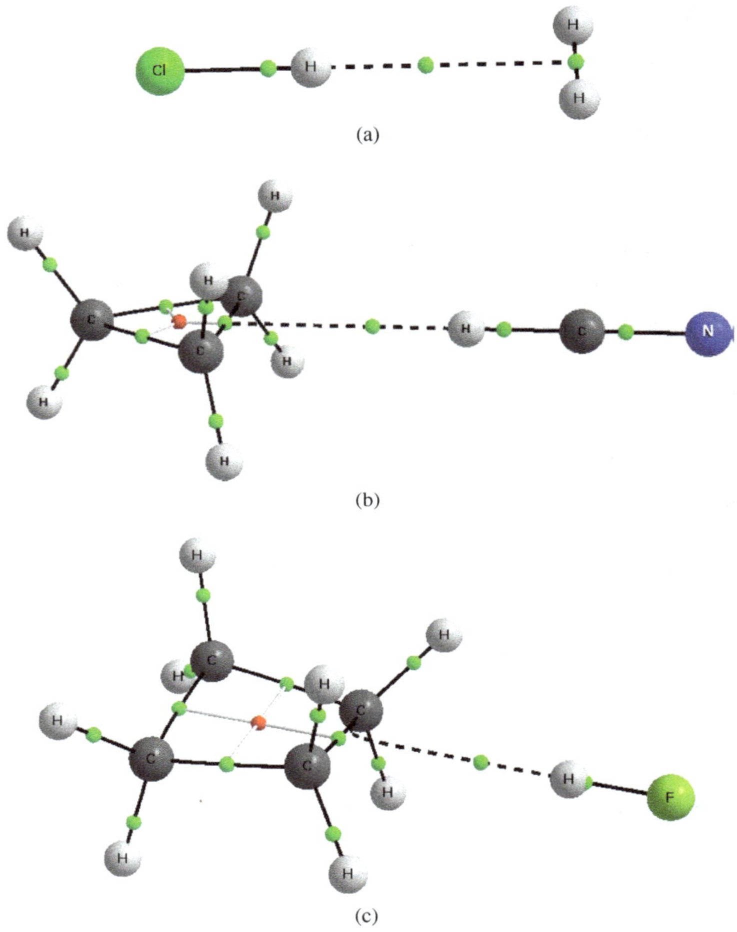

Fig. IV.12 The molecular graphs of complexes linked by A-H···σ hydrogen bonds: (a) H_2···HCl complex, (b) C_3H_6···HCN system, (c) C_4H_8···HF complex. Solid and broken lines correspond to bond paths, big circles to attractors and small green and red circles to BCPs and RCPs, respectively. Reprinted with permission from ref. 64. Copyright 2019 Wiley-VCH Verlag GmbH & Co.

corresponds to the elongation of the C-H bond as a result of complexation and such frequency decrease is observed for red-shifted hydrogen bonds. The H-H stretching frequency decrease in the complex, if compared with the corresponding frequency of the isolated molecular hydrogen, is related to the shift of the electron charge from the dihydrogen to the OCH^+ cation.[95] This type of electron charge shift, from the Lewis base to the Lewis acid unit, occurs for the hydrogen bonds as well as for other types of Lewis acid — Lewis base interactions.[7,13] QCISD(T)/6-311G(2df,2pd) calculations performed for this complex show that the T-shaped conformer energy minimum is 4.1 kcal/mol lower in energy than the sum of minima energies of isolated OCH^+ and H_2 monomers.[95]

There are theoretical studies on the A–H···H_2 complexes. MP2/aug-cc-pVTZ//MP2/6-311G(d,p) calculations were performed on the $NH_4^+\cdots H_2$ complex.[96] It means that the complexes were optimised at the MP2/6-311G(d,p) level and, furthermore, for such optimised structures the energetic single-point calculations were performed at the MP2/aug-cc-pVTZ level. The calculated binding energy for the latter complex is equal to −2.5 kcal/mol. The hydrogen molecule is perpendicular to the N–H bond of the NH_4^+ cation in this complex, thus it is the T-shaped configuration (Fig. IV.13a). The calculations on $NH_4^+\cdots(H_2)_n$ (n = 1–8) systems were also carried out, applying the same manner of calculations as for the $NH_4^+\cdots H_2$ complex.[96] It was found that for the $NH_4^+\cdots(H_2)_n$ clusters with n up to four dihydrogen molecules, each H_2 molecule is perpendicular, or nearly so, to the single N–H bond of the cation (Fig. IV.13b).[96] For n greater than four, all N-H bonds of the cation are saturated by the N–H···H_2 hydrogen bonds with four H_2 molecules, and next molecular hydrogens are in contact with the nitrogen centre by the H-N···H_2 pnicogen bonds (see Fig. IV.13c for the

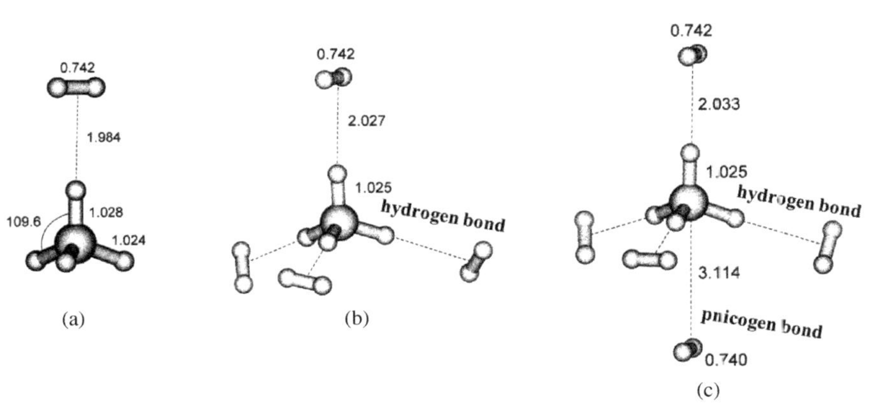

Fig. IV.13 Complexes of the ammonia cation linked by N-H···σ hydrogen bonds and the H-N···σ pnicogen bond: (a) $NH_4^+\cdots H_2$ complex, (b) $NH_4^+\cdots(H_2)_4$ cluster, (c) $NH_4^+\cdots(H_2)_5$ cluster. Modified with permission from ref. 96. Copyright 2001 Elsevier Science B.V.

$NH_4^+\cdots(H_2)_5$ cluster). However, the N-H···σ and H-N···σ arrangements were not analysed in terms of the hydrogen bonds and pnicogen bonds, respectively. It was pointed out that the molecular hydrogens may be vertex- or face-bonded to the ammonium cation.[96]

Figure IV.13a shows the $NH_4^+\cdots H_2$ complex linked by the N-H···σ hydrogen bond. One can see here that the N-H bond length is equal to 1.028 Å and the H···σ distance is equal to 1.984 Å (this is the distance between the H-centre of the proton-donating N-H bond and the mid-point of the molecular hydrogen). In the case of the $NH_4^+\cdots(H_2)_4$ cluster, all dihydrogens form hydrogen bonds with the N-H proton-donating bonds of the cation (Fig. IV.13b). The N-H bonds are shorter here (1.025 Å) than for the complex with one dihydrogen and the H···σ distances are longer, 2.027 Å. In other words, the hydrogen bonds are weaker in the cluster with four dihydrogens. It may be attributed to negative cooperativity, described in the previous section of this chapter. All dihydrogens deliver the electron charge to the cation, influencing the corresponding N-H bonds. In the case of clusters with more dihydrogens, these electron charge shifts are not so great per dihydrogens as it is for the complex that contains only one molecular hydrogen.

In the case of the $NH_4^+\cdots(H_2)_5$ cluster (Fig. IV.13c), all N-H bonds of the cation are saturated by the N-H···σ hydrogen bonds, and the fifth molecule is involved in the pnicogen bond. The H···σ distances of hydrogen bonds are equal to 2.033 Å, longer and consequently weaker than for the cluster containing four dihydrogen molecules. The distance is a rough measure of the hydrogen bond strength. The weakening of interactions with the increase of the number of ligands (dihydrogens) is attributed to the increase of negative cooperativity.

MP2/aug-cc-pVTZ calculations have been also performed in another study on the $AH_4^+\cdots H_2$ and $AH_4^+\cdots(H_2)_5$ (A = N, P, As, Sb, Bi) complexes.[97] In the $AH_4^+\cdots(H_2)_5$ clusters, four dihydrogen molecules form A–H···σ hydrogen bonds with the AH_4^+ ions, and the fifth dihydrogen molecule is linked by the pnicogen bond with the cation. Similarly, in the $AH_4^+\cdots H_2$ complexes the dihydrogen is linked with the cation by the hydrogen bond. This is the first study where A-H···σ interactions that link complexes were classified as hydrogen bonds.

The T-shaped configuration of the F-H···H_2 complex was a subject of experimental studies.[91–94] This complex was also analysed theoretically up to the MP2/6-311++G(3df,3pd) level, and it was pointed out that the H···σ linkage in the complex possesses characteristics typical of hydrogen bond interaction.[98] This is a weak interaction since the binding energy is equal to –0.8 kcal/mol. The elongation of both σ-bonds as a result of the complex formation, i.e., the H-F bond and the H-H bond of molecular hydrogen, by about 0.002 Å is observed. The bond elongations result from electron charge shifts from the dihydrogen Lewis base unit to the hydrogen fluoride Lewis acid unit. These electron charge shifts correspond to the $\sigma(H_2) \rightarrow \sigma^*(HF)$ orbital-orbital interaction.

Let us come back to systems where other σ-bonds act as proton acceptors in hydrogen bond interactions. Two such complexes are presented in Figs. IV.12b and IV.12c, namely C_3H_6···HCN and C_4H_8···HF complexes where the H-attractors of the proton-donating bonds, C-H and F-H, respectively, are connected by the bond paths with the BCPs of C-C bonds of cyclopropane and cyclobutane, respectively. These complexes are examples taken from the sample of moieties linked by H···σ interactions and calculated at the ωB97XD/aug-cc-pVTZ level.[64] These are complexes of species acting as Lewis base units: dihydrogen, cyclopropane, cyclobutane and cyclopentane, with Lewis acid units such as hydrogen fluoride, hydrogen chloride, water, hydrogen cyanide and acetylene. The strongest interaction in this sample is observed for the complex of cyclopropane with hydrogen fluoride, with the interaction energy equal to –5.0 kcal/mol. For the other complexes, the absolute value of the energy of interaction is lower than 3 kcal/mol, thus these are weak A-H···σ hydrogen bonds.[64]

There are other studies on systems linked by A-H···σ hydrogen bonds, for example, early studies on F-H···σ interactions of hydrogen fluoride with cyclopropane.[99,100] It is worth mentioning that the cyclopropane species is characterised by the accumulation of electron density at C-C bonds due to its molecular structure. Consequently these bonds have properties similar to CC double bonds and may act as the proton acceptor. One can also mention the theoretical study on complexes of an isomer of diborane, B_2H_4(4), with Lewis acid units such as HF, HNC, HCl, HCN, HCCH and H_2; the diborane acts here as the Lewis base through the B-B single σ-bond.[101]

There are experimental evidences of the existence of A-H···σ hydrogen bonds in the Cambridge Structural Database[102,103] where simple derivatives of cyclopropane, cyclobutane, and cyclopentane play the role of Lewis base centres through their CC σ-bonds.[64] About 40 accurately solved crystal structures were found where such types of hydrogen bonds occur.

IV.6 Hydrogen Bonds with Multi-Centre Proton Donors

Interactions described so far in the previous two sections of this chapter are commonly accepted as hydrogen bonds by the majority of definitions, including the IUPAC definition.[9,10] However, there is a group of interactions possessing characteristics of the hydrogen bond that are not taken into account by existing definitions. These are systems with multi-centre proton donors.[98,104–109] The proton-bound homodimer of acetylene is an example. There are different conformations, and one of them, probably corresponding to the global minimum, is characterised by the location of the proton between two acetylene molecules that are perpendicular to each other, at least approximately.[104] This protonated acetylene dimer was a subject of theoretical studies[104,107–109] as well as experimental analysis where the $C_2H_3^+$ cation and the proton-bound acetylene dimer cation $C_2H_2 \cdots H^+ \cdots C_2H_2$ were produced by a pulsed supersonic nozzle/pulsed electrical discharge cluster source.[106] Figure IV.14 presents the molecular graph and values of several parameters of this species, optimised at the MP2/aug-cc-pVTZ level. The proton is closer to one of the acetylene molecules, and due to the electron charge shifts from acetylene molecules its charge is lower than unity, +0.368 au. The distance between this H-centre and the C≡C bond mid-point of the closer acetylene is equal to 1.265 Å, and 1.726 Å for the further acetylene. However, the barrier height for the proton transfer between acetylene units was calculated to be only 0.04 kcal/mol. It means that the proton moves freely between the acetylene units in the larger region. The shorter contact with the closer acetylene possesses characteristics of a multi-centre covalent bond according to QTAIM parameters:[110,111] the electron density at H···σ BCP is equal to 0.152 au, the Laplacian of the electron density at BCP, $\nabla^2\rho_{BCP}$, is negative, thus the total electron energy density at BCP, H_{BCP}, is negative. It The longer H···σ contact in the complex, $C_2H_2 \cdots H^+ \cdots C_2H_2$, may be treated as a strong hydrogen bond since, despite the positive $\nabla^2\rho_{BCP}$ value, H_{BCP} is negative. It means that this hydrogen bond is strong and partly covalent in nature.

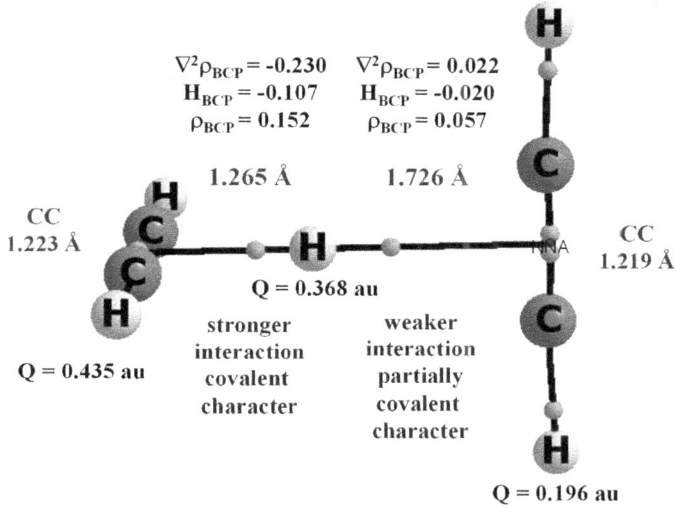

Fig. IV.14 The molecular graph of the proton-bound acetylene homodimer. Circles correspond to attractors, small circles to BCPs, and the small circle between two BCPs of acetylene corresponds to non-nuclear attractor. The solid lines correspond to bond paths. Values of various parameters are presented (they are described in the corresponding text).

The acetylene molecule that is closer to the proton in the $C_2H_2\cdots H^+\cdots C_2H_2$ complex is characterised by a slightly longer CC bond than in the farther acetylene molecule. This is because of electron charge transfer, which are greater from the closer molecule than from the farther one: their positive charges are equal to 0.435 au and 0.196 au, respectively. The electron charge outflow from the closer acetylene molecule is evidently greater. It is important that the H···σ link for this proton-bound homodimer, $C_2H_2\cdots H^+\cdots C_2H_2$, is classified as a hydrogen bond according to the 2sHB definition, with the other H···σ contact classified as a covalent bond. Hence the whole system may be considered as a complex of the $C_2H_3^+$ unit and the acetylene molecule, C_2H_2, i.e., a $C_2H_3^+\cdots C_2H_2$ complex.

In the T-shaped configuration of the free $C_2H_3^+$ cation, the proton is covalently bonded to the acetylene molecule. Figure IV.15a shows the molecular graph of this cation with the Laplacian of the electron density isolines. One can see that all attractors of this cation are in the area of the negative Laplacian values. This confirms the multicentre covalent bond in this cation. In contrast, in the $C_2H_2Li^+$ cation, the Li-centre is not bonded covalently with the acetylene species (see Fig. IV.15b). The bond path links

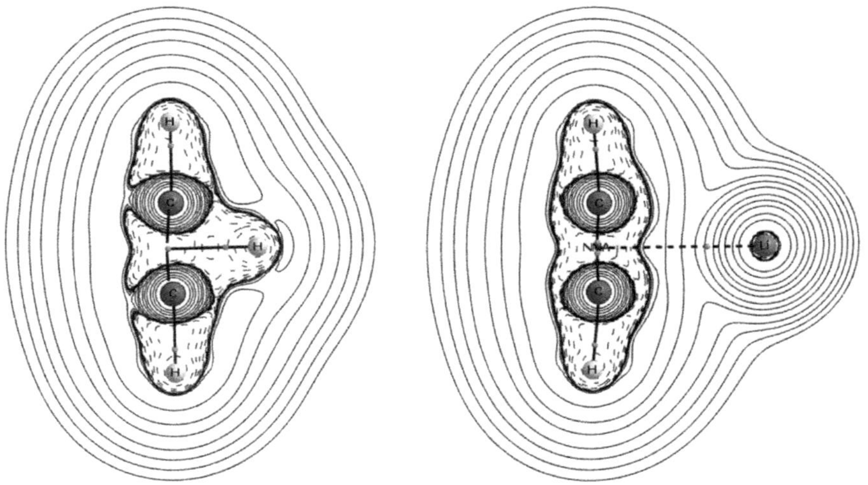

(a) (b)

Fig. IV.15 The contours of the Laplacian of electron density, $\nabla^2\rho$, for the (a) $C_2H_3^+$ cation, and (b) $C_2H_2Li^+$ cation. Continuous lines correspond to the positive values of $\nabla^2\rho$ and broken lines to the negative values. Reprinted with permission from ref. 108. Copyright 2021 MDPI.

here the Li-attractor with the non-nuclear attractor of the CC bond of acetylene. The BCP corresponding to this bond path is in the region of the positive values of the Laplacian of electron density.

There are numerous other interactions that possess multi-centre proton donors and that are classified as hydrogen bonds according to the 2sHB definition. These are $\pi\cdots H^+\cdots B$ and $\sigma\cdots H^+\cdots B$ interactions where there are multi-centre proton donors with the proton bound with π-electrons and σ-electrons, respectively, and there is the single-atom proton acceptor centre. There are also other systems classified as hydrogen bonds with multi-centre proton donors and multi-centre proton acceptors: $\pi\cdots H^+\cdots\pi$, $\sigma\cdots H^+\cdots\sigma$, $\pi\cdots H^+\cdots\sigma$, and $\sigma\cdots H^+\cdots\pi$. In these systems the multi-centre donors and acceptors are π or σ electrons. The proton-bound homodimer of acetylene discussed above is an example of a system linked by $\pi\cdots H^+\cdots\pi$ hydrogen bond.

IV.7 Hydride Bonds

It was described earlier that the majority of definitions of the hydrogen bond explain that the H-centre in the A-H$\cdots$B hydrogen-bonded bridge is characterised by a positive

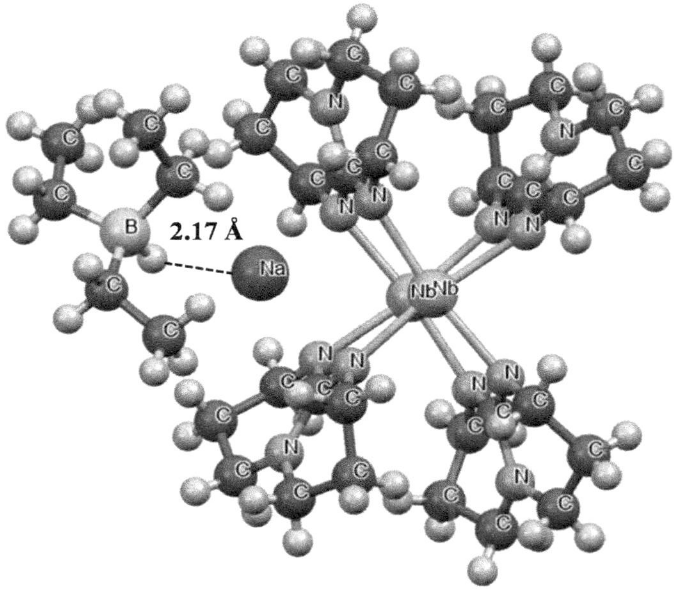

Fig. IV.16 A fragment of the crystal structure of tetrakis (μ_2-1,3,4,6,7,8-hexahydropyrimido (1,2-a) pyrimidinato)-diniobium bis (sodium triethylborate) (PAWDUM refcode, ref. 112).

charge; this is mentioned in the IUPAC definition[9,10] as well as in the 2sHB definition.[7,11] However, in some definitions, it is not mentioned which should be the sign of the H-atom charge; the definition of Pimentel and McClellan[6] presented earlier is an example. Hence the latter definition allows us to also classify as hydrogen bonds those systems that possess a negatively charged H-centre that connects two electrophilic sites. For example, B-H···Na^+ (B designates boron here) interactions were detected in crystal structures containing $NaEt_3BH$ adducts.[112] Figure IV.16 presents a fragment of the crystal structure of tetrakis (μ_2-1,3,4,6,7,8-hexahydropyrimido (1,2-a) pyrimidinato)-diniobium bis (sodium triethylborate) (PAWDUM refcode) where the B-H···Na^+ interaction is indicated.[112] The H···Na^+ distance is equal to 2.17 Å while the B-H-Na^+ angle amounts to 108° that is far from linearity, often observed for typical hydrogen bonds. In the case of the H···Na^+ contact, we can take into account the van der Waals radius of the sodium cation, estimated to be 1.27 Å,[113] while this radius for the H-centre is equal to 1.1 Å according to Rowland and Taylor (see Table I.1 in Chapter I).[114] The H-centre in this contact is negatively charged, so one can consider the van der Waals radius of the hydride anion as the other limit value, which is equal to 1.67 Å.[113] Hence the sum of van der Waals radii in the $H^{-\delta}\cdots Na^{+}$ contact is between 2.37 Å and 2.94 Å. Thus the $H\cdots Na^{+}$ distance of 2.17 Å in the crystal structure (Fig. IV.16) is lower than both limit values. This is the characteristic that is usually attributed to hydrogen bonds, that for the A-H···B system the H···B distance should be lower than the corresponding sum of van der Waals radii.[1,2,5,11]

The interactions with negatively charged H-centres were named as hydride bonds,[115] and henceforth this name is used in this book. However, other names were proposed in earlier studies, for example in the study of Rozas and co-workers,[116] where some of these complexes they analysed are presented in Figure IV.17. H···Li contacts occur for these complexes. If we take into account the calculated Li^{+} cation radius equal to 0.93 Å,[114] and the van der Waals radii mentioned above for the H-centre and the hydride ion, 1.1 Å and 1.67 Å, respectively, we have limit values for these contacts, 2.03 Å and 2.6 Å. All distances calculated at the MP2/6-311++G** level (Fig. IV.17) are shorter than both limit values. In the case of the $HBH_3^{-}\cdots LiH$ complex, the calculated contacts of 2.01 Å are

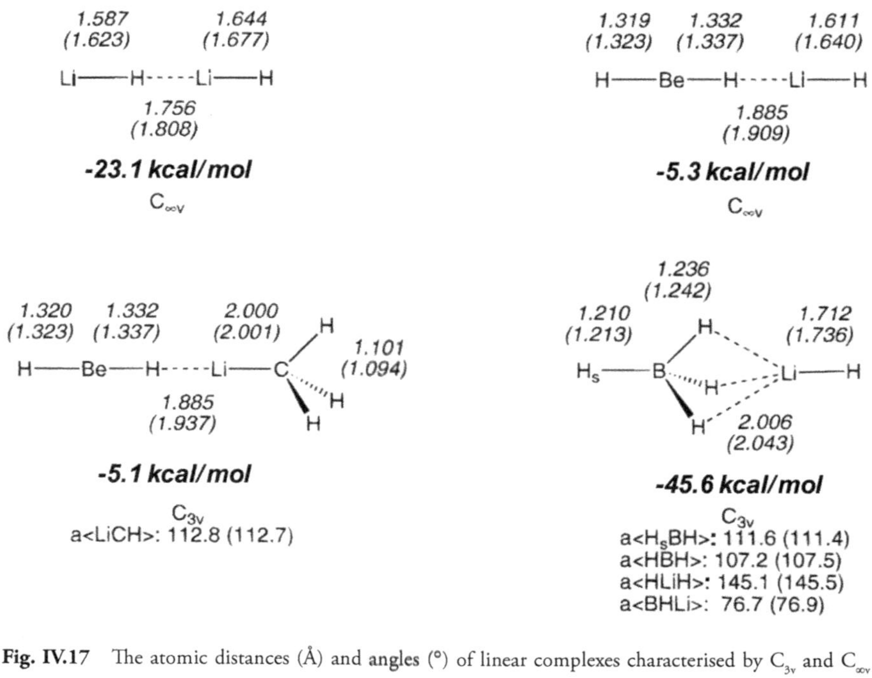

Fig. IV.17 The atomic distances (Å) and angles (°) of linear complexes characterised by C_{3v} and $C_{\infty v}$ symmetry; the MP2/6-311++G(d,p) level results (the HF/6-31G(d,p) results are given in parentheses) are presented. Reprinted with permission from ref. 116. Copyright 1997 American Chemical Society.

only slightly shorter than the lower limit of 2.03 Å. However, in this complex there are three H···Li contacts with three H-centres and one lithium site. Such interactions were described earlier in this chapter as weak interactions characterised by long contacts.

Figure IV.17 also presents the energies of interactions. One can see that the absolute value of the energy of interaction for two complexes linked by the Be-H···Li hydride bonds amounts to ~5 kcal/mol. In the case of complexes linked by the Li-H···Li and B-H···Li bonds (by three such interactions as it was mentioned above), these absolute values are 20 kcal/mol and 40 kcal/mol, respectively.

One can mention a few other theoretical studies on hydride bonds. For example, calculations up to the MP2/aug-cc-pVQZ level were performed on simple complexes of BeH_2 species with Li^+, Na^+ and Mg^{2+} cations.[115] The negatively charged hydridic H-atoms of the BeH_2 unit act here as the Lewis base centres while the cations are the Lewis acid centres. In this way the linear H-Be-H···cation (cation = Li^+, Na^+ and Mg^{2+}) complexes are formed. The MP2/aug-cc-pVQZ binding energies (BSSE correction included) for BeH_2 complexes with the cations are equal to –18.3 kcal/mol, –11.7 kcal/mol and –59.7 kcal/mol, respectively. The formation of these complexes is accompanied by electron charge transfer from the BeH_2 unit to the Li^+, Na^+, and Mg^{2+} cations, at 41 mē, 31 mē and 163 mē (mē — millielectron, i.e., 10^{-3} ē), respectively. These are great transfers since, for example, for the trans-linear conformation of the water dimer linked by the O-H···O hydrogen bond, the electron charge shift from the water unit acting as the Lewis base to the water acting as the Lewis acid unit is equal to only 13 mē (MP2/6-31+G(d,p) result).[117]

For the hydride-bonded complexes, the Laplacian of the electron density at the BCP, $\nabla^2\rho_{BCP}$, of the H···cation bond path is positive, thus these are not covalent bonds. For the H-Be-H··· Mg^{2+} complex, for the H···Mg^{2+} bond path the total electron energy density, H_{BCP}, is negative thus the latter interaction is partly covalent in nature according to the QTAIM approach. For the two remaining hydride-bonded complexes, the H_{BCP} is positive.

The decomposition of the energy of interaction performed for these complexes shows that the electrostatic interaction energy term and the delocalisation term are very important. The delocalisation term is related to the covalent character of interaction as well as to the electron charge shifts resulting from complexation. For the HBeH···Li^+ complex the electrostatic and delocalisation terms are almost equal to each other. For the

HBeH···Na$^+$ complex, among attractive terms, the electrostatic interaction energy is the most important. And for the HBeH···Mg$^+$ complex, the absolute value of the delocalisation term is almost two times greater than that of the electrostatic term. The results show that the hydride bonds are electrostatic interactions and only for very strong interactions, like for the HBeH···Mg$^+$ complex, partially covalent character is manifested.

It was mentioned above that there are early studies of Rozas and co-workers on hydride bonds,[116] but another name was proposed for such interactions, i.e., inverse hydrogen bonds. It was noted there that the dihydrogen bond, A-H$^{+\delta}$···$^{-\delta}$H-B, is a conventional A-H$^{+\delta}$···B hydrogen bond from one site and an inverse hydrogen bond (hydride bond), A-H$^{-\delta}$···B, from the other site. Scheme IV.5 presents this situation; the electron charge shifts are also indicated in this scheme.

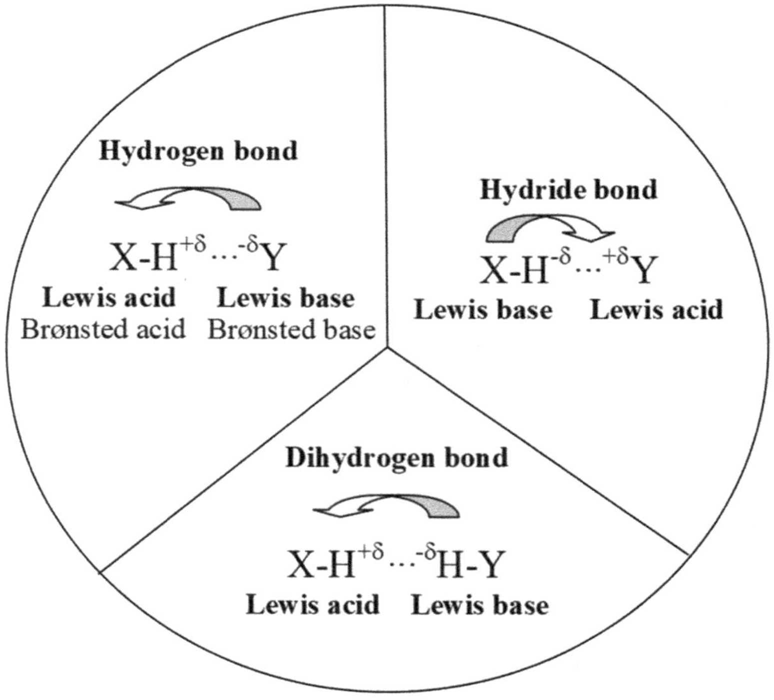

Scheme IV.5 The dihydrogen bond as an interaction that possesses characteristics of the conventional hydrogen bond and the hydride bond. The electron charge shifts for all these interactions are indicated. This is the modification of the scheme presented in ref. 116.

There are interactions that may be treated as a sub-class of hydride bonds, which are the charge-inverted hydrogen bonds (see Scheme IV.3). This term was proposed by Jabłoński to indicate interactions between the hydrogen centre characterised by a partial negative charge and the centre characterised by electron vacancy (the electron gap).[118–120] These interactions were designated as A-H···⊃B (A-H$^{-\delta}$···⊃$^{+\delta}$B). One can mention several studies where such interactions were analysed theoretically. For example, B3LYP/aug-cc-pVTZ and MP2/aug-cc-pVTZ calculations were performed on complexes of Lewis base units Li-H, H-Be-H, Cu-H, H_3Si-H and H_3Ge-H with Lewis acid species BH_3, AlH_3 and GaH_3.[118] In the Lewis acid species the electron vacancy at B, Al and Ga centres occurs. Great interaction energies were calculated for these complexes; for example, for all analysed complexes of LiH the absolute energies of interaction calculated by the MP2 method are over 50 kcal/mol. The characteristics of various kinds of charge-inverted hydrogen bonds were described in other studies.[119,120]

IV.8 Intramolecular Hydrogen Bonds

An intramolecular hydrogen bond, intraHB, occurs between a proton-donating bond and an electron-rich site in the same chemical unit.[1,2,11] In other words, intraHB is an intramolecular link that closes the chain of atoms connected by covalent bonds. In such a way the ring is formed, where among atom-atom links one of them is the hydrogen bond and the remaining ones are covalent bonds. The six-membered rings are often energetically stable, and this is a reason why they are analysed in various studies. The rings that contain less than six atoms are not stable usually due to geometric constraints, except for five-membered rings that may be stable sometimes. It was stated that in crystal structures the intraHBs in six-membered rings are often preferred to intermolecular hydrogen bonds. For example, Etter described that "six-membered rings with intramolecular hydrogen bonds are formed in preference to intermolecular hydrogen bonds."[121]

These six-membered rings are important in numerous chemical and biological processes. They are often observed in crystal structures, and LBHBs occur sometimes in such rings,[122,123] such as in some *ortho*-substituted aromatic species like phenols, *ortho* derivatives of benzoic acid, etc.[11,124]

Table IV.2 Classification of intramolecular hydrogen bonds. Modified with permission from ref. 11. Copyright 2021 The Royal Society of Chemistry.

Intramolecular hydrogen bond	Characterisation	Types	Examples
A-H···B	6-membered and 5-membered rings larger rings 3c-4e, RAHBs, LBHBs	N-H···O, O-H···O, C-H···O	malonaldehyde, o-chlorophenol
A-H···H-B	$H^{+\delta}\cdots{}^{-\delta}H$ contact B is often a transition metal	N-H···H-Ir, C-H···H-C	tetraphenylphosphonium squarate
A-H···π	π may mark two-centre acetylene or ethylene and multicentre aromatic acceptors	O-H···π, S/Se-H···π, C-H···π	o-phenylphenol, 3-butyne-1-thiol
A-H···σ	σ may mark two-centre CC single bonds	O-H···σ	cyclopropylmethanol

Table IV.2 presents different types of intraHBs. There are typical 3c-4e interactions where the single atom proton donor and proton acceptor occur, i.e., these are A-H···B intraHBs. There is the special sub-class where the negatively charged hydrogen centre is the proton acceptor; these are intramolecular dihydrogen bonds.[11] It should be mentioned that among the first experimental evidence of the occurrence of dihydrogen bonds were crystal structures with such intramolecular interactions.[125–128] There are also interactions where π-electrons play the role of the proton acceptor. One can mention very early studies on O-H···π intraHBs in structures of primary alcohols.[129,130] The CC double and triple bonds were proton acceptors in some of these structures. Also, O-H···π intraHB systems with aromatic fragments or cyclopropane rings as the proton acceptors were discussed.[130] In the case of cyclopropane, the formally single CC bonds as proton acceptors occur, but the authors claim that because of the electron charge accumulation in such rings, these interactions with cyclopropane as the proton acceptor may be treated as O-H···π intraHBs. On the other site, from the formal point of view these are O-H···σ intraHBs.

The types of intraHBs collected in Table IV.2 occur also for intermolecular hydrogen bonds. It was discussed earlier in this chapter that much more types of interactions occur for intermolecular links than for the intramolecular ones (look at Schemes IV.3 and IV.4 as well as at Table IV.2). The reason is that numerous types of intraHBs that have their intermolecular counterparts were not investigated, neither experimentally nor theoretically. The systems with multi-centre proton donors are examples of such a situation.

One of the main characteristics of intraHBs is that the A-H-B angle of the hydrogen bond bridge is far from linearity because of geometrical constraints, especially for systems that contain a small number of atoms, such as five- and six-membered rings. Thus it is worth mentioning here the categorisation proposed by Desiraju and Steiner[8] that concerns C-H···B (B is any single-atom proton acceptor) intraHBs, i.e., systems with a C-H proton-donating bond. Two types of such intraHBs may be indicated according to this categorisation. The first type are those in which a flexible and long chain of atoms linked by covalent bonds is closed by the intraHB. For such systems the geometrical constraints are not observed, thus the hydrogen bond bridges possess the same characteristics as for the intermolecular systems; consequently such bridges tend to be linear. The second category concerns the intraHBs with a small number of atoms in rings. For such hydrogen bond bridges the geometrical constraints are very important and the A-H-B angle is often far from linearity.

It was mentioned earlier that the resonance-assisted hydrogen bonds (RAHBs)[47,51,131–133] are related to the cooperativity phenomenon. Jeffrey described two types of cooperativity in the case of hydrogen-bonded systems in crystals.[2] The first type, the so-called σ-bond cooperativity, occurs in continuous chains or cycles linked by hydrogen bonds. The chain of hydrogen fluoride molecules connected through F-H···F hydrogen bonds or the chain of hydrogen cyanide molecules linked by C-H···N hydrogen bonds are simple systems where this phenomenon is observed.[4,47] The σ-bond cooperativity enhances the strength of the hydrogen bond. It relates to the increase of polarisation of the A-H proton-donating bond. Such an increase results from the formation of the hydrogen bond; additional hydrogen bond(s) lead to further and greater polarisation of this bond. The greater this polarisation, the stronger the corresponding hydrogen bond. The σ-bond cooperativity may also be understood in terms of the valence bond theory as the effect of the partial mixing of two pseudo-tautomeric resonance forms.

The second type of cooperativity, according to Jeffrey's descriptions, concerns hydrogen bonds between molecules with conjugated multiple π-bonds. There are numerous inter- and intramolecular hydrogen-bonded systems that may be classified as those where this second type of cooperativity occurs. For example, dimers of carboxylic acids or amides exhibit this cooperativity. This cooperativity is also referred to as resonance assistance and the corresponding hydrogen bonds are thus known as RAHBs. This term was introduced by Gilli and coworkers,[131,132] mainly for the intraHBs with conjugated bonds; however, intermolecular hydrogen bonds may also be classified as RAHBs.[51,133]

In the RAHBs systems, the formation of the hydrogen bond is accompanied by electron charge shifts that lead to an increase of the π-electron delocalisation within the conjugated system, and it is accompanied by a synergic enhancement of the hydrogen bond strength.[47] For example, this phenomenon is observed in six-membered rings with single and double bonds alternately that are closed by a hydrogen bond. O-H...O and N-H...O intramolecular RAHBs are analysed most often. Scheme IV.6 presents the electron charge shifts (designated by arrows) for intramolecular RAHBs closed by six-membered rings. This scheme corresponds to the malonaldehyde species if the R1, R2 and R3 substituents are H-atoms. The real structure is a mixture of the resonance forms presented in this scheme. Hence the real structure may be treated as an effect of π-electron delocalisation with respect to resonance forms since in the latter the single and double bonds occur while in the real structure the shortening and lengthening of single and double bonds, respectively, occur. The π-electron delocalisation leads also to shortening of the O···O distance of the hydrogen bond bridge and movement of the H-atom of this bridge to the mid-point. In the extreme case of equal contribution of both resonance forms, the H-atom is located in the mid-point of the O···O distance, which is very short, about 2.4 Å.[133]

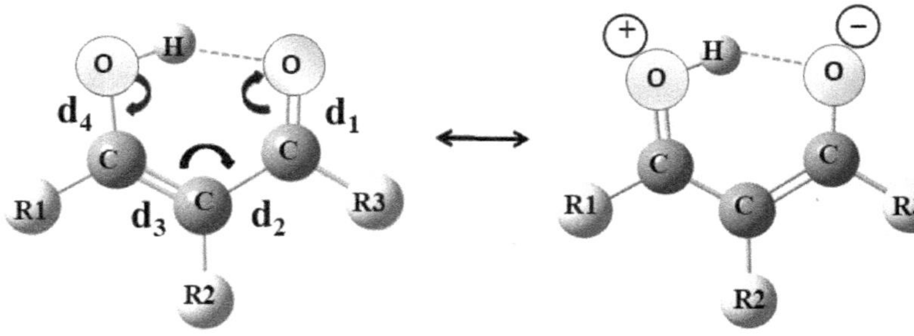

Scheme IV.6 Resonance forms of malonaldehyde and its derivatives. Reprinted with permission from ref. 11. Copyright 2021 The Royal Society of Chemistry.

The Q index was introduced to express the degree of π-electron delocalisation:[131]

$$Q = q_1 + q_2 \quad \text{and} \quad q_1 = d_4 - d_1 \text{ and } q_2 = d_2 - d_3 \qquad \text{(IV.7)}$$

Terms of the above equation are defined in Scheme IV.6 (the structure on the left). The Q-parameter was introduced for O-H···O hydrogen bonds in six-membered rings for derivatives of malonaldehyde, but it may be modified for other six-membered rings closed by other types of hydrogen bonds. The Q-parameter is equal to zero for "complete" equalisation of bonds, i.e., for the totally delocalised system, and it increases with the decrease of the degree of π-electron delocalisation. The greatest delocalisation is observed for the shortest O···O distances that correspond to the strongest O-H···O hydrogen bonds. It is worth mentioning that such strong intraHBs in six-membered rings are often classified as SSHBs and LBHBs. The RAHB model was contested in various studies.[134–136] It was proved that in systems with conjugated double bonds, there is no resonance-assisted stabilisation.[136] Other approaches have also been applied to analyse so-called RAHBs systems: the modern valence-bond theory,[137,138] the decomposition of the energy of interaction,[70,71] and the QTAIM approach,[110,111] and it was justified that the stabilisation of systems with intra- and intermolecular hydrogen bonds that are classified as RAHB results from the electron charge delocalisation and that the resonance stabilisation energies for these systems are negligible.[136] Thus the RAHB concept, which delivers numerous valuable findings, may be modified to be based mainly on the effects of π-electron delocalisation.

Another topic related to intraHBs concerns the evaluation of their strengths. In the case of intermolecular hydrogen bonding, the binding or interaction energy is calculated as the difference between the energy of the hydrogen-bonded complex and the sum of energies of monomers (eq. II.14 in Chapter II). In intraHB there is no reference monomers. Hence numerous other parameters are applied for the evaluation of the A-H···B intraHB strength:[11] A···B distance, electron density at the H···B BCP, etc. There are other, more accurate methods to calculate the interaction energy of intraHBs; the reader is referred to other studies and monographs since it is not a scope of this book.[139,140]

IV.9 Mechanisms of Hydrogen Bond Formation

The electron charge shift from the Lewis base unit to the Lewis acid unit is characteristic for the so-called Lewis acid-base interactions (see Chapter I). This is also in force for their sub-class hydrogen bonds where the $n_B \rightarrow \sigma_{AH}^*$ overlap is related to such an electron charge shift, and consequently A-H bond elongation.[19,63,141] However, this overlap does not explain the shortening of the A-H proton-donating bond that occurs in some hydrogen-bonded systems, known as blue-shifting hydrogen bonds.[18–21]

Alabugin and coworkers have explained that two effects are responsible for the hydrogen bond formation: hyperconjugative A-H bond weakening and rehybridisation-promoted A-H bond strengthening.[19] If the former effect prevails, A-H bond lengthening is observed; if the latter effect responsible for A-H bond shortening prevails, then the blue-shifting hydrogen bond is detected. It is worth mentioning that both effects occur for all hydrogen bonds, thus there are no significant differences between blue-shifting and red-shifting hydrogen bonds.[20]

The hyperconjugative effect is related to the orbital-orbital overlap that is characteristic for hydrogen bonds, even mentioned as obligatory in the definition of Weinhold and Klein (see Chapter II).[63] The rehybridisation process is connected with the increase of the s-character of the A-centre hybrid orbital in the A-H bond.[19] The σ_{AH} orbital corresponds to the latter bond (see Chapter II, eq. II.23). It was justified that "it is the common, fundamental characteristic related to the hydrogen bond formation — the increase of the s-character of the h_A hybrid orbital of the proton donating A-H bond."[11] The increase in s-character is a consequence of Bent's rule[142] which states that in atoms the s-character increases for hybrid orbitals directed to electropositive substituents, while it decreases with concomitant increase of p-character in atoms for hybrid orbitals aimed at electronegative substituents. The electropositive character of the H-centre of the A-H proton-donating bond increases because of the hydrogen bond formation since the positive charge of this centre increases. Thus, the hydrogen bond formation follows the Bent rule since the s-character increases for the h_A hybrid orbital of the A-H bond. There are numerous correlations between the hydrogen bond strength and the increase of the s-character of hybrid orbitals. The importance of s-character was described in the study of Alabugin and co-workers where various relationships between the s-character and other parameters were presented.[19]

For example, in the A-H···B hydrogen bond the s-character increases if the H···B and A···B distances decrease, the absolute values of interaction and binding energies

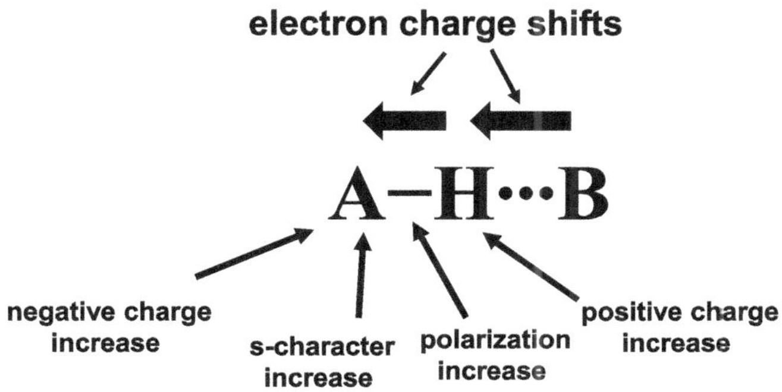

Scheme IV.7 The electron charge shifts and changes of parameters resulting from the hydrogen bond formation.

increase, the electron density at H···B BCP increases, the polarisation of the A-H bond increases, the electron charge shift from the B Lewis base unit to the A-H Lewis acid unit increases, the energy related to the $n_B \rightarrow \sigma_{AH}^*$ overlap increases, etc. All these relationships and properties result from the electron charge redistribution.[20] The formation of the hydrogen bond relates to the increase of positive charge of the H-centre and the increase of negative charge of the A-centre.[11,20,141] These changes accompanying the formation of the hydrogen bond are presented in Scheme IV.7.

There are other changes of parameters resulting from the hydrogen bond formation; for example, the decrease of the QTAIM H-atom radius and the increase of the QTAIM A-atom radius.[20] Both are components of the A-H bond path since they may be defined as distances between the BCP of the A-H bond path and the H and A attractors, respectively. If the decrease of the H-atom radius overwhelms the increase of the A-atom radius, then one can observe the blue-shifting hydrogen bond.[20]

Summary

There are different types of the hydrogen bond interaction; one can mention systems with single-atom proton acceptors and those where the multi-centre proton acceptors occur. In the latter species, π-electrons of ethylene and acetylene derivatives or aromatic moieties may play the role of the Lewis base sites. However, σ-electron systems may also play the role of the proton acceptor; this may be the molecular hydrogen or σ-bonds of other species. All links in systems mentioned above are classified as hydrogen bonds according to commonly known definitions, particularly according to the IUPAC definition.[9,10]

However, there are systems which are not covered by known definitions, which are complexes with multi-centre proton donors. The proton inserted between triple bonds of acetylene molecules is an example of a system where an untypical hydrogen bond occurs. The $H^{+}\cdots\pi$ links in the complex are classified as hydrogen bonds by the 2sHB definition.

Special attention may be paid to the intraHBs where the proton donor and the proton acceptor belong to the same chemical species. These links often possess specific characteristics, different from intermolecular analogues, since geometrical constraints may occur. This results, for example, in an A-H···B arrangement that is far from linearity.

The unique characteristics of the hydrogen bond may be specified since the formation of this interaction is connected with electron charge redistribution that is revealed by the increase of positive charge of the H-centre and the increase of negative charge of the A-centre, for example.

It is important that not only geometrical and energetic parameters of the A-H···B bridge are useful to describe properties of hydrogen bonds. These are also other parameters, such as QTAIM topological parameters or those that describe orbitals participating in the hydrogen bond formation. For example, the s-character of the hybrid atomic orbital of the A-centre always increases as a result of hydrogen bond formation. Such an increase may be useful to evaluate the strength of interaction.

Appendix

The following tools were applied to construct figures presented in this chapter.

Figures IV.10,14

AIMAll program — Reference: Keith, T.A. *AIMAll (Version 11.08.23)*; TK Gristmill Software: Overland Park, KS, USA, **2011**.

Figures IV.2,3,9,16

Graphical tools of the Cambridge Structural Database — References: (1) Groom, C.R.; Bruno, I.J.; Lightfoot, M.P.; Ward, S.C. The Cambridge structural database. *Acta Cryst.* **2016**, *B72*, 171–179; (2) Wong, R.; Allen, F.H.; Willett, P. The scientific impact of the Cambridge Structural Database: A citation-based study. *J. Appl. Cryst.* **2010**, *43*, 811–824.

Figures IV.4,5,7,8

Jmol program — Reference: Hanson, R.M. Jmol SMILES and Jmol SMARTS: specifications and applications. *J. Cheminform.* **2016**, *8*, 50.

Schemes IV.3,4,6 and Figures IV.1,6,11,12,13,15,17

Reprinted with permission.

Schemes IV.1,2,5,7

PowerPoint.

References

1. Jeffrey, G.A.; Saenger, W. *Hydrogen Bonding in Biological Structures*, Springer, Berlin, **1991**.
2. Jeffrey, G.A. *An Introduction to Hydrogen Bonding*, Oxford University Press, New York, **1997**.
3. Lewis, G.N. *Valence and the Structure of Atoms and Molecules*, American Chemical Society Monograph Series, New York, **1923**.
4. Pauling, L. *The Nature of the Chemical Bond: Third Edition*, Cornell University Press, New York, **1960**.
5. Desiraju, G.R. Hydrogen bridges in crystal engineering: interactions without borders. *Acc. Chem. Res.* **2002**, *35*, 565–573.
6. Pimentel, G.C.; McClellan, A.L. *The Hydrogen Bond*, W.H. Freeman and Company, San Francisco and London, **1960**, p. 6 and pp. 193–205.
7. Grabowski, S.J. Hydrogen bond types which do not fit accepted definitions. *Chem. Comm.* **2024**, 60, 6239–6255.
8. Desiraju, G.R.; Steiner, T. *The Weak Hydrogen Bond in Structural Chemistry and Biology*, Oxford University Press, New York, **1999**.

9. Arunan, E.; Desiraju, G.R.; Klein, R.A.; Sadlej, J.; Scheiner, S.; Alkorta, I.; Clary, D.C.; Crabtree, R.H.; Dannenberg, J.J.; Hobza, P.; Kjaergaard, H.G.; Legon, A.C.; Mennucci, B.; Nesbitt, D.J. Defining the hydrogen bond: An account (IUPAC Technical Report). *Pure Appl. Chem.* **2011**, *83*, 1619–1636.
10. Arunan, E.; Desiraju, G.R.; Klein, R.A.; Sadlej, J.; Scheiner, S.; Alkorta, I.; Clary, D.C.; Crabtree, R.H.; Dannenberg, J.J.; Hobza, P.; Kjaergaard, H.G.; Legon, A.C.; Mennucci, B.; Nesbitt, D.J. Definition of the hydrogen bond (IUPAC Recommendations 2011). *Pure Appl. Chem.* **2011**, *83*, 1637–1641.
11. Grabowski, S.J. *Understanding Hydrogen Bonds: Theoretical and Experimental Views,* The Royal Society of Chemistry, Cambridge, **2021.**
12. Politzer, P.; Murray, J.S.; Clark, T. Halogen bonding: An electrostatically-driven highly directional noncovalent interaction. *Phys. Chem. Chem. Phys.* **2010**, *12*, 7748–7758.
13. Grabowski, S.J. What is the covalency of hydrogen bonding? *Chem. Rev.* **2011**, *111*, 2597–2625.
14. Johnston, H.S. Large tunnelling corrections in chemical reactions rates. *Adv. Chem. Phys.* **1960**, *3*, 131–170.
15. Johnston, H.S.; Parr, Ch. Activation energies from bond energies. I. Hydrogen transfer reactions. *J. Am. Chem. Soc.* **1963**, *85*, 2544–2551.
16. Grabowski, S.J.; Krygowski, T.M. Molecular Geometry as a Source of Chemical Information — Applications of the Bond Valence — Bond Number Models, in Charton, M.; Charton, B.I. (Eds.), *Advances in Quantitative Structure Property Relationships*, Vol. 3, Elsevier, Amsterdam, **2002**, pp. 27–66.
17. This was named as the bond order conservation rule by Johnston and in latter studies the term bond number conservation rule was applied (see ref. 16).
18. Hobza, P.; Havlas, Z. Blue-shifting hydrogen bonds. *Chem. Rev.* **2000**, *100*, 4253–4264.
19. Alabugin, I.V.; Manoharan, M.; Peabody, S.; Weinhold, F. Electronic basis of improper hydrogen bonding: a subtle balance of hyperconjugation and rehybridization. *J. Am. Chem. Soc.* **2003**, *125*, 5973–5987.
20. Grabowski, S.J. Red- and blue-shifted hydrogen bonds: the Bent rule from quantum theory of atoms in molecules perspective. *J. Phys. Chem. A* **2011**, *115*, 12789–12799.
21. Joseph, J.; Jemmis, E.D. Red-, blue-, or no-shift in hydrogen bonds: a unified explanation. *J. Am. Chem. Soc.* **2007**, *129*, 4620–4632.
22. Roziere, J.; Williams, J.M. The hydrated proton $H^+(H_2O)_n$. IV. A high-precision neutron diffraction study of the diaquohydrogen ion, $(H_2O \cdot H \cdot OH_2)^+$, in *trans*-dichlorobis(ethylenediamine)cobalt(iii) chloride hydrochloride dihydrate. *Inorg. Chem.* **1976**, *15*, 1174–1178.

23. Zundel, G.; Metzger, H. Energy bands tunneling excess protons in liquid acids. IR spectroscopic study of the nature of $H_5O_2^+$ groups. *Z. Phys. Chem.* **1968**, *58*, 225–245.

24. Lammers, S.; Meuwly, M. Investigating the relationship between infrared spectra of shared protons in different chemical environments: a comparison of protonated diglyme and protonated water dimer. *J. Phys. Chem. A* **2007**, *111*, 1638–1647.

25. Pitsevich, G.; Malevich, A.; Kozlovskaya, E.; Mahnach, E.; Doroshenko, T.; Pogorelov, V.; Pettersson, L.G.M.; Sablinskas, V.; Balevicius, V. MP4 study of the anharmonic coupling of the shared proton stretching vibration of the protonated water dimer in equilibrium and transition states. *J. Phys. Chem. A* **2017**, *121*, 2151–2165.

26. Selenius, C.-O.; Lundgren, J.-O. Neutron diffraction study of 2-nitro-1,3-indandione dihydrate. *Acta Cryst.* **1980**, *B36*, 3172–3175.

27. Wenthold, P.G.; Squires, R.R. Bond dissociation energies of F_2^- and HF_2^-. A gas-phase experimental and G2 theoretical study. *J. Phys. Chem.* **1995**, *99*, 2002–2005.

28. Sode, O.; Hirata, S. Second-order many-body perturbation study of solid hydrogen fluoride under pressure. *Phys. Chem. Chem. Phys.* **2012**, *14*, 7765–7779.

29. Stein, C.; Oswald, R.; Sebald, P.; Botschwina, P.; Stoll, H.; Peterson, K.A. Accurate bond dissociation energies (D_0) for FHF^- isotopologues. *Mol. Phys.* **2013**, *111*, 2647–2652.

30. Golubev, N.S.; Melikova, Shchepkin, D.N.; Shenderovich, I.G.; Tolstoy, P.M.; Denisov, G.S. Interpretation of hydrogen/deuterium isotope effects on NMR chemical shifts of $[FHF]^-$ ion base on calculations of nuclear magnetic shielding tensor surface. *Z. Phys. Chem.* **2003**, *217*, 1549–1563.

31. Panich, A.M. NMR study of the F-H…F hydrogen bond. Relation between hydrogen atom position and F-H…F bond length. *Chem. Phys.* **1995**, *196*, 511–519.

32. Grabowski, S.J. $[FHF]^-$ — the strongest hydrogen bond under the influence of external interactions. *Crystals* **2016**, *6*, 3.

33. Silvestre, J-P.; Dao, N.Q.; Heger, G.; Cousson, A. Refinement by neutron diffraction of the crystal structure of hydroxyethylidene bisphosphonic acid monohydrate: $C(CH_3)(OH)(PO_3H_2)_2 \cdot H_2O$. *Phosphorus Sulfur Silicon Relat. Elem.* **2002**, *177*, 277–288.

34. Harkema, S.; Ter Brake, J.H.M. Structure of urea-oxalic acid (1/1), $CH_4N_2O.C_2H_2O_4$, determined by neutron diffraction. *Acta Cryst.* **1984**, *C40*, 1733–1734.

35. Chęcińska, L.; Grabowski, S.J. Partial hydrogen bonds — structural studies on thioureidoalkylphosphonates. *J. Phys. Chem. A* **2005**, *109*, 2942–2947.

36. Chęcińska, L.; Grabowski, S.J.; Małecka, M. An analysis of bifurcated H-bonds: crystal and molecular structures of O,O-diphenyl 1-(3-phenylthioureido)pentanephosphonate and O,O-diphenyl 1-(3-phenylthioureido)butanephosphonate. *J. Phys. Org. Chem. A* **2003**, *16*, 213–219.

37. Parthasarathi, R.; Amutha, R.; Subramanian, V.; Nair, B. U.; Ramasami, T. Bader's and reactivity descriptors' analysis of DNA base pairs. *J. Phys. Chem. A* **2004**, *108*, 3817–3828.
38. Espinosa, E.; Souhassou, M.; Lachekar, H.; Lecomte, C. Topological analysis of the electron density in hydrogen bonds. *Acta Crystallogr.* **1999**, *B55*, 563–572.
39. Alkorta, I.; Blanco, F.; Deyà, P.M.; Elguero, J.; Estarellas, C.; Frontera, A.; Quiñonero, D. Cooperativity in multiple unusual weak bonds. *Theor. Chem. Acc.* **2010**, *126*, 1–14.
40. Mó, O.; Yañez, M.; Del Bene, J.E.; Alkorta, I.; Elguero, J. Cooperativity and proton transfer in hydrogen-bonded triads. *ChemPhysChem* **2005**, *6*, 1411–1418.
41. Del Bene, J.; Pople, J.A. Intermolecular energies of small water polymers. *Chem. Phys. Lett.* **1969**, *4*, 426–428.
42. Dunitz, J.D. *X-Ray Analysis and the Structure of Organic Molecules*, Cornell University Press, Ithaca, **1979**.
43. Novoa, J.J.; Braga, D.; Addadi, L. (Eds.). *Engineering of Crystalline Materials Properties*, Springer, Dordrecht, **2007**.
44. Leiserowitz, L. Molecular packing modes. Carboxylic acids. *Acta Cryst.* **1976**, *B32*, 775–802.
45. Karothu, D.P.; Weston, J.; Desta, I.T.; Naumov, P. Shape-memory and self-healing effects in mechanosalient molecular crystals. *J. Am. Chem. Soc.* **2016**, *138*, 13298-13306.
46. Almenningen, A.; Bastiansen, O.; Motzfeldt, T. A reinvestigation of the structure of monomer and dimer formic acid by gas electron diffraction technique. *Acta Chem. Scand.* **1969**, *23*, 2848–2864.
47. Grabowski, S.J. Noncovalent Interactions in Crystal Structures: Quantifying Cooperativity in Hydrogen and Halogen Bond, in Novoa, J.J. (Ed.), *Intermolecular Interactions in Crystals: Fundamentals of Crystal Engineering*, The Royal Society of Chemistry, United Kingdom, **2018**, pp. 673–718.
48. Scheiner, S. *Hydrogen Bonding: A Theoretical Perspective*, Oxford University Press, New York, **1997**, pp. 230–290.
49. Ziółkowski, M.; Grabowski, S.J.; Leszczynski, J. Cooperativity in hydrogen-bonded interactions: ab initio and "atoms in molecules" analyses. *J. Phys. Chem. A* **2006**, *110*, 6514–6521.
50. Grabowski, S.J.; Ugalde, J.M. Ab initio calculations on $C_6H_6\cdots(HF)_n$ clusters — X–H···π hydrogen bond. *Can. J. Chem.* **2010**, *88*, 769–778.
51. Gilli, P.; Bertolasi, V.; Ferretti, V.; Gilli, G. Covalent nature of the strong homonuclear hydrogen bond. Study of the O-H···O system by crystal structure correlation methods. *J. Am. Chem. Soc.* **1994**, *116*, 909–915.
52. Grabowski, S.J. Hydrogen bonding strength — measures based on geometric and topological parameters. *J. Phys. Org. Chem.* **2004**, *17*, 18–31.

53. Calleja, M.; Mason, S.A.; Prince, P.D.; Steed, J.W.; Wilkinson, C. Single crystal neutron and X-ray diffraction studies of $(H_7O_3)[AuCl_4]$·15-crown-5. *New J. Chem.* **2001**, *25*, 1475–1478.

54. Luger, P. *Modern X-Ray Analysis on Single Crystals: Second Fully Revised and Extended Edition*, Walter de Gruyter, Berlin, **2014**.

55. Giacovazzo, C. (Ed.). *Fundamentals of Crystallography: Second Edition*, Oxford University Press, New York, **2002**.

56. Scheiner, S. Ab initio studies of hydrogen bonds: the water dimer paradigm. *Annu. Rev. Phys. Chem.* **1994**, *45*, 23–56.

57. Parra, R.D.; Bulusu, S.; Zeng, X.C. Cooperative effects in one-dimensional chains of three-center hydrogen bonding interactions. *J. Chem. Phys.* **2003**, *118*, 3499–3509.

58. Lamoureux, G.V.; Chaves-Carballo, K. Teaching hydrogen bridges: it is not FON anymore! *Chemistry Teacher International* **2024**. https://doi.org/10.1515/cti-2023–0027

59. Suttor, D.J. Evidence for the existence of C-H···O hydrogen bonds in crystals. *J. Chem. Soc.* **1963**, 1105–1110.

60. Donohue, J. Editorial, in Rich, A.; Davidson, N. (Eds.), *Structural Chemistry and Molecular Biology*, W. H. Freeman and Company, San Francisco, **1968**, pp. 459–463.

61. Taylor, R.; Kennard, O. Crystallographic evidence for the existence of C-H···O, C-H···N and C-H···Cl hydrogen bonds. *J. Am. Chem. Soc.* **1982**, *104*, 5063–5070.

62. Nishio, M.; Hirota, M.; Umezawa, Y. *The CH/π Interaction: Evidence, Nature, and Consequences*, Wiley-VCH, **1998**.

63. Weinhold, F.; Klein, R.A. What is a hydrogen bond? Mutually consistent theoretical and experimental criteria for characterizing H-bonding interactions. *Mol. Phys.* **2012**, *110*, 565–579.

64. Grabowski, S.J. A-H···σ hydrogen bonds: dihydrogen and cycloalkanes as proton acceptors. *ChemPhysChem* **2019**, *20*, 565–574.

65. Richardson, T.B.; de Gala, S.; Crabtree, R.H. Unconventional hydrogen bonds: intermolecular B-H···H-N interactions. *J. Am. Chem. Soc.* **1995**, *117*, 12875–12876.

66. Epstein, L.M.; Shubina, E.S. New types of hydrogen bonding in organometallic chemistry. *Coord. Chem. Rev.* **2002**, 231, 165–181.

67. Bakhmutov, V.I. *Dihydrogen Bonds: Principles, Experiments, and Applications*, John Wiley & Sons, Hoboken, NJ, **2008**.

68. Grabowski, S.J. Dihydrogen bond and X–H···σ interaction as sub-classes of hydrogen bond. Hydrogen bonding strength — measures based on geometric and topological parameters. *J. Phys. Org. Chem.* **2013**, *26*, 452–459.

69. Grabowski, S.J.; Sokalski, W.A.; Leszczynski, J. Wide spectrum of H···H interactions: van der Waals contacts, dihydrogen bonds and covalency. *Chem. Phys.* **2007**, *337*, 68–76.

70. Sokalski, W.A.; Roszak, S.; Pecul, K. An efficient procedure for decomposition of the SCF interaction energy into components with reduced basis set dependence. *Chem. Phys. Lett.* **1988**, *153*, 153–159.
71. Sokalski, W.A.; Roszak, S. Efficient techniques for the decomposition of intermolecular interaction energy at SCF level and beyond. *J. Mol. Struct.* **1991**, *234*, 387–400.
72. Kitaura, K.; Morokuma, K. A new energy decomposition scheme for molecular interactions within the Hartree-Fock approximation. *Int. J. Quantum Chem.* **1976**, *10*, 325–340.
73. Umeyama, H.; Morokuma,K.; Yamabe, S. Molecular orbital studies of electron donor-acceptor complexes. 4. Energy decomposition analysis of halo-complexes: ammonia-fluorine, -chlorine, -chlorine fluoride, methanamine-chlorine fluoride, formaldehyde-fluorine, hydrogen fluoride-chlorine fluoride, and fluorine-fluorine. *J. Am. Chem. Soc.* **1977**, *99*, 330–343.
74. Grabowski, S.J.; Lipkowski, P. Characteristics of X-H···π interactions: ab initio and QTAIM studies. *J. Phys. Chem. A* **2011**, *115*, 4765–4773.
75. Alder, R.W. Strain effects on amine basicities. *Chem. Rev.* **1989**, *89*, 1215–1223.
76. Staab, H.A.; Saupe, T. “Proton sponges” and the geometry of hydrogen bonds: aromatic nitrogen bases with exceptional basicities. *Angew. Chem. Int. Ed. Engl.* **1988**, *27*, 865–879.
77. Stoyanov, E.S.; Reed, C.A. IR spectrum of the $H_5O_2^+$ cation in the context of proton disolvates $L–H^+–L$. *J. Phys. Chem. A* **2006**, *110*, 12992–13002.
78. Mildvan, A.S.; Massiah, M.A.; Harris, T.K.; Marks, G.T.; Harrison, D.H.T.; Viragh, C.; Reddy, P.M.; Kovach, I.M. Short, strong hydrogen bonds in enzymes: NMR and mechanistic studies. *J. Mol. Struct.* **2002**, *615*,163–175.
79. Chen, J.; McAllister, M.A.; Lee, J.K.; Houk, K.N. Short, strong hydrogen bonds in the gas phase and in solution: theoretical exploration of pK_a matching and environmental effects on the strengths of hydrogen bonds and their potential roles in enzymatic catalysis. *J. Org. Chem.* **1998**, *163*, 4611–4619.
80. Warshel, A.; Papazyan, A. Energy considerations show that low-barrier hydrogen bonds do not offer a catalytic advantage over ordinary hydrogen bonds. *Proc. Natl. Acad. Sci. USA* **1996**, *93*, 13665–13670.
81. Cleland, W.W.; Frey, P.A.; Gerlt, J.A. The low barrier hydrogen bond in enzymatic catalysis. *J. Biol. Chem.* **1998**, *273*, 25529–25532.
82. Cremer, D.; Kraka, E. A description of the chemical bond in terms of local properties of electron density and energy. *Croat. Chem. Acta* **1984**, *57*, 1259–1281.
83. Jenkins, S.; Morrison, I. The chemical character of the intermolecular bonds of seven phases of ice as revealed by AB initio calculation of electron densities. *Chem. Phys. Lett.* **2000**, *317*, 97–102.

84. Arnold, W.D.; Oldfield, E. The chemical nature of hydrogen bonding in proteins via NMR: J-couplings, chemical shifts, and AIM theory. *J. Am. Chem. Soc.* **2000**, *122*, 12835–12841.

85. Rozas, I.; Alkorta, I.; Elguero, J. Behavior of ylides containing N, O, and C atoms as hydrogen bond acceptors. *J. Am. Chem. Soc.* **2000**, *122*, 1154–11161.

86. Nygren, C.L.; Wilson, C.C.; Turner, J.F.C. On the solid state structure of 4-iodobenzoic acid. *J. Phys. Chem. A* **2005**, *109*, 2586–2593.

87. Ward, M.D.; Huang, H-T.; Zhu, L.; Biswas, A.; Popov, D.; Badding, J.V.; Strobel, T.A. Chemistry through cocrystals: pressure-induced polymerization of $C_2H_2{\cdot}C_6H_6$ to an extended crystalline hydrocarbon. *Phys. Chem. Chem. Phys.* **2018**, *20*, 7282–7294.

88. Gholami, S.; Aarabi, M.; Grabowski, S.J. Coexistence of intra- and intermolecular hydrogen bonds: salicylic acid and salicylamide and their thiol counterparts. *J. Phys. Chem. A* **2021**, *125*, 1526–1539.

89. Grabowski, S.J. What Can Be Learnt from a Location of Bond Paths and from Electron Density Distribution, in Chauvin, R.; Lepetit, C.; Silvi, B.; Alikhani, E. (Eds.), *Applications of Topological Methods in Molecular Chemistry*, Springer, **2016**, pp. 399–433.

90. Grabowski, S.J. Unpublished results.

91. Jucks, KJ.W.; Miller, R.E. Infrared Stark spectroscopy of the hydrogen–HF binary complex. *J. Chem. Phys.* **1987**, *87*, 5629–5633.

92. Moore, D.T.; Miller, R.E. Dynamics of hydrogen–HF complexes in helium nanodroplets. *J. Chem. Phys.* **2003**, *118*, 9629–9636.

93. Moore, D.T.; Miller, R.E. Solvation of HF by molecular hydrogen: helium nanodroplet vibrational spectroscopy. *J. Phys. Chem. A* **2003**, *107*, 10805–10812.

94. Moore, D.T.; Miller, R.E. Rotationally resolved infrared laser spectroscopy of $(H_2)n$-HF and $(D_2)n$-HF (n = 2–6) in helium nanodroplets. *J. Phys. Chem. A* **2004**, *108*, 1930–1937.

95. Bieske, E.J.; Nizkorodov, S.A.; Bennett, F.R.; Maier, J.P. The infrared spectrum of the H_2–HCO^+ complex. *J. Chem. Phys.* **1995**, *102*, 5152–5164.

96. Urban, J.; Roszak, S.; Leszczynski, J. Shellvation of the ammonium cation by molecular hydrogen: a theoretical study. *Chem. Phys. Lett.* **2001**, *346*, 512–518.

97. Szymczak, J.J.; Grabowski, S.J.; Roszak, S.; Leszczynski, J. H···σ interactions — an ab initio and 'atoms in molecules' study. *Chem. Phys. Lett.* **2004**, *393*, 81–86.

98. Grabowski, S.J.; Sokalski, W.A.; Leszczynski, J. Can H···σ, π···H^+···σ and σ···H^+···σ interactions be classified as H-bonded? *Chem. Phys. Lett.* **2006**, *432*, 33–39.

99. Rozas, I.; Alkorta, I.; Elguero, J. Unusual hydrogen bonds: H···π interactions. *J. Phys. Chem. A* **1997**, *101*, 9457–9463.

100. Galano, A.; Alvarez-Idaboy, J.R.; Vivier-Bunge, A. Non-alkane behavior of cyclopropane and its derivatives: characterization of unconventional hydrogen bond interactions. *Theor. Chem. Acc.* **2007**, *118*, 597–606 and references therein.
101. Alkorta, I.; Soteras, I.; Elguero, J.; Del Bene, J.E. The boron–boron single bond in diborane(4) as a non-classical electron donor for hydrogen bonding. *Phys. Chem. Chem. Phys.*, **2011**, *13*, 14026–14032.
102. Groom, C.R.; Bruno, I.J.; Lightfoot, M.P.; Ward, S.C. The Cambridge structural database. *Acta Cryst.* **2016**, *B72*, 171–179.
103. Wong, R.; Allen, F.H.; Willett, P. The scientific impact of the Cambridge Structural Database: A citation-based study. *J. Appl. Cryst.* **2010**, *43*, 811–824.
104. Grabowski, S.J.; Sokalski, W.A.; Leszczynski, J. Is a $\pi\cdots H^{+}\cdots\pi$ complex hydrogen bonded? *J. Phys. Chem. A* **2004**, *108*, 1806–1812.
105. Grabowski, S.J. π-H···O hydrogen bonds: multicenter covalent π-H interaction acts as the proton-donating system. *J. Phys. Chem. A* **2007**, *111*, 13537–13543.
106. Douberly, G.E.; Ricks, A.M.; Ticknow, B.W.; McKee, W.C.; Schleyer, P. v. R.; Duncan, M.A. Infrared photodissociation spectroscopy of protonated acetylene and its clusters. *J. Phys. Chem. A* **2008**, *112*, 1897–1906.
107. Grabowski, S.J.; Ruipérez, F. $\pi\cdots H^{+}\cdots\pi$ hydrogen bonds and their lithium and gold analogues: MP2 and CASPT2 calculations. *ChemPhysChem* **2017**, *18*, 2409–2417.
108. Aarabi, M.; Gholami, S.; Grabowski, S.J. Hydrogen and lithium bonds—Lewis acid units possessing multi-center covalent bonds. *Molecules* **2021**, *26*, 6939.
109. Gholami, S.; Aarabi, M.; Grabowski, S.J. Proton and lithium cations linked to π-electron and σ-electron systems: are such interactions beyond or within the definition of hydrogen/lithium bond? *ChemPhysChem*, **2022**, *23*, e202200273.
110. Bader, R.F.W. *Atoms in Molecules: A Quantum Theory*, Oxford University Press, Oxford, **1990**.
111. Matta, C.J.; Boyd, R.J. (Eds.). *Quantum Theory of Atoms in Molecules: Recent Progress in Theory and Application*, Wiley-VCH, **2007**.
112. Cotton, F.A.; Matonic, J.H.; Murillo, C.A. Triply bonded Nb_2^{4+} tetragonal lantern compounds: some accompanied by novel B–H···Na^{+} interactions. *J. Am. Chem. Soc.* **1998**, *120*, 6047–6052.
113. Smith, M.; Li, Z.; Landry, L.; Merz, K.M. Jr.; Li, P. Consequences of overfitting the van der Waals radii of ions. *J. Chem. Theory Comput.* **2023**, *19*, 2064–2074.
114. Rowland, R.S.; Taylor, R. Intermolecular nonbonded contact distances in organic crystal structures: comparison with distances expected from van der Waals radii. *J. Phys. Chem.* **1996**, *100*, 7384–7391.

115. Grabowski, S.J.; Sokalski, W.A.; Leszczynski, J. Hydride bonding — ab initio studies of $BeH_2 \cdots Li^+$, $BeH_2 \cdots Na^+$ and $BeH_2 \cdots Mg^{2+}$ model systems. *Chem. Phys. Lett.* **2006**, *422*, 334–339.

116. Rozas, I.; Alkorta, L.; Elguero, J. Inverse hydrogen-bonded complexes. *J. Phys. Chem. A* **1997**, *101*, 4236–4244.

117. Gu, Y.; Kar, T.; Scheiner, S. Fundamental properties of the CH···O interaction: Is it a true hydrogen bond? *J. Am. Chem. Soc.* **1999**, *121*, 9411–9422.

118. Jabłoński, M. Binding of X–H to the lone-pair vacancy: Charge-inverted hydrogen bond. *Chem. Phys. Lett.* **2009**, *477*, 374–376.

119. Jabłoński, M. Charge-inverted hydrogen bond vs. other interactions possessing a hydridic hydrogen atom. *Chem. Phys.* **2014**, *433*, 76–84.

120. Jabłoński, M. Ten years of charge-inverted hydrogen bonds. *Struct. Chem.* **2020**, *31*, 61–80.

121. Etter, M.C. Encoding and decoding hydrogen-bond patterns of organic compounds. *Acc. Chem. Res.* **1990**, *23*, 120–126.

122. Madsen, G.K.H., McIntyre, G.J.; Schiøtt, B.; Larsen, F.K. The low-barrier hydrogen bond of deuterated benzoylacetone probed by very low temperature neutron and X-ray diffraction studies and theoretical calculations. *Chem. Eur. J.* **2007**, *13*, 5539–5547.

123. Madsen, G.K.H.; Iversen, B.B.; Larsen, F.K.; Kapon, M.; Reisner, G.M.; Herbstein, F.H. Topological analysis of the charge density in short intramolecular C–H···O hydrogen bonds. Very low temperature X-ray and neutron diffraction study of benzoylacetone. *J. Am. Chem. Soc.* **1998**, *120*, 10040–10045.

124. Sidgwick, N.V.; Callow, R.K. LXVII. — abnormal benzene derivatives. *J. Chem. Soc.* **1924**, *125*, 527–538.

125. Park, S.; Ramachandran, R.; Lough, A.J.; Morris, R.H. A new type of intramolecular H ··· H ··· H interaction involving N–H ··· H(Ir)··· H–N atoms. Crystal and molecular structure of $[IrH(\eta^1\text{-}SC_5H_4NH)_2(\eta^2\text{-}SC_5H_4N)(PCy_3)]BF_4 \cdot 0.72CH_2Cl_2$. *J. Chem. Soc. Chem. Commun.* **1994**, 2201–2202.

126. Lough, A.J.; Park, S.; Ramachandran, R.; Morris, R.H. Switching on and off a new intramolecular hydrogen-hydrogen interaction and the heterolytic splitting of dihydrogen. *J. Am. Chem. Soc.* **1994**, *116*, 8356–8357.

127. Peris, E.; Lee, J.C. Jr.; Rambo, J.R.; Eisenstein, O.; Crabtree, R.H. Factors affecting the strength of X-H···H-M hydrogen bonds. *J. Am. Chem. Soc.* **1995**, *117*, 3485–3491.

128. Park, S.H.; Lough, A.J.; Yap, G.P.A.; Morris, R.H. The effect of ancillary ligands on intramolecular proton-hydride (NH···HIr) bonding in complexes of iridium(III). *J. Organomet. Chem.* **2000**, *609*, 110–122.

129. Baker, A.W.; Shulgin, A.T. Intramolecular hydrogen bonds to π-electrons and other weakly basic groups. *J. Am. Chem. Soc.* **1958**, *80*, 5358–5363.
130. Schleyer, P. v. R.; Trifan, D.S.; Bacskai, R. Intramolecular hydrogen bonding involving double bonds, triple bonds and cyclopropane rings as proton acceptors. *J. Am. Chem. Soc.* **1958**, *80*, 6691–6692.
131. Gilli, G.; Bellucci, F.; Ferretti, V.; Bertolasi, V. Evidence for resonance-assisted hydrogen bonding from crystal-structure correlations on the enol form of the β-diketone fragment. *J. Am. Chem. Soc.* **1989**, *111*, 1023–1028.
132. Bertolasi, V.; Gilli, P.; Ferretti, V.; Gilli, G. Evidence for resonance-assisted hydrogen bonding. 2. Intercorrelation between crystal structure and spectroscopic parameters in eight intramolecularly hydrogen bonded 1,3-diaryl-1,3-propanedione enols. *J. Am. Chem. Soc.* **1991**, *113*, 4917–4925.
133. Gilli, G.; Gilli, P. *The Nature of the Hydrogen Bond*, Oxford University Press, Oxford, **2009**.
134. Alkorta, I.; Elguero, J.; Mó, O.; Yañez, M.; Del Bene, J.E. Do coupling constants and chemical shifts provide evidence for the existence of resonance-assisted hydrogen bonds? *Mol. Phys.* **2004**, *102*, 2563–2574.
135. Alkorta, I.; Elguero, J.; Mó, O.; Yañez, M.; Del Bene, J.E. Are resonance-assisted hydrogen bonds 'resonance assisted'? A theoretical NMR study. *Chem. Phys. Lett.* **2005**, *411*, 411–415.
136. Góra, R.W.; Maj, M.; Grabowski, S.J. Resonance-assisted hydrogen bonds revisited. Resonance stabilization vs. charge delocalization. *Phys. Chem. Chem. Phys.* **2013**, *15*, 2514–2522.
137. Thorsteinsson, T.; Cooper, D.L.; Gerratt, J.; Karadakov, P.B.; Raimondi, M. Modern valence bond representations of CASSCF wavefunctions. *Theor. Chim. Acta* **1996**, *93*, 343–366.
138. Thorsteinsson, T.; Cooper, D.L. Nonorthogonal weights of modern VB wavefunctions. Implementation and applications within CASVB. *J. Math. Chem.* **1998**, *23*, 105–126.
139. Jabłoński, M. A critical overview of current theoretical methods of estimating the energy of intramolecular interactions. *Molecules* **2020**, *25*, 5512.
140. Grabowski, S.J. Intramolecular hydrogen bond energy and its decomposition — O–H···O interactions. *Crystals* **2021**, *11*, 5.
141. Weinhold, F.; Landis, C. *Valency and Bonding: A Natural Bond Orbital Donor — Acceptor Perspective*, Cambridge University Press, Cambridge, **2005**.
142. Bent, H.A. An appraisal of valence-bond structures and hybridization in compounds of the first-row elements. *Chem. Rev.* **1961**, *61*, 275–311.

CHAPTER V

Comparison and First Classifications of Selected Lewis Acid — Lewis Base Interactions

Abstract: Interactions categorised as σ-hole bonds and π-hole bonds are compared with the hydrogen bond. The classifications and names are presented, a majority of which is based on the name of the atom, or the name of the group of the periodic table to which the interacting atom belongs; this atom would be the centre of the link between the Lewis acid and Lewis base units. The characteristics of selected interactions are described, and the tetrel bond is analysed in particular. The change of configuration of the Lewis acid centres that result from interactions are discussed; particular attention is paid to changes of the tetrahedral structures.

Keywords: σ-hole bond, π-hole bond, classification of interactions, hydrogen bond, tetrahedral configuration

V.1 First Classifications of Lewis Acid – Lewis Base Interactions

Among various interactions, the hydrogen bond is analysed most often. This was described in detail in the previous chapter, where its unique properties and importance in numerous chemical, physical and biological processes were mentioned.[1–3] Even in early studies, the similarities between the hydrogen bond bridge and other molecular arrangements were sought.

For example, the lithium bond was analysed as an analogous interaction to the hydrogen bond by Kollman very early on, in the seventies of the last century.[4] The lithium bond was briefly described in Chapter I. The most important is that di-coordinated lithium, i.e., the A-Li···B arrangement, occurs rarely,[4] in contrast to the hydrogen bond where such di-coordination is common and greater coordination occurs from time to time only, such as the bifurcated proton donor hydrogen bonds described in the previous chapter. It is worth mentioning here a study where systems containing di-coordinated and multi-coordinated proton, H^+, and H^- anion were discussed; one of the main characteristics of these ions is the large electron charge shifts that lead the charges of these ions to be far from +1 or –1 au.[5]

There are other differences between the hydrogen bond and the lithium bond that were briefly discussed in Chapter I.[6–8] The main characteristic of the A-Li···B lithium bond is that the Li-centre charge is very close to +1 au, while in the case of A-H···B hydrogen bonds the H-centre charge is positive but often far from +1 au due to the electron charge shifts related to the polarisation of the covalent A-H bond. Additionally, hydrogen bond formation usually leads to enhancement of this polarisation since the increase of positive charge of the H-atom is observed, in comparison with its charge in the A-H unit not involved in any interaction.

The halogen bond, A-X···B (X designates the halogen centre), seems to be a counterpart of the hydrogen bond since surprisingly it possesses more common characteristics with the hydrogen bond than does the lithium bond. Despite X being an electronegative atom, despite X usually possessing the dual character (i.e., it may act simultaneously as Lewis acid and Lewis base), and despite the electron structure of X being much more complicated than that of the H-centre, there are a lot of similarities between the hydrogen bond and the halogen bond. One can mention the linearity of the A-H/X···B bridge and the coordination of the X-centre, often two since X is covalently bonded to the A-centre and it may form, similar to the hydrogen in hydrogen bonds, weaker interaction with the B Lewis base site. It was justified that the hydrogen and halogen bonds are ruled by the same mechanisms.[9,10] Scheme V.1 presents results of the formation of both interactions; examples shown are the water dimer HOH···OH_2 linked by the O-H···O hydrogen bond, and the F_3CCl···OCH_2 complex linked by the C-Cl···O halogen bond. One can see that for both, the electron charge shift from the Lewis base unit to the Lewis acid is observed. And for both interactions, due to the rehybridisation process, further shifts of the electron charge from the H and Cl centres to the remaining parts of the Lewis acid units are observed. This leads to the increase of charge of the H and Cl centres that are in contact with the Lewis base sites. It means that the positive charge of the H-atom of the water molecule acting as the Lewis acid

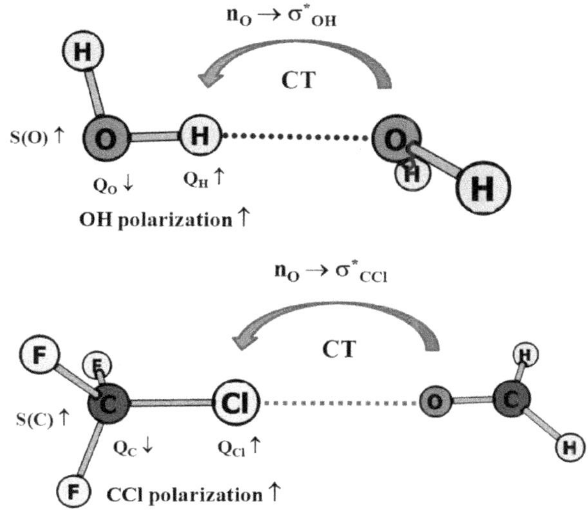

Scheme V.1 Changes in parameters resulting from hydrogen bond formation for the water dimer, $(H_2O)_2$, and halogen bond formation for the $CF_3Cl–OCH_2$ complex; Q denotes the atomic charge. Reprinted with permission from ref. 10. Copyright 2017 The Royal Society of Chemistry.

unit increases after hydrogen bond formation and the negative charge of the O-centre bonded to this hydrogen increases due to the electron charge shifts within the Lewis acid unit. It This seems to be the common process for hydrogen-bonded systems, although exceptions occur.[3] In the case of the A-X···B halogen-bonded system (as presented for the $F_3CCl···OCH_2$ complex in Scheme V.1), the increase of positive charge of the halogen centre (or decrease of its negative charge) occurs that is accompanied by the increase of negative charge of the A centre bound to X. For both interactions the most important overlaps are the n → σ interactions. And for both A-centres of the halogen bond and the hydrogen bond, the s-character of the hybrid atomic orbital of the A-centre increases. It agrees with the Bent rule, as was discussed for hydrogen bonds in the previous chapter.[11] The electron charge shifts described above lead to the increase of polarisation of the O-H and C-Cl bonds. It seems that the changes resulting from halogen bond or hydrogen bond formation are the same or at least similar for a majority of other types of σ-hole bonds.[10]

The properties of the halogen bond were discussed in Chapter I; however, one additional topic should be discussed. It concerns the question of whether the halogen bond may be considered, similar to the hydrogen bond, as the preliminary stage of a chemical process. The hydrogen bond often initialises the proton transfer reaction.[12] Is it possible in the case of the halogen bond that the halonium cation, which is much heavier than a proton, may transfer between two electronegative centres? Several recent studies are related to this topic. One can mention experimental studies supported by theoretical calculations where NMR, spectroscopic, and diffraction methods were used to analyse complexes containing halonium cation and other cations that are located between nitrogen centres.[13] These systems may be designated as $[NZN]^+$ where $Z^+ = H^+, Li^+, Na^+, F^+, Cl^+, Br^+, I^+, Ag^+$ and Au^+.[13] Two series of complexes were considered, one with two pyridine structures and another containing the (1,2-bis(pyridin-2-ylethynyl)benzene) structure. In almost all complexes the symmetrical positions of Z^+ cations are observed, except systems where $Z^+ = H^+$ and F^+ where asymmetric systems occur. For the fluorine species only theoretical calculations were performed without experimental confirmations. There are examples of asymmetric linear silver and halonium complexes characterised by $[NAgN]^+$ and $[NIN]^+$ bridges, respectively.[14] These interactions were analysed by ^1H and ^1H-^{15}N HMBC NMR spectroscopy, as well as by X-ray diffraction results in crystals. The latter study supports the possibility of halonium cation transfer and of the transfer of other cations mentioned above. One can see that the halonium cation transfer in halogen bonds is not as common as proton transfer in hydrogen bonds.[3,12] However, the fact that this transfer occurs indicates that halogen bonds possess numerous similar properties to hydrogen bonds.

Scheme V.2 shows interrelations between the halogen and hydrogen bonds on one hand and those of the hydride and halide bonds on the other hand. For the former interactions the central atom possesses properties of the Lewis acid centre, while for the latter interactions the central atom possesses properties of the Lewis base centre. It was described earlier in Chapter I that Lewis acid properties are revealed by positive electrostatic potential (EP) and not by positive charge; this means that the depletion of electron charge may lead to positive EP areas even if the total atomic charge is negative. In contrast, the accumulation of negative charge may lead to negative EP areas. Thus, contacts between these centres are possible if they are between sites characterised by opposite EP. In such a way the combination of the hydrogen bond and the hydride bond leads to the hydrogen-hydride bond, i.e., to the dihydrogen

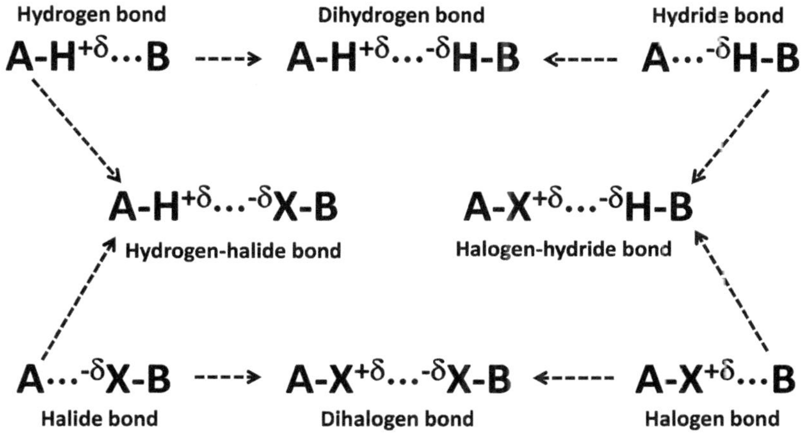

Scheme V.2 Categorisation of hydrogen and halogen bonds. Modified from ref. 10.

bond, an interaction where two hydrogen atoms are in contact. One may say that they are characterised by opposite charges. However, the more proper indication is that they possess opposite EPs, despite the fact that in the case of the light hydrogen centre its positive or negative EP is usually related to its positive or negative charge, respectively.

A similar situation occurs for the halogen bond where the site of the halogen centre possessing positive EP interacts with the site possessing negative EP. The latter site may be of another halogen centre. Hence two halogen centres may be in contact, and this is the combination of the halogen and halide bonds — the halogen-halide bond, i.e., the dihalogen bond.

However, there are meaningful differences between the dihalogen bond and its counterpart, the dihydrogen bond. Firstly, the hydrogen is either the Lewis acid centre or Lewis base while the halogen centre may possess dual character, i.e., it may act as Lewis acid and base simultaneously. Hence the dihalogen bond may be formed between the same centres of the same species, although contact between different species is also possible. Even more, the halogen-halogen contact concerns not only the same kind of elements. Hence not only chlorine-chlorine or bromine-bromine dihalogen bonds are observed, for example, but also mixed ones as bromine-iodine dihalogen bonds. Thus, there is a huge variety of dihalogen bonds in comparison with dihydrogen bonds.

Besides, according to the anisotropic distribution of electron charge of the halogen centre (see Scheme I.1 in Chapter 1), which is more manifested for heavier halogen elements, two types of halogen-halogen contacts are possible (Scheme V.3). For the first type, the R_1-X_1-X_2 and X_1-X_2-R_2 angles are equal to each other, at least approximately, and the $X_1 \cdots X_2$ contact may be treated as composed of two Lewis acid – Lewis base links. For the second type, only one Lewis acid – Lewis base link is observed and the R_1-X_1-X_2 and X_1-X_2-R_2 angles are equal to 90° and 180° (or nearly so), respectively. This situation occurs for the crystal structure of 4-iodobenzoic acid that was discussed in Chapter I (Fig. I.5);[15] the iodine centres act as the Lewis acid and Lewis base sites simultaneously. It is worth mentioning that the two types of dihalogen bonds presented in Scheme V.3 do not exhaust all cases. This is because the electrophilic site of the halogen centre, i.e., the σ-hole, may be in contact with the halogen anion (halide ion). However, it was mentioned earlier that according to the author of this book the dihalogen bond occurs if both halogen centres being in contact possess dual character.

For example, ωB97XD/aug-cc-pVTZ calculations were performed on complexes containing one or two diiodotetrafluorobenzene molecules and one halide ion,[16] and the C-I$\cdots$X$^-$ and C-I$\cdots$X$^-\cdots$I-C arrangements were analysed (X = F, Cl, Br and I). The

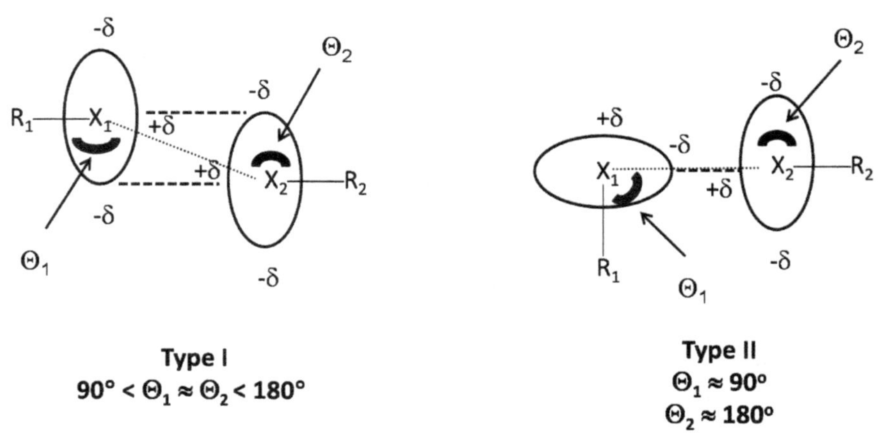

Scheme V.3 Two types of halogen-halogen interactions where mutual arrangements of halogen-halogen contacts depend on the electron charge distributions at halogen centres. Modified with permission from ref. 3. Copyright 2021 The Royal Society of Chemistry.

latter arrangements may be classified as the electron donor bifurcated halogen bond. Figure V.1 presents the complex of iodide anion with two molecules of 1,2,4,5-tetrafluoro-3,6-diiodobenzene. This is the case of the C-I···I⁻···I-C bifurcated interaction. One can see that the I-I-I angle is far from linearity; it is close to 90°. It is worth mentioning that the σ-hole for the iodine centre in 1,2,4,5-tetrafluoro-3,6-diiodobenzene molecule is located in the extension of the C-I bond. The non-linearity of the I-I-I system may be explained in the following way. For the iodide anion, the negative EP is equal for the whole surface characterised by the same electron density, say 0.001 au. For the complex of the iodide anion with the single 1,2,4,5-tetrafluoro-3,6-diiodobenzene molecule, the C-I···I^- angle is equal to 180° and less negative EP at the iodide anion is observed in the elongation of the C-I···I^- direction than at this anion in the direction perpendicular to the C-I···I^- line. Hence the next tetrafluoro-3,6-diiodobenzene molecule is attached to the complex in such a way that the I-I-I angle is close to 90°. This attachment is through the σ-hole at the iodine centre of the 1,2,4,5-tetrafluoro-3,6-diiodobenzene molecule and the more negative EP area in the iodide anion linked with another 1,2,4,5-tetrafluoro-3,6-diiodobenzene molecule.

It is interesting that for complexes analysed in this study two attractive interaction energy terms are the most important ones, electrostatic and orbital-orbital contributions;

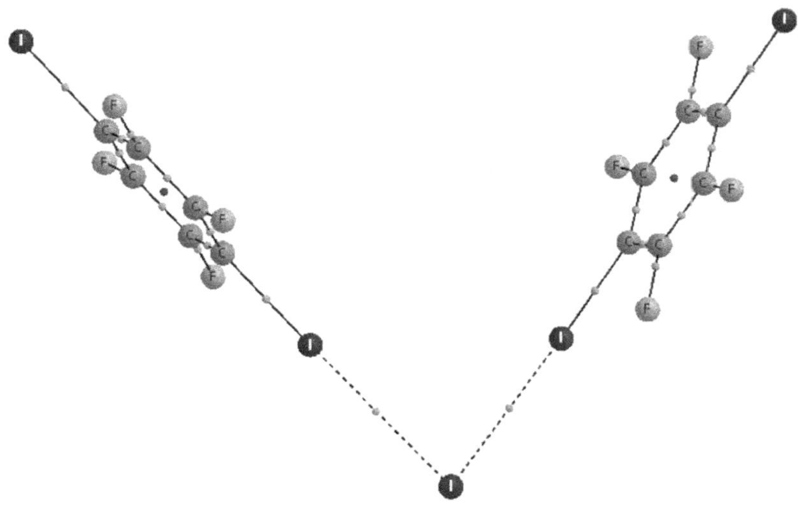

Figure V.1 The complex of iodide anion with two molecules of 1,2,4,5-tetrafluoro-3,6-diiodobenzene.

the latter is related to electron charge shifts. They are approximately equal to each other, and the dispersion term is negligible; in a majority of cases it is less than 1% of the sum of all attractive terms.[16]

Let us look again at interactions presented in Scheme V.2. The combination of the hydrogen bond with the halide bond leads to the hydrogen-halide interaction that is simply the hydrogen bond with the electronegative halogen, acting here by its site of the negative EP as the proton acceptor. In the case of another combination there is a sub-class of the halogen bond, the halogen-hydride bond. The latter interactions were analysed theoretically and also the experimental evidences of their occurrence in crystal structures were described.[17] The complexes of F_3CCl and F_3SiCl with hydrides such as LiH, BeH_2 and HBeF were analysed by calculations at different levels of approximation up to MP2/6-311++G(3df,3pd) and MP2/aug-cc-pVTZ. The chlorine centre acts as the Lewis acid site through the σ-hole situated in the elongation of the C-Cl or Si-Cl bond. The hydridic, negatively charged hydrogen in the hydrides mentioned above is the Lewis base centre. The complexes of F_3CCl form stronger links than the corresponding complexes of F_3SiCl. The complexes of LiH connected with the Lewis acid units are stronger than those of BeH_2 and HBeF. The strongest interaction occurs for the $F_3CCl\cdots HLi$ complex where the binding energy amounts to –3.5 kcal/mol. For the remaining complexes weaker interactions occur, where absolute values of the binding energy do not exceed 2.2 kcal/mol.

Scheme V.2 shows the halogen and hydrogen bonds, their counterparts with the Lewis base centre (halide and hydride bonds), and their combinations. Nevertheless, numerous combinations may be created for any other pair of interactions. For example, the Si–H···Se interaction in 1-mesitylselanyl-8-(dimethylsilyl)naphthalene was analysed and the X-ray crystal structure was determined accompanied by results of different theoretical approaches.[18] This Si–H···Se interaction was classified as a chalcogen-hydride bond.[18]

Table V.1 presents different types of interactions; their names that often appear in various studies are given. A similar classification of interactions was presented in an early study, however not all names given in Table V.1 were presented therein.[10] One can see (Table V.1) that names of interactions are mainly related to the Lewis acid

Table V.1 Types of interactions that most often appear in literature; acidic central atoms are not marked while grey areas correspond to basic central atoms. Bolded interactions correspond to those where the Lewis acid centre is the group element.

Group	Name of interaction	Central atom
1	Hydrogen bond	H
1	Hydride bond	H
1*	Dihydrogen bond	H...H
1	**Alkali bonds**	**Li, Na, K, Rb, Cs**
1	Lithium bond	Li
1	Sodium bond	Na
2	**Alkaline-earth bond**	**Be, Mg, Ca, Sr, Ba**
2	Beryllium bond	Be
2	Magnesium bond	Mg
10	**Regium bond**	**Ni, Pd, Pt**
11	**Regium bond**	**Cu, Ag, Au**
11	Gold bond	Au
12	**Spodium bond**	**Zn, Cd, Hg**
13	**Triel bond**	**B, Al, Ga, In, Tl**
13	Boron bond	B
14	**Tetrel bond**	**C, Si, Ge, Sn, Pb**
14	Carbon bond	C
15	**Pnicogen bond**	**N, P, As, Sb, Bi**
16	**Chalcogen bond**	**O, S, Se, Te, Po**
17	**Halogen bond**	**F, Cl, Br, I, At**
17	Fluorine bond	F
17**	Dihalogen bond	X···X (X: halogen)
17	**Halide bond**	**F, Cl, Br, I, At**
18	**Aerogen bond**	**Ne, Ar, Kr, Xe, Rn**

*In other words, it is the hydrogen-hydride bond.

**Halogen-halide bond is an example.

properties of the central atom. In the case of the A-X···B halogen bond or the A-H···B hydrogen bond, the X-halogen or H-hydrogen centre possesses Lewis acid properties and interacts with the B electron donor. The names are related to the groups the element belongs to; for example, the tetrel bond or the alkaline-earth bond concerns the Lewis acid properties of the 14th and 2nd group of elements, respectively. However, there are also names related to the single elements; the carbon bond, beryllium bond or hydrogen bond are examples of such interactions. The names related to groups are bolded in Table V.1 while those concerning single elements are not.

In the grey areas, interactions with names coming from the Lewis base centre are presented. These are the hydride and halide bonds, for example. The "mixed" interactions are also presented in this table, i.e., the dihydrogen and dihalogen bonds. Figure V.2 presents a classification similar to that presented in an earlier study[10] but for better clarity, interactions are presented with the periodic system.[19] The names of interactions related to groups are shown, but a few interactions that concern single elements are also marked: the hydrogen bond — HB, the lithium bond — LiB and the beryllium bond — BeH. There are no interactions related to Lewis base properties of elements in this figure, i.e., the halide and hydride bonds.

It is worth to refer here to the study of Joy and Jemmis,[20] who analysed in detail the Lewis acid properties of transition metal centres. They justified that elements of the 12th group may participate in Lewis acid — Lewis base interactions, playing the role of electrophiles. Such interactions are named as spodium bonds (Table V.1 and

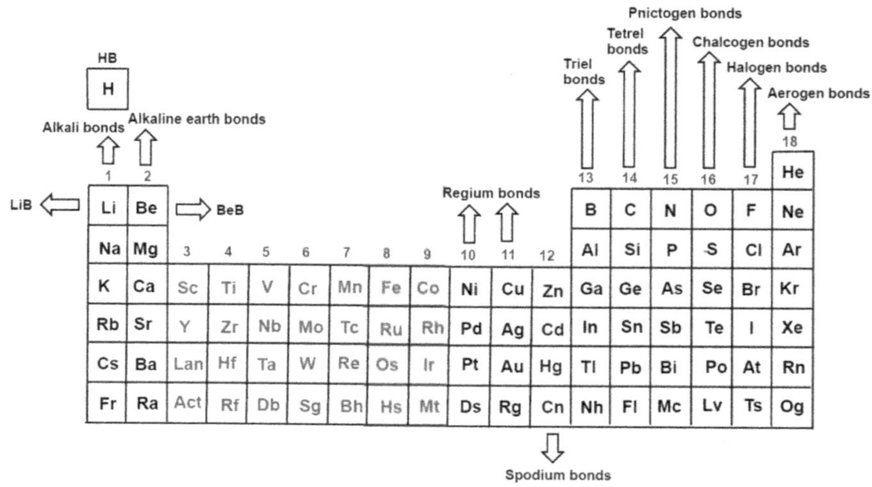

Figure V.2 Classification of interactions based on Lewis acid centres. Reprinted with permission from ref. 19. Copyright 2020 MDPI.

Fig. V.2). However, the elements of groups 3–10 rarely participate as Lewis acid centres in interactions. The latter was justified by results of searches through the Cambridge Structural Database.[21,22]

V.2 Counterparts of Hydrogen, Halogen and Triel Bonds

Table V.2 presents examples of interactions which are analysed very often in various experimental and theoretical studies. These are hydrogen, halogen and triel bonds — they are compared among themselves. For these three types of interactions, the electrostatic and polarisation forces are the most important attractive components (see eq. I.13 in Chapter I). The polarisation forces are understood here as those related to electron charge shifts; in different decomposition schemes other names are applied, such as induction, charge transfer, delocalisation, etc.[27] Polarisation is related in various studies to the covalent character of interaction; this term is very important for the triel bonds where very often the electron-deficient species, such as boron trihalide, interacts with the unit that supplies the missing electron pair. Such interaction often possesses characteristics of a covalent bond. The polarisation/electrostatic ratio of forces increases with the increase of the strength of interaction, consequently often with the decrease of the length of intermolecular contact. It is important that these relationships related to the interaction energy terms, strength of interaction and the intermolecular distance are fulfilled for other σ-hole bonds such as the pnicogen, tetrel and chalcogen bonds.[28] The dispersion attractive forces are not as significant; however, they may play an important role for weaker interactions.

The hydrogen bond may be classified as a σ-hole bond.[29] The halogen bonds are usually classified as σ-hole bonds, but π-hole halogen bonds are also possible.[30] In the case of triel bonds, most often π-hole interactions are observed, but σ-hole triel bonds are also possible.[31,32] For these three interactions, the mechanisms of formation are the same or at least similar. The hyperconjugation is related to the n(B) → σ^*_{AH} and n(B) → σ^*_{AX} orbital-orbital interactions for the A-H···B hydrogen bond and the A-X···B halogen bond, respectively. In the case of the Z-T···B triel bond, where T marks the triel Lewis acid centre while Z is a centre connected with the triel atom, the n(B) → σ^*_{ZT} orbital-orbital interaction occurs. However, the situation is slightly different here than for the other two interactions. The T centre is not monovalent; usually it is trivalent, thus three such orbital-orbital interactions occur and additionally

Table V.2 Comparison of the hydrogen bond with the halogen and triel bonds.

Properties	**A-H···B hydrogen bond**	**A-X…B halogen bond**	**T…B**
Forces	Electrostatic and polarisation (covalent) terms are accompanied by dispersion.		
Nature	Importance of covalency; the polarisation/electrostatic ratio increases with the increase of strength of interaction.		
Type	σ-hole bond	σ-hole bond most often π-hole bond sometimes	π-hole bond most often σ-hole bond sometimes
Mechanism	hyperconjugation and rehybridisation		
Orbital-orbital interaction	$n(B) \rightarrow \sigma^*_{AH}$	$n(B) \rightarrow \sigma^*_{AX}$	$n(B) \rightarrow \sigma^*_{TZ}$ and $n(B) \rightarrow n(T)^*$
Absolute value of energy of interaction (kcal/mol)	2–40	Strength increases in the order: Cl < Br < I, comparable with hydrogen bonds.	1–120
Directionality	Directional	Directional, more than hydrogen bond	Directionality not so strictly observed
Influence of interaction on the A-H/X/T bond length	Usually increases — red shift, sometimes decreases — blue shift.	Increases or decreases; blue shift occurs more often than in hydrogen bonds.	Concerns a few bonds of the triel centre; systematic changes are not observed.
H/X/T electronegativity[23]	H – 2.2	F – 3.98; Cl – 3.16; Br – 2.96; I – 2.66	B – 2.04; Al – 1.61; Ga – 1.81; In – 1.78; Tl – 1.62
van der Waals radius (Å)[24–26]	H – 1.1	F – 1.47; Cl – 1.75; Br – 1.83; I – 1.98	B – 1.92; Al – 1.84; Ga – 1.87; In – 1.93; Tl – 1.96
H/X/T charge	Positive	Usually negative	Positive
H/X/T electrostatic potential	Positive for the whole hemisphere.	Usually positive in the extension of A-X bond; negative in the orthogonal direction to A-X bond.	Positive for the T π-hole region.
Acidity/basicity of H/X/T	Acidic	Usually dual acidic/basis character of the same X centre	Acidic and acidic/basic for heavier triel atoms
H/X/T valency	Monovalent	Monovalent or multivalent	Multivalent
Cooperativity	Occurs		
Bifurcation	Occurs		
Preliminary stage of reaction	Proton transfer	X^+ transfer, very rare process	Release of molecular hydrogen

the n(B) → n(T)* overlap occurs where n(T)* is the antibonding orbital of the lone electron pair of the triel centre. For these three interactions, the rehybridisation process that is related to further electron charge shifts in the Lewis acid units occurs.

It is interesting to analyse the strength of interactions. In the case of the hydrogen bond, the absolute value of the energy of interaction is in the range of 2 kcal/mol and 40 kcal/mol as often presented in various studies.[1–3] In the case of the halogen bond, fluorine as the Lewis acid centre is rather controversial, but in several studies halogen bonds with the fluorine centre were investigated.[33–36] In the case of chlorine centres, the interactions are usually weaker than their hydrogen-bonded counterparts, while halogen bonds with bromine and iodine centres are usually stronger than the corresponding hydrogen bonds. It seems that the range of strength of halogen bonds is similar to that of hydrogen bonds. It was found for the broad spectrum of various halogen bonds analysed that the absolute value of the binding energy is between 1 kcal/mol and 45 kcal/mol.[37] It was found in another study that the halogen bonds are characterised by energies up to approximately 79 kcal/mol.[38] These ranges depend on the level of calculation and on the samples of complexes that are investigated.

The strength of the triel bonds is discussed in several studies; one may mention the study on these interactions in the gas phase and in crystals.[31] It was discussed that the complexes of boron trihydride and boron trihalides with ammonia and other electron-donating units are linked by strong triel bonds. The boron in these isolated Lewis acid units is characterised by the electron vacancy, thus it acts as the electrophilic centre. Figure I.1a (Chapter I) presents the $F_3B\cdots NH_3$ complex as an example. The absolute values of energies of interaction for such complexes are often over 100 kcal/mol that is typical for covalent bonds. The other characteristics of the B···N links also indicate their covalent character.

The A-H···B hydrogen bonds and A-X···B halogen bonds are both directional interactions, which means that the A-H-B and A-X-B angles, respectively, tend to linearity. It was analysed in studies of Politzer and coworkers in terms of the σ-hole and π-hole concept[39–42] that the halogen bond is often closer to linearity than the hydrogen bond.[39] The reason is that for the latter interaction, the EP is positive for the hydrogen hemisphere. In the case of the halogen bond, the part of the halogen sphere that is usually characterised by positive EP is often only a small area at the elongation of the A-X bond. Hence, in comparison with the whole hemisphere of hydrogen, linearity for the halogen bonds occur more often. In the case of the triel centres, they are often multivalent and the π-holes possessing positive EP are observed for the areas above and below the plane of the triel centre and neighbouring atoms. The electron-donating unit is directed to such areas. Thus, it is difficult to discuss about the linearity of the triatomic system in the case of triel bonds. The same occurs for the halogen bonds if a multivalent centre is involved in such interaction. In such a case, very often, π-hole halogen bonds are observed. Information on the areas of positive EP as well as the valency of hydrogen, halogen and triel centres are included in Table V.2.

Table V.2 presents the van der Waals radii of the hydrogen, halogen and triel atoms.[24–26] It is worth mentioning that for the corresponding interactions, at least for the strong and medium interactions, contacts of the Lewis acid — Lewis base centres are shorter than the corresponding sum of van der Waals radii.[10,31] These interactions are also treated as the preliminary stages of chemical reactions. In the case of the hydrogen and halogen bonds, these are the proton transfer[3,43] and the halonium cation transfer,[13,14] respectively. The situation in the case of the triel bonds is much more complicated. The reactions of adducts linked by the triel bonds may lead to the release of molecular hydrogen.[44] This may be the reaction of ammonia borane, or molecular hydrogen may be released from ammonium borohydride, diammoniate of diborane and polyamino borane.[45]

Table V.2 indicates that the hydrogen bond formation leads to lengthening of the A-H proton-donating bond that causes a red shift of the corresponding stretching mode. Sometimes the shortening of the A-H bond is observed as a result of complexation, resulting in a blue shift of the stretching mode.[46] In the case of the halogen bond, the elongation and shortening of the A-X bond occur but the shortening is observed more often than in the case of the hydrogen bond.[47]

Table V.2 indicates that the cooperativity and bifurcated effects occur not only in the case of the hydrogen bond but also for the halogen bond[48,49] and the triel bond.[31] For example, cooperativity in the $H_2CO\cdots ClF\cdots ClF$ cluster as well as in systems containing a greater number of the ClF units was analysed in an early study.[49] Figure V.3a shows the molecular graph of the $H_2CO\cdots ClF\cdots ClF$ cluster; the isolines of Laplacian of

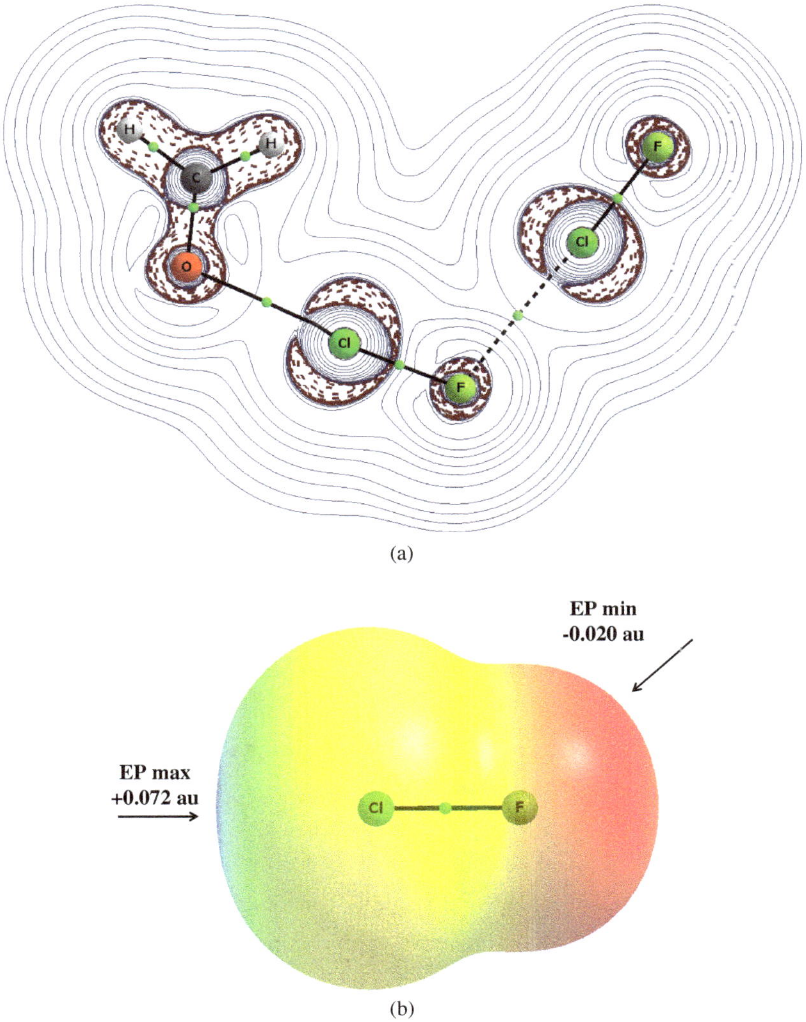

Figure V.3 (a) The molecular graph of $H_2CO\cdots ClF\cdots ClF$ cluster. The isolines of Laplacian of electron density are also presented; continuous lines correspond to positive Laplacian values and broken lines to negative values. (b) The molecular graph of the ClF unit. The EP map is presented here for the molecular surface of the 0.001 au electron density. Reprinted with permission from ref. 48. Copyright 2018 The Royal Society of Chemistry.

electron density are also given. Figure V.3b displays the molecular graph of the ClF unit; the EP map is presented here for the molecular surface characterised by the electron density of 0.001 au. The latter figure shows the polarisation of the Cl-F bond due to the difference between electronegativities of chlorine and fluorine centres. The difference leads to the greater shift of electron density to the F-centre and to the negative EP at this centre. Positive EP at chlorine is observed for this ClF unit. Consequently, this unit may act as the Lewis base through the fluorine and as the Lewis acid through the chlorine by the σ-hole located in the elongation of the F-Cl bond. This is why the chlorine is attached to the oxygen of the H_2CO species and the ClF molecule may be attached to the fluorine of $H_2CO\cdots ClF$ complex by the Cl-centre.

Figure V.3a shows that the positive EP at Cl-centre in ClF units is accompanied by the depletion of electron charge (in extension of the F-Cl bond) since positive Laplacian is observed in this area. MP2/aug-cc-pVTZ calculations were performed for these systems and O···Cl distances equal to 2.558 Å and 2.481 Å were found for the $H_2CO\cdots ClF$ and $H_2CO\cdots ClF\cdots ClF$ complexes, respectively.[49] For these systems full optimisations were performed without any geometrical constraints. The interaction energies for these complexes were calculated to be equal to –4.14 kcal/mol and –5.20 kcal/mol, respectively (these energies were corrected for the basis set superposition error). It is worth mentioning that for both complexes the interaction energies were calculated in such a way that the H_2CO species is one unit and the remaining ClF molecule/molecules is/are treated as the second unit. Hence one can see the strengthening of the O···Cl interaction for the system containing two ClF molecules if it is compared with the complex with one ClF molecule. This is the positive cooperativity effect.

A similar enhancement of interaction is observed for numerous systems linked by hydrogen bonds (see Chapter IV). For example, for the $H_2CO\cdots(HF)_n$ clusters the strength of the O···H interaction increases with the increase of the number of HF units in the system.[50] The cooperativity effects for systems linked by different types of interactions were analysed with the use of various theoretical approaches.[51] Thus it is a phenomenon occurring not only for the hydrogen-bonded systems.

V.3 Tetrahedral Centres as Lewis Acid Sites

Numerous studies show that interactions often lead to changes of configurations.[10,52–55] The only condition of the occurrence of such changes is that interactions are sufficiently strong. For example, the triel centres characterised by trigonal configurations may change their configurations into tetrahedral ones if interacting with Lewis base units.[31] However, the configuration in the complex is usually intermediate between trigonal, which corresponds to the isolated Lewis acid moiety, and tetrahedral, which occurs in the complex linked by a very strong and often covalent interaction, i.e., the triel bond. For example, the latter situation occurs for the boron trihydrides and trihalides interacting with strong Lewis base units,[31,56] such as fluoride and chloride anions.

It is interesting to analyse changes of configurations of tetrel, pnicogen, chalcogen and halogen centres that may possess tetrahedral structures. The following characteristics of tetrahedral centres may be mentioned: the tetrel centres possess four covalent bonds, pnicogen centres have three bonds and one electron pair, chalcogen centres possess two bonds and two electron pairs and for the halogen centres one bond and three electron pairs are observed (see Scheme V.4).[55] One may assume that these centres are characterised approximately by a tetrahedral structure and sp^3 hybridisation.

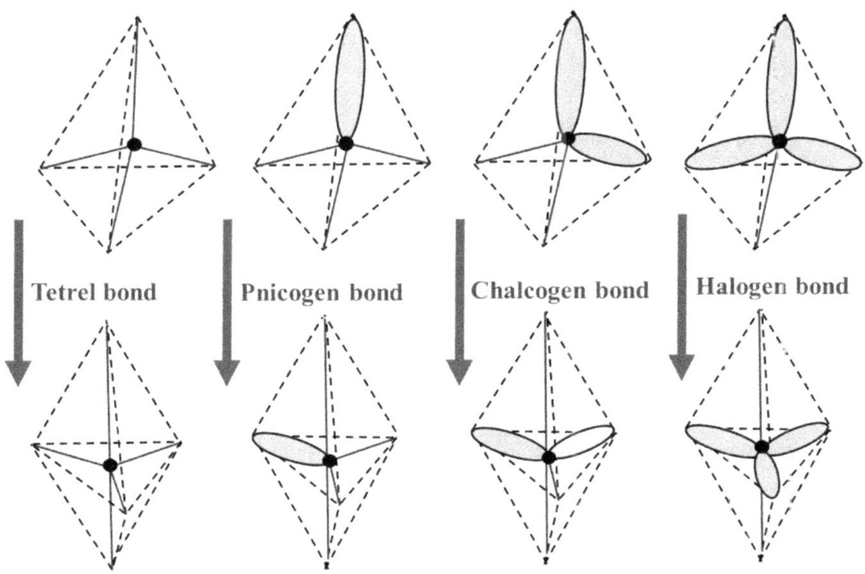

Scheme V.4 Changes of the tetrahedral configuration into the trigonal bipyramid resulting from the formation of tetrel, pnicogen, chalcogen and halogen bonds. Continuous lines correspond schematically to covalent bonds (or strong interactions) and grey ellipsoids to lone electron pairs. Reprinted with permission from ref. 55. Copyright 2022 MDPI.

However, this is only an approximate and model assumption. The ideal tetrahedral structure is observed for the tetrel centre that is bound with four equivalent substituents; the methane species is the simplest example of such a situation. Different substituents attached to the tetrel centre cause disturbance of the tetrahedral structure. The same occurs for the pnicogen, chalcogen and halogen centres where the lone electron pairs disturb the ideal tetrahedral structure.[28] For example, the water molecule in the gas phase is characterised by C_{2v} symmetry if positions of atoms are considered. For the water structure two O-H covalent bonds and two lone electron pairs occur. In the case of sp^3 hybridisation of the oxygen centre of water, two equivalent O-H bonds and two equivalent lone electron pairs are directed to vertices of a tetrahedron (see Scheme V.4 — the case of the chalcogen centre) that is deformed but the C_{2v} symmetry is still preserved even if one takes into account not only the positions of atoms but also the electron structure of water. Such a picture of water is often presented in numerous monographs and scholarly books. However, for the isolated water molecule the lone electron pairs are not equivalent, and they are not fitted into sp^3 hybridisation.[57] On the other hand, the arrangement of molecular species around water molecules agree approximately with sp^3 hybridisation of water oxygen. That is observed in numerous crystal structures.

It is worth mentioning that in the hydrogen and triel bonds, the hydrogen and triel centres are usually positively charged. A different situation is observed for charges of halogen centres where they are usually negatively charged, but halogen centres of positive charge are also observed sometimes. However, it was pointed out that the charge is not informative of Lewis acid or Lewis base properties of the centre considered.[29,39,40] A similar situation occurs for electronegativities (Table V.2),[23] where the hydrogen and triel centres possess low values of electronegativity while the halogen centres are characterised by high electronegativities. The latter may confirm the Lewis base properties of halogen atoms. However, they usually possess duel character. They act as Lewis acids due to the occurrence of σ-holes and, sometimes for multivalent halogens, due to the π-holes. On the other hand, the halogens also act as Lewis base centres due to the lone electron pairs. The hydrogen and triel centres possess acidic character; heavier triel centres reveal basic character sometimes.[31]

Similarly, sp^3 hybridisation does not occur in halogen centres. However, the arrangement of molecules is often in agreement with this hybridisation.[52,54] For example, the HF dimer possesses a structure that corresponds to this hybridisation for the fluorine centre; it is not a linear complex.[52]

Scheme V.4 shows the tetrahedral centres that may be transformed into trigonal bipyramids due to interactions, particularly σ-hole bonds. These interactions may be often treated as preliminary stages of S_N2 reactions; mainly it is observed for the tetrel bonds where the tetrahedral tetrel centres are characterised by a lack of lone electron pairs.[58] However, the number of lone electron pairs may be reduced if they are in contact with electrophiles. For example, the species containing tetrahedral pnicogen centre, with one electron pair, may interact with a cation that leads to a system similar to the tetrel tetrahedral centre that does not contain lone electron pairs. Hence, neutral tetrel centres may occur as well as positively charged pnicogen tetrahedral centres; both are characterised by four covalent bonds. The NH_4^+ cation is an example, and other cations containing tetrahedral pnicogen centres possess similar characteristics. The tetrahedral chalcogen and halogen centres possess two and three lone electron pairs, respectively. Hence, corresponding tetrahedral cations containing only covalent bonds are rather rare. One can refer to the H_4O^{2+} cation that was analysed theoretically as well as its counterparts that were investigated with the use of experimental tools.[59]

Figure V.4 presents examples of neutral species containing the tetrel tetrahedral centre and cations containing the tetrahedral pnicogen centre that interact with nucleophiles, NCH and NCLi moieties. The latter Lewis base units interact with tetrel or pnicogen centres by the lone electron pair of the nitrogen. The selected geometrical and energetic parameters of complexes linked by the pnicogen and tetrel bonds are presented in this figure. These are examples of the greater sample of complexes that were optimised at the MP2/aug-cc-pVTZ level and analysed in detail.[53]

As mentioned above, the pnicogen and tetrel interactions presented in Fig. V.4 may be treated as preliminary stages of reactions that lead to the change of tetrahedral configuration into trigonal bipyramid (Scheme V.4). The $NH_4^+\cdots NCH$ complex linked by the N-H···N positively charge-assisted hydrogen bond, CAHB(+), is also presented in Fig. V.4 for comparison (Fig. V.4a). The H···N distance in the linear N-H···N system is equal to 1.742 Å, much less than the corresponding sum of van der Waals radii, 2.65 Å (see Table I.1 in Chapter I). Hence it is a rather strong hydrogen bond that is also confirmed by other parameters of this complex since the energy of

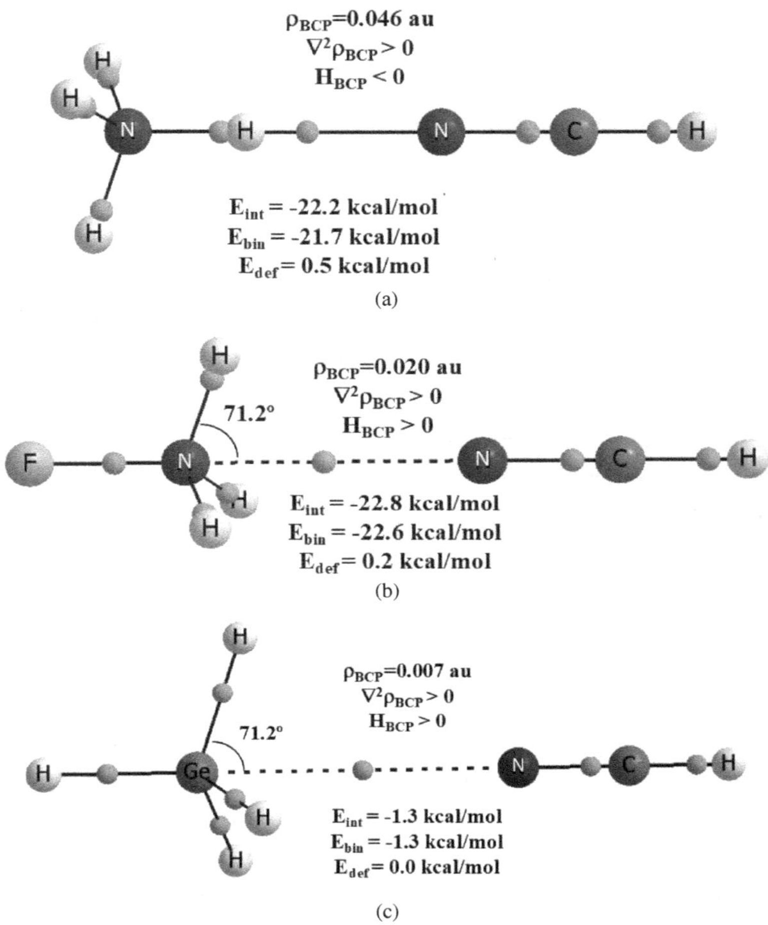

Figure V.4 Examples of molecular graphs of the tetrel-, pnicogen- and hydrogen-bonded complexes; big circles correspond to attractors, small circles to BCPs. The QTAIM characteristics of BCPs are included. Reprinted with permission from ref. 53. Copyright 2019 Springer.

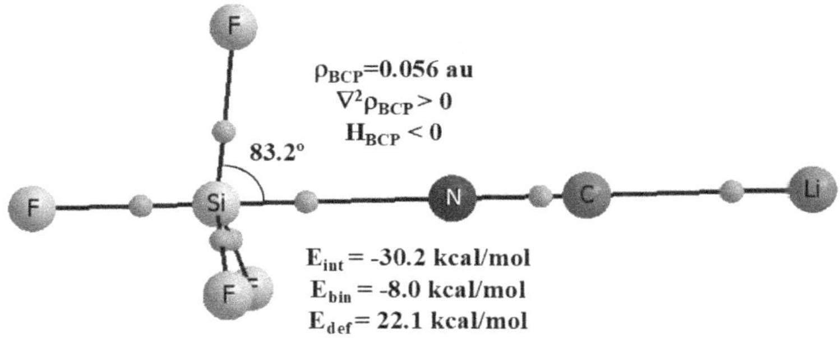

(d)

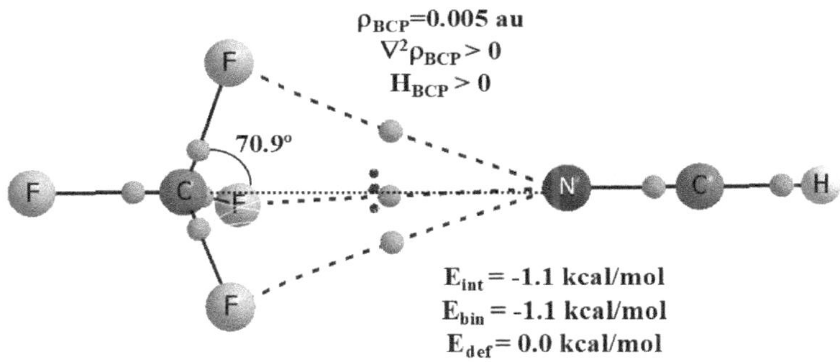

(e)

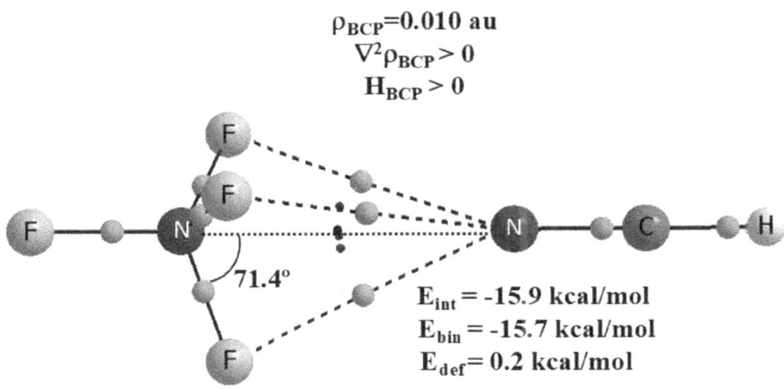

(f)

Figure V.4 *(Continued)*

interaction, E_{int}, amounts to -22.2 kcal/mol, and the binding energy, E_{bin}, is equal to -21.7 kcal/mol that indicates a rather small value of the deformation energy, E_{def}, of 0.5 kcal/mol. This energy is mainly the result of elongation of the N-H bond of the ammonium cation because of hydrogen bond formation. The topological parameters of the bond critical point (BCP) of the H···N bond path also show that the corresponding hydrogen bond is strong: the electron density at BCP, ρ_{BCP}, is equal to 0.046 au, the Laplacian of the electron density, $\nabla^2\rho_{BCP}$, is positive that means it is not a covalent bond but the total electron energy density at BCP, H_{BCP}, is negative that means it is partly covalent in nature.

The NFH_3^+···NCH complex (Fig. V.4b) is linked by the N···N pnicogen bond. The nitrogen centre of the NFH_3^+ cation acts as the Lewis acid site through the σ-hole situated in the extension of the F-N bond while the nitrogen of the hydrogen cyanide unit is the Lewis base site due to the occurrence of the lone electron pair. This electron pair at the nitrogen centre occurs for both NCH and NCLi Lewis base units that participate in all interactions presented in Fig. V.4, and also in the hydrogen-bonded system analysed previously. The EP at the N-centre of the NFH_3^+ unit, at the σ-hole located in the extension of the F-N bond, is equal to +0.308 Hartree for the surface of 0.001 au. The latter confirms the Lewis acid properties of this nitrogen centre. This EP was calculated for the NFH_3^+ unit optimised separately. In the same way, for the other complexes shown in Fig. V.4 and discussed later, the EP values for separated Lewis acid units and for the surfaces of 0.001 au will be presented.

The linear F-N···N system is observed here for the NFH_3^+···NCH complex with the N···N distance equal to 2.557 Å, less than the corresponding sum of van der Waals radii, 3.10 Å (Table I.1). All F/H-Z···N systems linked by pnicogen and tetrel bonds that are presented in Fig. V.4 are linear arrangements (Z designates the pnicogen or tetrel centre), thus for the next complexes this information will not be repeated. For the NFH_3^+···NCH complex the E_{int} and E_{bin} values are equal to –22.8 kcal/mol and –22.6 kcal/mol, respectively, thus a negligible E_{def} of 0.2 kcal/mol occurs here. It corresponds to negligible geometrical deformations resulting from the complex formation and it is reflected by the low value of the angle, 71.2°.

This angle for the NFH_3^+···NCH complex and for complexes considered further here is between the line containing the F/H-Z bond that crosses the σ-hole at the Z-centre participating in the pnicogen or tetrel bond and one of the remaining F/H-Z bonds of the Lewis acid unit; it is designated as the α-angle henceforth. The α angle for the ideal tetrahedron not disturbed by additional interactions is equal to 70.5°. Very strong interactions lead to trigonal bipyramids characterised by the α angle equal to 90°.

The topological parameters of the N···N BCP for the NFH_3^+···NCH complex indicate non-covalent character of this pnicogen bond: $\nabla^2\rho_{BCP}$ and H_{BCP} at BCP are positive, and ρ_{BCP} is equal to 0.020 au. This is slightly not in agreement with the energetic parameters that indicate a rather strong pnicogen bond.

In the case of the GeH_4···NCH complex (Fig. V.4c), all parameters reveal the occurrence of a weak Ge···N tetrel bond. The α angle is equal to 71.2° that is far from the 90.0° of the structure of a trigonal bipyramid. And practically there are no deformations of species participating in this interaction; E_{int} and E_{bin} are both equal to –1.3 kcal/mol. The GeH_4 species acts as the Lewis acid unit through the σ-hole located in the elongation of the H-Ge bond. The EP for this σ-hole is equal to +0.026 Hartree. The Ge···N distance amounts to 3.276 Å, slightly greater than the corresponding sum of van der Waals radii, 3.21 Å. It means that the GeH_4···NCH complex is stabilised by very weak forces, mainly dispersion forces. The parameters of the Ge···N BCP also show that this tetrel bond is very weak: $\nabla^2\rho_{BCP}$ and H_{BCP} at BCP are positive, and ρ_{BCP} is equal to only 0.007 au.

The SiF_4···NCLi system presented in Fig. V.4d is an interesting example of a complex linked by a strong Si···N tetrel bond. The EP at the σ-hole of the Si-centre located in the extension of the F-Si bond in the SiF_4 species is equal to 0.073 Hartree. The E_{int} and E_{bin} values are equal to –30.2 kcal/mol and –8.0 kcal/mol, respectively, thus a huge E_{def} of 22.1 kcal/mol occurs here. The deformation of units in this complex concern mainly the Lewis acid SiF_4 species since the α angle is equal to 83.2°, close to the trigonal bipyramid. It is worth mentioning that for the larger sample analysed in the study on tetrahedral systems,[53] a few systems where the α angle is closer to 90° were analysed. For example, for the AsF_4^+···NCLi system, the E_{int} and E_{bin} values are equal to –112.0 kcal/mol and –78.6 kcal/mol, respectively, thus the E_{def} is equal to 33.4 kcal/mol. The huge deformations are mainly related to the Lewis acid AsF_4^+ cation since the α angle is equal to 86.7°.

Returning to the SiF_4···NCLi complex, the Si···N distance is equal to 2.065 Å while the corresponding sum of van der Waals radii amounts to 3.2 Å. The topological parameters confirm that this tetrel bond is a strong interaction since the $\nabla^2\rho_{BCP}$ is positive but the H_{BCP} value at the Si···N BCP is negative indicating partly covalent interaction; ρ_{BCP} is equal to 0.056 au. The strong Lewis acid properties of the Si-centre in this complex result from the electron-withdrawing properties of the fluorine substituents that cause the silicon centre to be largely devoid of electron charge.

Let us look at the CF_4···NCH complex (Fig. V.4e) where the fluorine substituents withdraw the electron charge from the carbon centre, similar to the SiF_4 species. However, the carbon centre is smaller, thus it is covered to a greater extent by substituents than the silicon centre. Hence there is no direct interaction and consequently no direct bond path between the carbon centre of the CF_4 unit and the nitrogen centre of the NCH species. However, three F···N bond paths between pairs of negatively charged centres are observed here that may seem to be controversial. This may be one of the strange examples where the bond paths do not correspond to the local stabilising interactions.[60–63] The characteristics of F···N BCPs show that the corresponding local interactions are very weak: the $\nabla^2\rho_{BCP}$ and H_{BCP} values are positive, and ρ_{BCP} is equal to 0.005 au for all these three BCPs. It is worth mentioning that the EP at fluorine centres is equal to –0.002 Hartree that indicates weak electrostatic F···N repulsions. On the other hand, attractive dispersion forces may occur for these contacts. There is no C···N bond path; however, positive EP at the C-centre is observed that is equal to +0.034 Hartree. This indicates attractive C···N electrostatic interaction. The α angle is equal to 70.9° that corresponds to the lack of deformations of the interacting units. The E_{int} and E_{bin} energies are equal to –1.1 kcal/mol for this weakly bound complex. The C···N intermolecular distance is equal to 3.363 Å, more than the corresponding sum of van der Waals radii, 3.25 Å.

At first glance, a similar situation occurs for the NF_4^+···NCH system (Fig. V.4f). However, for the whole surface of the NF_4^+ cation, positive EP values are observed, +0.205 Hartree for fluorine centres and +0.289 Hartree for the N-centre in elongations of F-N bonds. It means that the F···N bond paths correspond here to stabilising electrostatic interactions; the $\nabla^2\rho_{BCP}$ and H_{BCP} at the F···N BCPs are positive, but the ρ_{BCP} value is equal to 0.010 au, more than in the case of the CF_4···NCH complex. The α-angle is equal to 71.4° that means geometrical deformations are not significant, and the E_{def} energy is equal to only 0.2 kcal/mol. The E_{int} and E_{bin} energies amount to −15.9 kcal/mol and −15.7 kcal/mol, respectively. The N···N distance is equal to 2.933 Å, less than the corresponding sum of van der Waals radii, 3.10 Å. Thus, the replacement of the CF_4 species by the NF_4^+ cation in the complex of the NCH species does not change the molecular graph. However, for the complex with the cation the local stabilising interactions are observed without any doubt, while for the complex with the neutral Lewis acid unit the local stabilising interactions are controversial.

V.4 Tetrel Bond — From Tetrahedron to Trigonal Bipyramid and Octahedron Structures

The tetrel bond is one of the most important interactions that is analysed in numerous studies. It seems that the most often analysed tetrel bonds are those where the tetrel tetrahedral centre acts as the Lewis acid site interacting with the electron-donating systems through the σ-hole. In general, the tetrel bond is crucial in many chemical, physical and biological processes.[64] It may be considered, for example, as a preliminary stage of the S_N2 reaction,[58] and it is often also responsible for arrangements of molecules and ions in crystals,[65] similar to other interactions that influence on such arrangements.[66,67] The hydrogen bond[1–3] or halogen bond[47,68] are examples of such influences.

The number of studies about the tetrel bond has increased in recent years where it was described in terms of the σ-hole and π-hole concept[69] and where the term "tetrel bond" was assigned to this interaction.[58,70] The tetrel bond may be defined briefly as a Lewis acid — Lewis base interaction where the tetrel centre (an atom of the 14th group of elements) plays the role of electron acceptor that is in contact with an electron-rich site playing the role of the Lewis base centre. It is worth mentioning that special attention was put on the carbon centre as the Lewis acid that may interact with electron donors; the term "carbon bond" was assigned for such interactions that were compared with hydrogen and halogen bonds.[71]

The tetrel bond is analysed in detail in this section because of its unique properties. First, the tetrel centre is often covered by substituents and its contacts with electron-rich sites seem to be problematic very often, especially for smaller tetrel atoms like carbon or silicon. The other interesting matter is that the properties of different centres of the 14th group differ significantly among themselves. Besides, the strong tetrel bond may change the tetrahedral structure of the tetrel centre into a trigonal bipyramid, or to a structure whose geometry is between tetrahedron and trigonal bipyramid structures. Tetrel centres possessing trigonal bipyramid structure may further interact with electron-rich species that leads to changes towards octahedron geometry.

It is worth mentioning that studies on the tetrel bond were performed early on despite this interaction not being named, described or characterised in detail. One can mention very early studies of Pfeiffer who analysed addition compounds of tetrel tetrahalides with ethers; he found that the ability to form such complexes decreases with the increase of the size of the halide.[72,73] These findings were confirmed later by studies of Sisler and coworkers who analysed complexes of carbon, silicon, germanium and tin tetrachlorides with ethers.[74–76] They found that the ability to form such connections decreases with the decrease of atomic number of the tetrel centre. They concluded that one of the reasons is that tetrel centres of lower atomic number "leave insufficient space for an oxygen atom" of ethers.[75] The authors claimed that "the oxygen atoms in these ethers probably do not readily share electron pairs with the silicon and germanium atoms in the respective tetrachlorides".[74]

It was stated by Sisler and coworkers that the formation of the tin tetrachloride complexes with two ether molecules results "from Lewis acid-base reactions in which the tin atom expands its valence shell by accepting a pair of electrons from each of two molecules of ether to form compounds of the type of $SnCl_4 \cdot 2R_2O$".[76] The authors also explained that the coordination of the tetrel centres should not surpass six. One can see that the last quote may be treated practically as the characterisation of the tetrel bond or even as the definition of this interaction, i.e., the sharing of electrons of Lewis base species with the tetrel centre that is the Lewis acid site. These early studies indicate that the tetrel centre may be coordinated by up to six ligands that lead to the octahedron structure.

One can mention other early studies related to tetrel bond interaction; the study on the crystal structure of *trans*-dipyridenetetrachlorogermanium and its silicon counterpart,[77] the infrared spectra analysis of systems of tetrafluoro silicon that are coordinated by Lewis base ligands,[78] the crystal structure of the complex of silicon tetrafluoride with hydrazine,[79] the crystal structure of tetrafluorobispyridinesilicon,[80] the vibrational spectra analysis of complexes of tetrahalides with trimethylamine and trimethylphosphine,[81] or the infrared spectra of $SiF_4 \cdot NH_3$ complex.[82]

One of the first theoretical studies concern the $SiF_4 \cdot NH_3$ and $SiF_4 \cdot 2NH_3$ complexes[83] while another concerns silicon tetrafluoride complexes with amines.[84] There is one important study[69] where complexes linked by tetrel bonds were analysed for the first time in terms of the σ-hole concept.[39,40] There are numerous studies on systems linked by the tetrel bonds in crystal structures,[64,65] on the role of tetrel bonds in S_N2 reactions,[58,85,86] and on deformation of species linked by tetrel bonds.[87,88] Special attention should be paid to the π-hole tetrel bonds that are also analysed from time to time.[89–91]

However, the emphasis is put here on studies related to changes of structures of tetrel centres due to their interactions with electron-rich species. Let us describe in detail the transformation of the tetrel bond into the trigonal bipyramid. Scheme V.5 presents the $F^- \cdots H_3C\text{-}Cl \rightarrow F\text{-}CH_3 \cdots Cl^-$ reaction that is the S_N2-type process.[58] The $F^- \cdots H_3C\text{-}Cl$ configuration linked by the F···C tetrel bond is the preliminary stage of this reaction. For this stage, electron charge transfer from the fluoride anion to the $H_3C\text{-}Cl$ unit occurs, followed by redistribution of electron charge in this unit that leads to the accumulation of electron charge at the chlorine centre. The transition state is characterised by the planar CH_3 system and two halogen units possessing

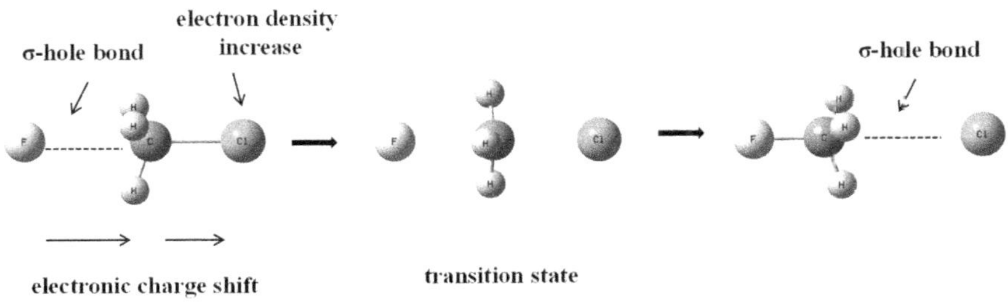

Scheme V.5 S_N2 reaction mechanism for the $F^- + CH_3Cl \rightarrow Cl^- + CH_3F$ reaction. Reprinted with permission from ref. 58. Copyright 2014 The Royal Society of Chemistry.

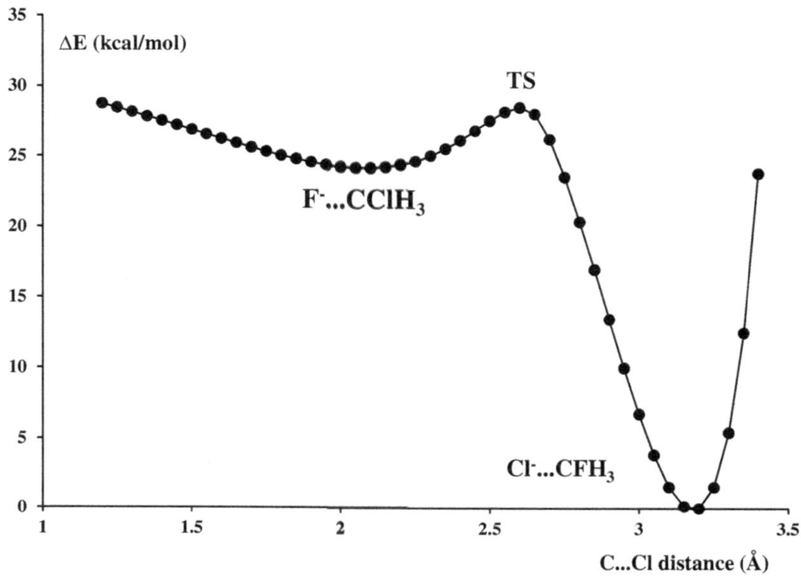

Figure V.5 The dependence between C…Cl^- distance and the energy of the system related to the global minimum) for the $F^- + CH_3Cl \rightarrow H_3FC...Cl^-$ reaction. Reprinted with permission from ref. 58. Copyright 2014 The Royal Society of Chemistry.

partial negative charges, $F^{-\delta}$ and $Cl^{-\delta}$. In other words, the chloride anion is not completely released from the CH_3 system and the fluoride anion is not completely attached. In the case of two equivalent substituents, like in the $Cl^-\cdots H_3C$-Cl process, the transition state is an ideal trigonal bipyramid. For the S_N2 reaction discussed here, this bipyramid may be slightly disturbed due to the different halogen substituents. This reaction leads to the F-$CH_3\cdots Cl^-$ complex where the units are linked by the C···Cl tetrel bond.

Figure V.5 presents the reaction path of this S_N2 reaction. This is the relationship between the C···Cl distance and the energy of the whole system, ΔE, which is related to the global minimum, i.e., to the energy of $H_3FC...Cl^-$ complex. The $F^-\cdots H_3C$-Cl system corresponds to the local shallow energetic minimum. The calculations were performed at the MP2/aug-cc-pVTZ level (ZPVE correction is included). The energy difference between the transition state and the global minimum is equal to

27.8 kcal/mol and the difference between the transition state and the local minimum of $H_3ClC...F^-$ is equal to 4.4 kcal/mol. The latter value is the potential barrier height for this S_N2 process. These energetic values are roughly in agreement with the other CCSD(T) level of calculations performed for the same S_N2 reaction, 26.5 kcal/mol and 3.1 kcal/mol, respectively.[92]

The transition state of the $F^- \cdots H_3C\text{-}Cl \rightarrow F\text{-}CH_3 \cdots Cl^-$ S_N2 reaction was presented here as one where the trigonal bipyramid occurs as a result of the σ-hole tetrel bond. However, the latter structure may occur because of the tetrel bond, not only for the transition state of the S_N2 process. A few examples were mentioned in the previous section, such as the $SiF_4 \cdots NCLi$ complex (Fig. V.4d). The α-angle is equal to 83.2° that is very close to the angle of 90°, corresponding to the transition state structure and consequently to the ideal trigonal bipyramid. Such changes were analysed for light[93] and heavy[94] tetrel centres. Figure V.6 presents molecular graphs of complexes of the germanium tetrafluoride with one and two arsine ligands (Figs. V.6a and V.6b, respectively). In the case of the $GeF_4 \cdots AsH_3$ complex, a weak Ge···As tetrel bond is observed where the energy of interaction, E_{int}, is equal here to −2.8 kcal/mol. The other parameters confirm this observation although the Ge···As distance amounts to 3.399 Å, which is less than the corresponding sum of van der Waals radii, 3.96 Å (Table I.1). However, the low value of the electron density at the Ge···As BCP is observed, 0.012 au. The Laplacian of electron density at the BCP is positive, but the total electron energy density at BCP, H_{BCP}, is close to zero according to the precision of calculations. The α angle is equal to 73.5° that is far from the 90° typical for the trigonal bipyramid, as was observed for the $SiF_4 \cdots NCLi$ complex, for example.

In the case of the $GeF_4 \cdots (AsH_3)_2$ complex (Fig. V.6b), much stronger Ge···As interactions are observed. The structure is an octahedron since the α angle is equal to 90° and lengths of all Ge-F are equal to each other. Two Ge···As distances are equal and they amount to 2.543 Å. That is less than the corresponding sum of van der Waals radii and less than this distance for the complex with one arsine ligand. The E_{int} is equal to −22.3 kcal/mol that indicates much stronger interaction than for the complex with one ligand. Hence one can see that the strength of interactions discussed here for complexes presented in Fig. V.6 concern the Ge···As contacts. For the $GeF_4 \cdots (AsH_3)_2$ complex, the ρ_{BCP} for the Ge···As bond path is equal to 0.052 au and H_{BCP} is negative

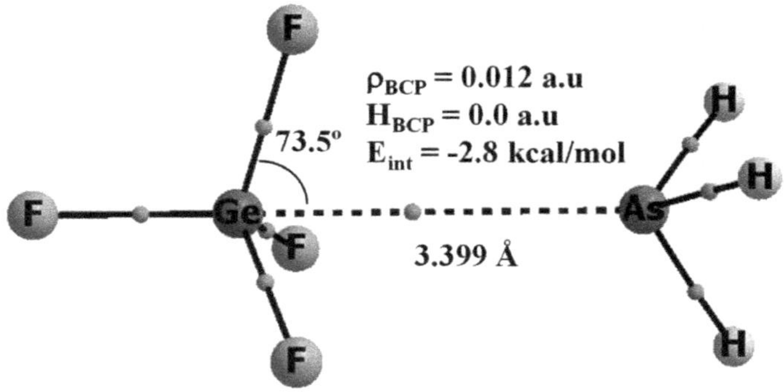

$GeF_4 \cdots AsH_3$

(a)

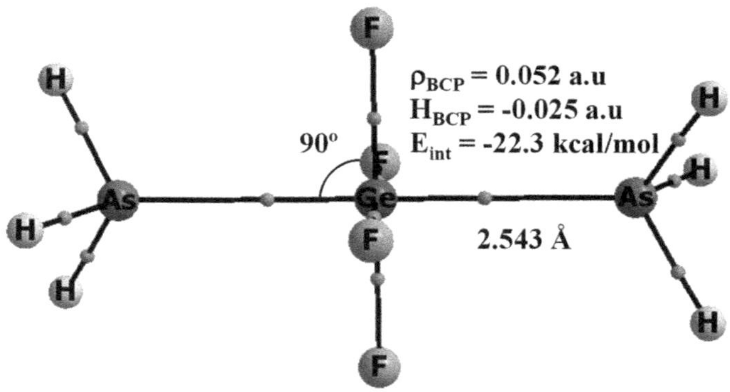

$GeF_4 \cdots (AsH_3)_2$

(b)

Figure V.6 Complexes of the germanium tetrafluoride with (a) one and (b) two arsine ligands. The molecular graphs are presented; big circles correspond to attractors amd small circles to BCPs, while the continuous lines correspond to bond paths. Modified with permission from ref. 93. Copyright 2017 MDPI.

that indicates partial covalent interaction. The two Ge···As contacts may be treated here as the electron acceptor bifurcated tetrel bonds; for both contacts the corresponding local interactions are stronger than the tetrel bond which is observed for the $GeF_4 \cdots AsH_3$ complex. Hence interactions in the complex with two AsH_3 ligands may be treated as a case of positive cooperativity.

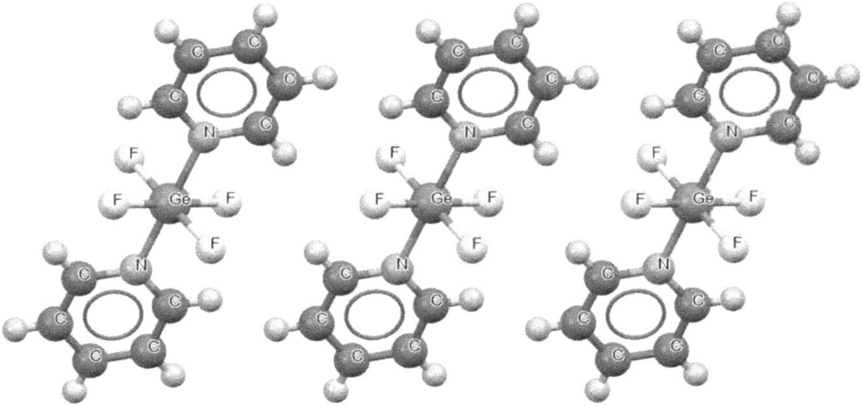

Figure V.7 A fragment of the crystal structure of *trans*-tetrafluoro-bis(pyridine)-germanium(IV).

As mentioned above, there are numerous studies on crystal structures where species linked by tetrel bonds occur.[94] One can mention the structures containing complexes of germanium tetrafluoride and tin tetrafluoride with thioethers[95] or phosphanes.[96] The octahedron structure is observed in such structures very often. For example, the Cambridge Structural Database[21,22] searches were performed to find structures containing complexes of the CF_4, SiF_4 and GeF_4 species connected through the tetrel centre with two Lewis base ligands where the basic site is located at the centre of the 15th or 16th group element.[93] The following search criteria were applied: structures without disorder, no structures with unresolved errors, no powder structures, no polymeric structures, and $R \leq 7.5\%$. Structures containing CF_4 were not found, nine structures with SiF_4 species and 13 that contain the GeF_4 unit were detected.[93] Figure V.7 presents a fragment of the crystal structure of *trans*-tetrafluoro-bis(pyridine)-germanium(IV) that may be treated as the structure where complexes of the germanium tetrafluoride molecules with two pyridine ligands occur.[97] In such a way the germanium centre is characterised by the octahedron structure. The Ge···N distances are equal to 2.011 Å (both GeF_4^- ligand contacts are equivalent due to the crystal symmetry; the Ge-centres are located at inversion centres). The other intermolecular contacts are rather weak, for example the H···F contacts between the $GeF_4\cdots(C_5H_5N)_2$ species are equal to 2.5 Å that may be classified as very weak hydrogen bonds, and the H···F low precision X-ray diffraction distance is close to the sum of van der Waals radii, 2.57 Å (Table I.1).

Summary

The early classifications of interactions were based on similarities between them. For example, the halogen bond may be treated as a counterpart of the hydrogen bond. There is the dihalogen bond — the special case of the halogen bond, similar to how the dihydrogen bond is a sub-class of the hydrogen bond interactions. However, a greater diversity of dihalogen bonds is observed in comparison with the dihydrogen bonds. The hydride bond is an interaction where the Lewis base hydrogen centre that is negatively charged is inserted between centres characterised by low values of electronegativity. The connection of this interaction with the halogen bond leads to the halogen-hydride bond where the halogen centre is the Lewis acid site and the negatively charged H-atom is the Lewis base centre. The reverse situation, the interaction between the protonic positively charged H-centre with the area of halogen of negative EP is the typical hydrogen bond interaction. The consideration of other pairs of interactions may lead to the generation of next interactions; the chalcogen-hydride bond is an example.[18]

It seems that the informative classification based on change of structures of centres involved in interactions is the promising one;[54] this classification is discussed in the next chapter. In the present chapter such changes are only described roughly. The strong intermolecular interactions lead to the change of configurations of centres involved. The triel centres characterised by the occurrence of Lewis acid properties of the π-hole when interacting with electron-rich species change the triel trigonal structure into tetrahedral. Other examples concern the tetrahedral tetrel, pnicogen, chalcogen and halogen centres that are involved in σ-hole interactions which change the structures of the Lewis acid sites into the trigonal bipyramid configuration. Special attention may be paid to the tetrahedral tetrel Lewis acid structures whose geometry depends on links with ligands; there is no lone electron pairs here, contrary to the pnicogen, chalcogen and halogen centres where one, two and three lone electron pairs are observed, respectively, for the tetrahedral centres. The tetrahedral tetrel centre interacting with one or two Lewis base ligands may change its configuration into the trigonal bipyramid or octahedron, respectively. There are numerous examples of such structures that occur in crystal structures.

Appendix

The following tools were applied to construct figures presented in this chapter.

Figure V.1

AIMAll program — Reference: Keith, T.A. *AIMAll (Version 11.08.23)*; TK Gristmill Software: Overland Park, KS, **2011**.

Figure V.7

Graphical tools of the Cambridge Structural Database — References: (1) Groom, C.R.; Bruno, I.J.; Lightfoot, M.P.; Ward, S.C. The Cambridge structural database. *Acta Cryst.* **2016**, *B72*, 171-179. and (2) Wong, R.; Allen, F.H.; Willett, P. The scientific impact of the Cambridge Structural Database: A citation-based study. *J. Appl. Cryst.* **2010**, *43*, 811-824.

Schemes V.1,3,4,5 and Figures V.2,3,4,5,6

For any modifications of these schemes and figures reprinted with permissions (for cases where they were made), PowerPoint was applied.

Scheme V.2

PowerPoint.

References

1. Jeffrey, G.A.; Saenger, W. *Hydrogen Bonding in Biological Structures*, Springer, Berlin, **1991**.
2. Jeffrey, G.A. *An Introduction to Hydrogen Bonding*, Oxford University Press, New York, **1997**.
3. Grabowski, S.J. *Understanding Hydrogen Bonds: Theoretical and Experimental Views*, The Royal Society of Chemistry, Cambridge, **2021.**
4. Kollman, P.A.; Liebman, J.F.; Allen, L.C. The Lithium Bond. *J. Am. Chem. Soc.* **1970**, *92*, 1142–1150.
5. Grabowski, S.J.; Hoffmann, R. Stabilizng H_3^-: or are we stabilizing a proton? *ChemPhysChem* **2021**, *13*, 2286–2288.
6. Sannigrahi, A.B.; Kar, T.; Niyogi, B.G.; Hobza, P.; von Schleyer, P.R. The lithium bond reexamined. *Chem. Rev.* **1990**, *90*, 1061–1076.
7. Vila, A.; Vila, E.; Mosquera, R.A. Topological characterisation of intermolecular lithium bonding. *Chem. Phys.* **2006**, *326*, 401–408.
8. Lipkowski, P.; Grabowski, S.J. Could the lithium bond be classified as the σ-hole bond? — QTAIM and NBO analysis. *Chem. Phys. Lett.* **2014**, *591*, 113–118.

9. Grabowski, S.J. Hydrogen and halogen bonds are ruled by the same mechanisms. *Phys. Chem. Chem. Phys.* **2013**, *15*, 7249–7259.
10. Grabowski, S.J. Hydrogen bonds, and σ-hole and π-hole bonds — mechanisms protecting doublet and octet electron structures. *Phys. Chem. Chem. Phys.* **2017**, *19*, 29742–29759.
11. Bent, H.A. An appraisal of valence-bond structures and hybridization in compounds of the first-row elements. *Chem. Rev.* **1961**, *61*, 275–311.
12. Sobczyk, L.; Grabowski, S.J.; Krygowski, T.M. Interrelation between H-bond and pi-electron delocalization. *Chem. Rev.* **2005**, *105*, 3513–3560.
13. Reiersølmoen, A.C.; Battaglia, S.; Øien-Ødegaard, S.; Gupta, A.K.; Fiksdahl, A.; Lindh, R.; Erdélyi, M. Symmetry of three-center, four-electron bonds. *Chem. Sci.* **2020**, *11*, 7979–7990.
14. Ward, J.S.; Fiorini, G.; Frontera, A.; Rissanen, K. Asymmetric $[N–I–N]^+$ halonium complexes. *Chem. Commun.* **2020**, *56*, 8428–8431.
15. Nygren, C.L.; Wilson, C.C.; Turner, J.F.C. On the solid state structure of 4-iodobenzoic acid. *J. Phys. Chem. A* **2005**, *109*, 2586–2593.
16. Grabowski, S.J. Halogen bonds between diiodotetrafluorobenzenes and halide anions: theoretical analysis. *Cryst. Growth Des.* **2023**, *23*, 489–500.
17. Lipkowski, P.; Grabowski, S.J.; Leszczynski, J. Properties of the halogen-hydride interaction: an ab initio and "atoms in molecules" analysis. *J. Phys. Chem. A* **2006**, *110*, 10296–10302.
18. Keil, H.; Herbst-Irmer, R.; Rathjen, S.; Girschik, C.; Muller, T.; Stalke, D. Si–H···Se chalcogen–hydride bond quantified by diffraction and topological analyses. *Inorg. Chem.* **2022**, *61*, 6319–6325.
19. Alkorta, I.; Elguero, J.; Frontera, A. Not only hydrogen bonds: other noncovalent interactions. *Crystals* **2020**, *10*, 180.
20. Joy, J.; Jemmis, E.D. Contrasting behavior of the Z bonds in X–Z···Y weak interactions: Z = main group elements versus the transition metals. *Inorg. Chem.* **2017**, *56*, 1132–1143.
21. Groom, C.R.; Bruno, I.J.; Lightfoot, M.P.; Ward, S.C. The Cambridge structural database. *Acta Cryst.* **2016**, *B72*, 171–179.
22. Wong, R.; Allen, F.H.; Willett, P. The scientific impact of the Cambridge Structural Database: A citation-based study. *J. Appl. Cryst.* **2010**, *43*, 811–824.
23. Pauling, L. *The Nature of the Chemical Bond: Third Edition*, Cornell University Press, New York, **1960**, pp. 88–102.
24. Mantina, M.; Chamberlin, A.C.; Valero, S.; Cramer, C.J.; Truhlar, D.G. Consistent van der Waals radii for the whole main group. *J. Phys. Chem. A* **2009**, *113*, 5806–5812.
25. Rowland, R.S.; Taylor, R. Intermolecular nonbonded contact distances in organic crystal structures: comparison with distances expected from van der Waals radii. *J. Phys. Chem.* **1996**, *100*, 7384–7391.

26. Bondi, J. Van der Waals volumes and radii. *J. Phys. Chem.* **1964**, *68*, 441–451.
27. Pendas, A.M.; Blanco, M.A.; Francisco, E. The nature of the hydrogen bond A synthesis from the interacting quantum atoms picture. *J. Chem. Phys.* **2006**, *125*, 184112.
28. Grabowski, S.J.; Sokalski, W.A. Are various σ-hole bonds steered by the same mechanisms? *ChemPhysChem* **2017**, *18*, 1569–1577.
29. Clark, T. σ-Holes. *WIREs Comput. Mol. Sci.* **2013**, *3*, 13–20.
30. Varadwaj, P.R. Halogen bond via an electrophilic π-hole on halogen in molecules: does it exist? *Int. J. Mol. Sci.* **2024**, *25*, 4587.
31. Grabowski, S.J. Triel bond and coordination of triel centres–Comparison with hydrogen bond interaction. *Coord. Chem. Rev.* **2020**, *407*, 213171.
32. Tabernero, V.; Muñoz, M.T.; Palenzuela, M.; Gomila, R.M.; Frontera, A.; Mosquera, M.E.G. σ-Hole triel bonds in aluminium derivatives. *Dalton Trans.* **2023**, *52*, 551–555.
33. Pavan, M.S.; Prasad, K.D.; Guru Row, T.N. Halogen bonding in fluorine: experimental charge density study on intermolecular F···F and F···S donor–acceptor contacts. *Chem. Commun.* **2013**, *49*, 7558–7560.
34. Eskandari, K.; Lesani, M. Does fluorine participate in halogen bonding? *Chem. Eur. J.* **2015**, *21*, 4739–4746.
35. Scheiner, S. F-Halogen bond: conditions for its existence. *J. Phys. Chem. A* **2020**, *124*, 7290–7299.
36. Cole, J.C.; Taylor, R. Intermolecular interactions of organic fluorine seen in perspective. *Cryst. Growth Des.* **2022**, *22*, 1352–1364.
37. Oliveira, V.; Kraka, E.; Cremer, D. The intrinsic strength of the halogen bond: electrostatic and covalent contributions described by coupled cluster theory. *Phys. Chem. Chem. Phys.* **2016**, *18*, 33031–33046.
38. Riley, K.E.; Tran, K-A. Strength, character, and directionality of halogen bonds involving cationic halogen bond donors. *Faraday Discuss.* **2017**, *203*, 47–60.
39. Politzer, P.; Murray, J.S.; Clark, T. Halogen bonding: an electrostatically-driven highly directional noncovalent interaction. *Phys. Chem. Chem. Phys.* **2010**, *12*, 7748–7757.
40. Politzer, P.; Murray, J.S.; Clark, T. Halogen bonding and other σ-hole interactions: a perspective. *Phys. Chem. Chem. Phys.* **2013**, *15*, 11178–11189.
41. Clark, T.; Murray, J.S.; Politzer, P. A perspective on quantum mechanics and chemical concepts in describing noncovalent interactions. *Phys. Chem. Chem. Phys.* **2018**, *20*, 30076–30082.
42. Murray, J.S.; Politzer, P. Interaction and polarization energy relationships in σ-hole and π-hole bonding. *Crystals* **2020**, *10*, 76.
43. Grabowski, S.J. What is the covalency of hydrogen bonding? *Chem. Rev.* **2011**, *111*, 2597–2625.

44. Kumar, R.; Karkamkar, A.; Bowden, M.; Autrey, T. Solid-state hydrogen rich boron–nitrogen compounds for energy storage. *Chem. Soc. Rev.* **2019**, *48*, 5350–5380.
45. Karkamkar, A.; Kathmann, S. M.; Schenter, G. K.; Heldebrant, D. J.; Hess, N.; Gutowski, M.; Autrey, T. Thermodynamic and structural investigations of ammonium borohydride, a solid with a highest content of thermodynamically and kinetically accessible hydrogen. *Chem. Mater.* **2009**, *21*, 4356–4358.
46. Hobza, P.; Havlas, Z. Blue-shifting hydrogen bonds. *Chem. Rev.* **2000**, *100*, 4253–4264.
47. Cavallo, G.; Metrangolo, P.; Milani, R.; Pilati, T.; Priimagi, A.; Resnati, G.; Terraneo, G. The halogen bond. *Chem. Rev.* **2016**, *116*, 2478–2601.
48. Grabowski, S.J. Noncovalent Interactions in Crystal Structures: Quantifying Cooperativity in Hydrogen and Halogen Bonds, in Novoa, J.J. (Ed.), *Intermolecular Interactions in Crystals: Fundamentals of Crystal Engineering*, The Royal Society of Chemistry, **2018**, pp. 673–718.
49. Grabowski, S.J.; Bilewicz, E. Cooperativity halogen bonding effect — Ab initio calculations on $H_2CO\cdots(ClF)n$ complexes. *Chem. Phys. Lett.* **2006**, 427, 51–55.
50. Ziółkowski, M.; Grabowski, S.J.; Leszczynski, J. Cooperativity in hydrogen-bonded interactions: ab initio and "atoms in molecules" analyses. *J. Phys. Chem. A* **2006**, *110*, 6514–6521.
51. Alkorta, I.; Blanco, F.; Deyà, P.M.; Elguero, J.; Estarellas, C.; Frontera, A.; Quiñonero, D. Cooperativity in multiple unusual weak bonds. *Theor. Chem. Acc.* **2010**, *126*, 1–14.
52. Gillespie, R.J.; Hargittai, I. *The VSEPR Model of Molecular Geometry*. Allyn & Bacon, Boston, **1991**. Reprinted: Gillespie, R.J.; Hargittai, I. *The VSEPR Model of Molecular Geometry*. Dover Publications, New York, **2012**.
53. Grabowski, S.J. Pnicogen and tetrel bonds — tetrahedral Lewis acid centres. *Struct. Chem.* **2019**, *30*, 1141–1152.
54. Grabowski, S.J. Classification of so-called non-covalent interactions based on VSEPR model. *Molecules* **2021**, *26*, 4939.
55. Grabowski, S.J. σ-Hole bonds and the VSEPR model — from the tetrahedral structure to the trigonal bipyramid. *Sci* **2022**, *4*, 17.
56. Grabowski, S.J. Boron and other triel Lewis acid centers: From hypovalency to hypervalency. *Chem. Phys. Chem.* **2014**, *15*, 2985–2993.
57. Clauss, A.D.; Nelsen, S.F.; Ayoub, M.; Moore, J.W.; Landis, C.R.; Weinhold, F. Rabbit-ears hybrids, VSEPR sterics, and other orbital anachronisms. *Chem. Educ. Res. Pract.* **2014**, 15, 417–434.
58. Grabowski, S.J. Tetrel bond — σ-hole bond as a preliminary stage of the S_N2 reaction. *Phys. Chem. Chem. Phys.* **2014**, 16, 1824–1834.

59. Grabowski, S.J.; Casanova, D.; Formoso, E.; Ugalde, J.M. Tetravalent oxygen and sulphur centres mediated by carborane superacid: theoretical analysis. *ChemPhysChem* **2019**, *20*, 2443–2450.
60. Cioslowski, J.; Mixon, S.T. Topological properties of electron density in search of steric interactions in molecules: electronic structure calculations on ortho-substituted biphenyls. *J. Am. Chem. Soc.* **1992**, *114*, 4382–4387.
61. Cioslowski, J.; Mixon, S.T. Universality among topological properties of electron density associated with the hydrogen–hydrogen nonbonding interactions. *Can. J. Chem.* **1992**, *70*, 443–449.
62. Jabłoński, M. Bond paths between distant atoms do not necessarily indicate dominant interactions. *J. Comput. Chem.* **2018**, *39*, 2183–2195.
63. Jabłoński, M. On the uselessness of bond paths linking distant atoms and on the violation of the concept of privileged exchange channels. *ChemistryOpen* **2019**, *8*, 497–507.
64. Scheiner, S. (Ed.). *Tetrel Bonds*, MDPI, Basel, **2023**, printed edition of the special issue published in *Molecules*.
65. Bauzá, A.; Mooibroek, T.J.; Frontera, A. Tetrel bonding interactions. *Chem. Rec.* **2016**, *16*, 473–487.
66. Scheiner, S. (Ed.). *Noncovalent Forces*, Springer Cham, Heidelberg, **2015**.
67. Novoa, J.J. (Ed.). *Intermolecular Interactions in Crystals: Fundamentals of Crystal Engineering*, The Royal Society of Chemistry, **2018**.
68. Parisini, E.; Metrangolo, P.; Pilati, T.; Resnati, G.; Terraneo, G. Halogen bonding in halocarbon–protein complexes: a structural survey. *Chem. Soc. Rev.* **2011**, *40*, 2267–2278.
69. Bundhun, A.; Ramasami, P.; Murray, J.S.; Politzer, P. Trends in σ-hole strengths and interactions of F_3MX molecules (M = C, Si, Ge and X = F, Cl, Br, I). *J. Mol. Model.* **2013**, *19*, 2739–2746.
70. Bauzá, A.; Mooibroek, T.J.; Frontera, A. Tetrel-bonding interaction: rediscovered supramolecular force? *Angew. Chem. Int. Ed.* **2013**, *52*, 12317–12321.
71. Mani, D.; Arunan, E. The X–C···Y (X = O/F, Y= O/S/F/Cl/Br/N/P) carbon bond and hydrophobic interactions. *Phys. Chem. Chem. Phys.* **2013**, *15*, 14377–14383.
72. Pfeiffer, P. Zur Hydrolyse des Zinnchlorids und Zinnbromids. *Z. anorg. Chem.* **1914**, *87*, 235–247.
73. Pfeiffer. P.; Halperin, O. Über die Wechselnde Affinität der Nebenvalenzen des Zinns. *Z. anorg. Chem.* **1914**, *87*, 335–352.
74. Sisler, H.H.; Wilson, W.J.; Gibbins, B.J.; Batey, H.H.; Pfahler, B.; Mattair, R. Phase studies of some group 4a tetrachloride-aromatic ether systems. *J. Am. Chem. Soc.* **1948**, *70*, 3818–3821.

75. Sisler, H.H.; Batey, H.H.; Pfahler, B.; Mattair, R. Phase studies of some group 4a tetrachloride-aliphatic ether systems. *J. Am. Chem. Soc.* **1948**, *70*, 3821–3824.
76. Sisler, H.H.; Schilling, E.E.; Groves, W.O. Phase studies on the binary systems of tin tetrabromide, tin tetraiodide and trichlorosilane with some aliphatic ethers. *J. Am. Chem. Soc.* **1951**, *73*, 426–429.
77. Hulme, R.; Leigh, G.J.; Beattie, I.R. The crystal structure of trans-dipyridinetetrachlorogermanium (IV) and of its silicon and titanium analogues. *J. Chem. Soc.* **1960**, 366–371.
78. Gutmann, V.; Heilmayer, P.; Utvary, K. Zur Kenntnis von Koordinationsverbindungen des Silicium(IV)-fluorids. *Monatsh. für Chem.* **1961**, *92*, 322–329.
79. Aggarwal, R.C.; Onyszchuk, M. Coordination compounds of hydrazine part III. The interaction of silicon tetrafluoride with hydrazine. *Can. J. Chem.* **1963**, *41*, 876–882.
80. Bain, V.A.; Killean, R.C.G.; Webster, M. The crystal and molecular structure of tetrafluorobispirydinesilicon (IV). *Acta Cryst.* **1969**, *B25*, 156–159.
81. Beattie, I.R.; Ozin, G.A. Vibrational spectra, vibrational analysis, and shapes of some 1:1 and 1:2 addition compounds of group IV tetrahalides with trimethylamine and trimethylphosphine. *J. Chem. Soc. A* **1970**, 370–377.
82. Ault, B.S. Matrix-isolation studies of Lewis acid/base interactions: infrared spectra of the 1:1 adduct of $SiF_4 \cdot NH_3$. *Inorg. Chem.* **1981**, *20*, 2817–2822.
83. Chehayber, J.M.; Nagy, S.T.; Lin, C.S. Ab initio studies of complexes between SiF_4 and ammonia. *Can. J. Chem.* **1984**, *62*, 27–31.
84. Politzer, P.; Murray, J.S.; Lane, P.; Concha, M.C. Electrostatically driven complexes of SiF_4 with amines. *Int. J. Quantum Chem.* **2009**, *109*, 3773–3780.
85. Liu, M.; Li, Q.; Cheng, J.; Li, W.; Li, H.-B. Tetrel bond of pseudohalide anions with XH_3F (X = C, Si, Ge, and Sn) and its role in S_N2 reaction. *J. Chem. Phys.* **2016**, *145*, 224310.
86. Kubelka, J.; Bickelhaupt, F.M. Activation strain analysis of S_N2 reactions at C, N, O, and F Centers. *J. Phys. Chem. A* **2017**, *121*, 885–891.
87. Zierkiewicz, W.; Michalczyk, M.; Scheiner, S. Implications of monomer deformation for tetrel and pnicogen bonds. *Phys. Chem. Chem. Phys.* **2018**, *20*, 8832–8841.
88. Scheiner, S. Systematic elucidation of factors that influence the strength of tetrel bonds. *J. Phys. Chem. A* **2017**, *121*, 5561–5568.
89. Wei, Y.; Li, Q.; Scheiner, S. The π-tetrel bond and its influence on hydrogen bonding and proton transfer. *ChemPhysChem* **2018**, *19*, 736–743.
90. Politzer, P.; Murray, J.S.; Clark, T. The π-hole revisited. *Phys. Chem. Chem. Phys.* **2021**, *23*, 16458–16468.

91. Shen, S.; Zeng, Y.; Li, X.; Meng, L.; Zhang, X. Insight into the π-hole···π-electrons tetrel bonds between F_2ZO (Z=C, Si, Ge) and unsaturated hydrocarbons. *Int. J. Quantum Chem.* **2018**, *118*, e25521.
92. Mugnai, M.; Cardini, G.; Schettino, V. An *ab initio* molecular dynamics study of the S_N2 Reaction. *J. Chem. Phys.* **2003**, *118*, 2767–2774.
93. Grabowski, S.J. Lewis acid properties of tetrel tetrafluorides — the coincidence of the σ-hole concept with the QTAIM approach. *Crystals* **2017**, *7*, 43.
94. Grabowski, S.J. Tetrel bonds, penta- and hexa-coordinated tin and lead centres. *Appl. Organometal. Chem.* **2017**, *31*, e3727.
95. Davis, M.F.; Levason, W.; Reid, G.; Webster, M.; Zhang, W. The first examples of germanium tetrafluoride and tin tetrafluoride complexes with soft thioether coordination — Synthesis, properties and crystal structures. *Dalton Trans.* **2008**, 533–538.
96. Davis, M.F.; Levason, W.; Reid, G.; Webster, M. Complexes of germanium (IV) fluoride with phosphane ligands: Structural and spectroscopic authentication of germanium (IV) phosphane complexes. *Dalton Trans.* **2008**, *17*, 2261–2269.
97. Cheng, F.; Davis, M.F.; Hector, A.L.; Levason, W.; Reid, G.; Webster, M.; Zhang, W. Synthesis, spectroscopic and structural systematics of complexes of germanium(IV) halides (GeX_4, X = F, Cl, Br or I) with mono-, bi- and tri-dentate and macrocyclic nitrogen donor ligands. *Eur. J. Inorg. Chem.* **2007**, 4897–4905.

Chapter VI

Interactions and the VSEPR Model

Abstract: The formation of attractive Lewis acid — Lewis base interactions leads to numerous geometrical, energetic and electron charge distribution changes in units that are linked. However, it seems that all those changes work against Pauli repulsion that increases with the decrease of the distance between the acidic and basic sites in complexes formed. These changes try to protect the octet (or doublet in the case of hydrogen) electron structure of the Lewis acid centre being in contact with the electron-rich site of the Lewis base unit. If this centre is characterised by hypervalency, electron charge shifts occur to maintain the octet or doublet rule. There are several configurations typical for Lewis acid centres, such as trigonal, tetrahedral, trigonal bipyramid and octahedron that are changed in the case of strong interactions with electron donors. These changes follow the valence shell electron pair repulsion model, VSEPR.

Keywords: Pauli repulsion, electron donor, octet (doublet) rule, configurations of interacting centres, valence shell electron pair repulsion model, VSEPR

VI.1 The Pauli Repulsion that Steers Interactions

The variety of interactions that was described so far in this book presents a clear picture of mechanisms of formation as well as of processes that accompany them. The electron charge shifts and distributions were presented for the hydrogen bonds in Scheme IV.7 (Chapter IV), and particularly for the water dimer, $(H_2O)_2$, linked by the O-H···O hydrogen bond in Scheme V.1 (Chapter V). The changes resulting from the formation of the $CF_3Cl \cdots OCH_2$ complex that is linked by the C-Cl···O halogen bond are presented in the same Scheme V.1. Scheme VI.1, in this section, shows selected changes that are typical in the case of formation of a greater number of interactions, which concerns mainly the σ-hole bonds.

The whole process of the formation of interaction may be roughly described in the following way.[1,2] It seems that two units may be linked as Lewis acid — Lewis base pair if they possess at their molecular surfaces sites characterised by opposite electrostatic potentials (EPs). The contact between these sites, i.e., the Lewis acid — Lewis base link, may be classified as σ-hole or π-hole bond. In this way, for example, the water dimer is formed (Scheme V.1) through hydrogen bonding since the hydrogen centre, by its positive EP, is connected with the negative EP site of the oxygen centre of another water molecule. It was explained, in early studies, that such connections are formed by pairs of centres characterised by opposite charges. However, in later studies it was justified that charges are not good characteristics to explain connections between centres, and the analysis of EPs is better.[3–5]

One may conclude that such analysis of EP surfaces indicates the electrostatic character of interaction. However, for the links mentioned above, electron charge shifts are observed. This was described in Chapter I, that the electron charge shifts are attributed to the Lewis acid — Lewis base interactions. Stronger electrostatic interactions lead to shorter contacts between the Lewis acid and Lewis base units and to greater electron charge shifts. That is presented in Scheme VI.1, and it is also shown that such shifts are more complex. The electron charge is shifted from the Lewis base to the Z-centre, i.e., to the H-atom in the case of the hydrogen bond or to the X atom (F, Cl, Br or I) in the case of the halogen bond, for example. However, the positive charge of Z (H or X)

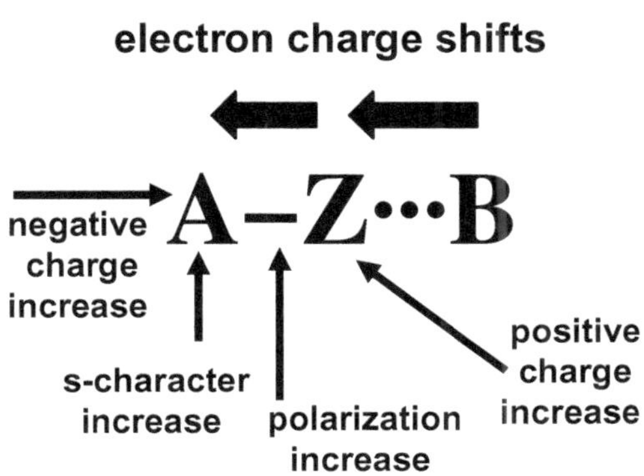

Scheme VI.1 Changes of parameters resulting from σ-hole bond formation. Z is the Lewis acid unit centre that is in contact with the electron-donating site, B, of the Lewis base unit. The cases of monovalent Z-centres are presented here.

does not decrease, and in fact usually increases. Why? Because there are further electron charge shifts from the Z-centre to the remaining parts of the Lewis acid unit (designated by A in Scheme VI.1).[6,7] One may ask why such further electron charge shifts occur, why they are not stopped at the Z-centre but lead to the strange and surprising increase of its positive charge? It is explained later in this chapter.

Scheme VI.1 presents the situation for the monovalent Z-centre. That often occurs for hydrogen-bonded systems as well as for halogen bonds. However, for the latter the regions of multivalent halogen centres may also play the role of Lewis acid sites.[8–12] The same occurs for triel, tetrel, chalcogen and aerogen centres, and for atoms of the first and second groups, which are usually multivalent. Additionally, for numerous centres, dual character is observed because they may possess lone electron pairs apart from σ-holes characterised by positive EPs. For such centres that possess dual character and/or are multivalent, the processes that are shown at Scheme VI.1 may also occur, both for the σ-hole and π-hole bonds.

Let us further discuss Scheme VI.1. The Z-centre becomes more positive and the A-centre (or centres in the case of multivalency of Z) becomes more negative due to the electron charge shifts resulting from the σ-hole bond formation in comparison with these charges in the isolated A-Z Lewis acid unit not involved in any interaction. This leads to increase of the polarisation of the A-Z bond. Such polarisation increases due to the increase of the difference between the values of charges of Z and A centres. Besides, this polarisation increase relates to increase of the s-character of the atomic hybrid orbital of the A-centre; this atomic hybrid orbital is a component of the A-Z σ-bond orbital (increase of the s-character follows the increase of A-Z bond polarisation).[1]

The energy decomposition analysis (EDA) schemes allow us to deepen understanding of mechanisms that accompany the formation of numerous σ-hole bonds and π-hole bonds.[13–17] Such schemes were presented in Chapter II (eqs. II.19 and II.20), and also the SAPT approach[18] (eqs. II.21 and II.22) was described that formally is not the EDA scheme since the partitioning of the energy of interaction is not performed here but all terms of energy are calculated separately. The repulsion interaction energy term appears in all schemes, and very often it is named as the Pauli repulsion contribution.

It is important to note that this repulsion is balanced by the attractive terms, electrostatic, orbital and dispersion if one refers to some of the EDA approaches where such names appear. Orbital contribution is often replaced by other names corresponding to terms that have different physical meanings; these are induction, delocalisation, polarisation, charge transfer, etc. As mentioned in Chapter I, in this book the term polarisation interaction energy is applied if any specific decomposition scheme is not referred to. The EDA scheme may be related approximately to the equation given below:

$$E_{int} = E_{rep} + E_{es} + E_{pol} + E_{disp} \tag{VI.1}$$

The following designations were applied in the above equation: E_{int} — the total interaction energy, E_{rep} — the Pauli repulsion interaction energy, E_{es} — the electrostatic interaction energy term, E_{pol} — the polarisation term and E_{disp} — the dispersion interaction energy contribution.

Coming back to the Pauli repulsion (most studies establish conventionally that repulsion interaction energy is positive while attractive contributions are negative), it increases if the distance between the interacting species decreases.[1] However, the absolute values of attractive terms also increase with the shortening of this distance.[1] The total interaction energy is negative for energetically stable complexes and while it is positive for unstable systems, at least it is an approximate relation since to be more precise the thermodynamic conditions should be taken into account (see eq. I.4 in Chapter I). For the complex that is in the energetic minimum, balance between all energetic terms occurs, or to say it better – balance between the processes accompanying the formation of the complex occurs.

One may say that for the stable system, if there is contact between the Lewis acid — base units, the repulsion between these units appears and the attractive terms overcompensate this repulsion. Figure VI.1 presents the relationship between the Pauli repulsion term and the sum of all attractive terms on one hand, and the relationship between this repulsion and the attractive term corresponding to the electron charge shifts on the other hand. The latter term is named as polarisation; however, in the EDA scheme applied, the Hybrid Variation Perturbation Theory approach,[16,17] for results of this figure, it is named as the delocalisation contribution. These results concern the sample of complexes linked by various σ-hole bonds,[19] i.e., halogen, chalcogen, pnicogen

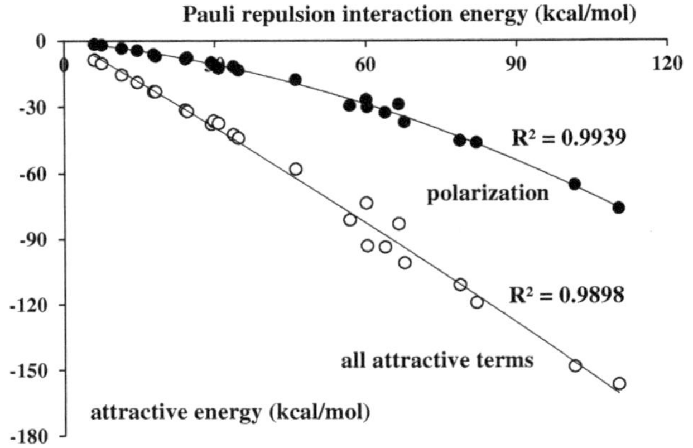

Figure VI.1 Relationships (second-order polynomials) between the Pauli repulsion interaction energy (kcal/mol) and the attractive interaction energy (kcal/mol); open circles — the sum of all attractive terms is considered (electrostatic, polarisation and correlation), full circles — only polarisation is considered. Reprinted with permission from ref. 1. Copyright 2017 The Royal Society of Chemistry.

and tetrel bonds. The following species play the role of the Lewis base unit in these complexes: NH_3 through the lone electron pair at the N-centre, the Cl^- anion, and the C_2H_2 species through the π-electrons of the CC bond. The selected complexes of this sample were presented in Chapter II and Fig. II.5 presents their molecular graphs. The binding, interaction and deformation energies of several complexes of this sample are also presented in Table II.4. The second-order polynomial relationships in Fig. VI.1 show that the attractive interaction energy terms overwhelm the Pauli repulsion term; besides, one can see that good correlation between polarisation and Pauli repulsion terms occurs. In other words, the shortening of the distance between Lewis acid and Lewis base units in the complex increases absolute values of all terms of interaction energy. The significance of the polarisation term corresponding to the electron charge shifts indicates that the formation of the intermolecular link relates to mechanisms that try to balance or even overcompensate the repulsion interaction energy.

Figure VI.2 shows the exponential relationship between the Pauli repulsion interaction energy term and the polarisation of the F-Z bond for the same sample. The F-Z bond participates in the σ-hole bond, F-Z···N, F-Z···Cl or F-Z···π. The Z-centre participates in these links through the σ-hole located in extension of the F-Z bond approximately.

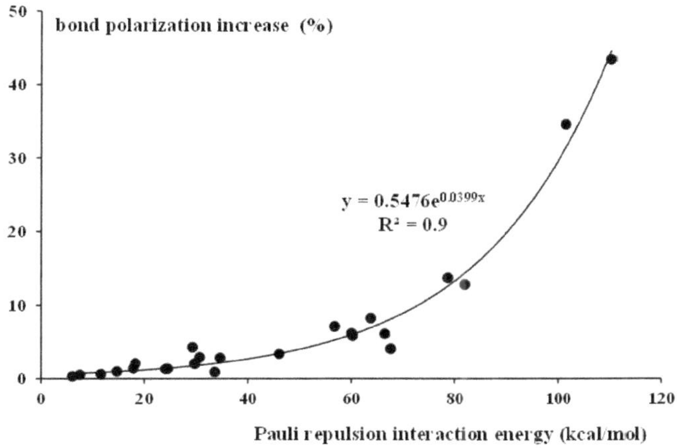

Figure VI.2 Exponential relationship between the Pauli repulsion interaction energy (kcal/mol) and the increase of polarisation (%) of the Z-F bond as an effect of complexation. Reprinted with permission from ref. 1. Copyright 2017 The Royal Society of Chemistry.

The increase of the Pauli repulsion presented in Fig. VI.2 is related to the increase of the strength of interaction (the shortening of the Lewis acid — base distance), which in turn increases because of the increase of electron charge shifts, in part from the Z centre to the F centre in the F-Z bond. This is connected directly with the increase of polarisation of the F-Z bond. In other words, the formation of the complex (creation of the σ-hole or π-hole bond) leads to the appearance of Pauli repulsion that next activates various mechanisms, mainly related to the electron charge shifts, which try to compensate it.

VI.2 Interactions as the Preliminary Stages of Chemical Reactions

It was mentioned in the previous chapter that interactions are often preliminary stages of chemical reactions.[20] The hydrogen bond may be the preliminary stage of the proton transfer (PT) process.[21] This process is more probable for stronger hydrogen bonds. Special attention should be paid to the so-called low barrier hydrogen bonds (LBHBs), where the barrier height for the PT is not so high.[22,23] That happens if the pK_a of A and

B centres of the A-H···B hydrogen bond are very close to each other such that the proton may occur with close to equal probability at each of both centres. Furthermore, in numerous studies it is stated that LBHBs occur if the barrier height is energetically located below the ground state of vibrational levels of A and B centres.[23] In such a way the proton moves "freely" between these centres, and in the case of the same centres (for A-H···A systems) and the same environments (often resulting from symmetry conditions in crystals), the mean position of the proton is exactly at the mid-point of the A···A distance. Such a situation occurs usually for O-H···O hydrogen bonds and for very short O···O distances, below 2.5 Å but close to 2.4 Å. However, this is not the final and fixed mid-point position of the proton, which occurs for systems characterised by the single-well potential energy curve. For the majority of LBHB systems, it is a statistical mean position of the proton that moves between oxygen centres.

Scheme VI.2 presents the classification of symmetrical O-H···O hydrogen bonds that is based on the O···O distance. For typical hydrogen bonds (Scheme VI.2A) with distances greater than ~2.5–2.55 Å, the potential barrier height is above the ground state vibrational levels and the proton is located at one of the oxygen centres. However, the PT process is possible, particularly if additional conditions are fulfilled; temperature and solvent may affect this process, for example. It seems that the double PT process occurs often for centrosymmetric dimers of carboxylic acids in crystals, particularly for the O···O distances in the range of ~2.55–2.65 Å.[24,25] Several crystal structures of carboxylic acids were presented so far in this book; for example, Figure III.10 (Chapter III) presents the crystal structure of terephthalic acid.[26]

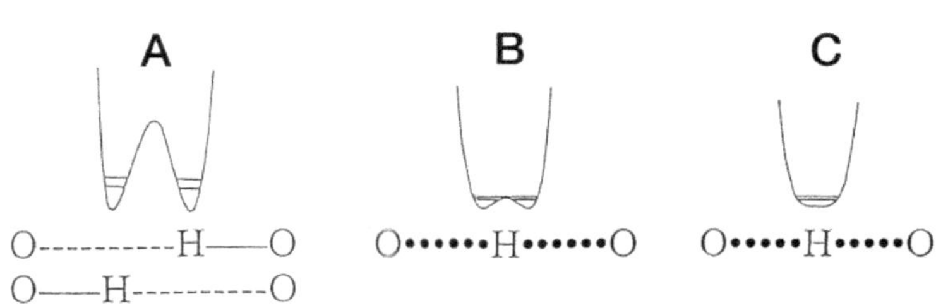

Scheme VI.2 Classification of O-H···O hydrogen bonds based on the donor–acceptor distances, from > 2.5–2.55 Å (VI.2A) to about ~2.4 Å (VI.2C). The horizontal lines within the wells represent zero-point vibration energies for the proton and deuteron. The LBHB (VI.2B) has the PT barrier height below the ground state vibrational levels of the donor and acceptor. Reprinted with permission from ref. 23. Copyright 1998 American Society for Biochemistry and Molecular Biology, Inc.

LBHB structures occur often for O···O distances lower than 2.5–2.55 Å (Scheme VI.2B). And for O···O distances equal to about ~2.4 Å and less, the proton may be located at the mid-point of the O···O distance since this is a case of structures characterised by the single-well potential energy (Scheme VI.2C). Such a situation may occur for the $H_5O_2^+$ cation.[27] However, a majority of experimental and theoretical results shows that this is the LBHB system where the central proton location is only the mean value (see also Chapter IV). It is worth mentioning that the $(FHF)^-$ anion is an example of a structure corresponding to the single-well potential energy curve without any doubt.[28,29]

There are neutron diffraction crystal structures with such mid-point proton position in O-H···O bridges, but it is difficult to say if these are single-well potential energy systems or if they correspond to LBHBs with a statistical mean proton position. Nevertheless, these neutron diffraction results may be useful to model the PT process. The structure-correlation method introduced in early studies[30–33] may be applied for modelling this process. According to this method, a fragment of a molecular species taken from a crystal structure may be treated as a "frozen stage" of a chemical reaction that is analysed. The other related fragments that differ by geometry and that come from different crystal structures may be treated as geometrical fragments which reflect structural changes occurring during this reaction.[30–33] The structure-correlation method was applied to analyse different reactions such as electrocyclic ring closure of polyenes, nucleophilic addition to a carbonyl group or nucleophilic substitution at tetrahedrally coordinated atoms (S_N1 and S_N2 reactions were considered).[30–33]

The PT process may also be analysed using the structure-correlation method. For example, –C=O···H-O-C- geometrical fragments were taken from different crystal structures for comparison and for construction of the reaction path of the corresponding –C=O···H-O-C- ⇔ –C-O-H···O=C- PT process.[1,20,34,35] These geometrical fragments were taken from the Cambridge Structural Database (CSD).[36,37] In one of these studies,[20] the following search criteria were applied: accurate crystal structures with estimated standard deviation ≤ 0.005 Å, R ≤ 7.5%, error-free structures, structures without disorder, no polymers and no powder diffraction results. Besides, only neutron diffraction crystal structures were taken into account because such structures are characterised by precisely determined positions of H-atoms.[38] It was described in Chapter III that neutron diffraction results are characterised by high precision of positions of H-atoms while for the X-ray results the refinement of crystal structures is most often based on spherical approximation of the atomic electron densities that is connected, roughly speaking, with inaccurate positions. For some crystal structures the H-atom is located in the mid-point of the O···O distance, therefore the –C=O···H^+···O=C- geometrical fragments were also used to model the PT process.

Figure VI.3 presents the modelled PT reaction that is based on neutron diffraction results of 56 geometrical fragments. Similar fragments where the H-atom is replaced by deuterium, i.e., –C=O···D-O-C- fragments, are included in this figure; however, in this case only four structures were found in the CSD (CSD updates up to May 2020 were taken into account).[20]

Figure VI.3 shows the relationship between the Δr parameter and the O···O distance. Δr is defined as the distance of the H-atom of the O-H···O bridge from the O···O mid-point. It is worth mentioning that results of this figure are symmetrical about the

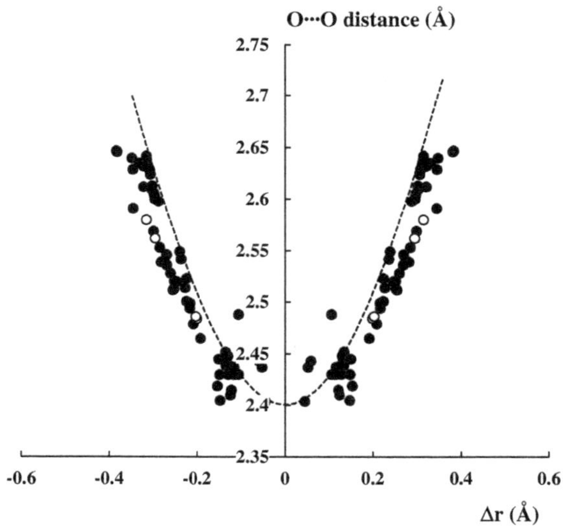

Figure VI.3 Dependence between Δr (Å), the displacement of the proton (or the deuteron) position from the O···O mid-point, and the O···O distance (Å), for the O-H···O (black circles) and O-D···O systems (white circles). The broken line corresponds to the bond number conservation rule (eq. VI.2). Reprinted with permission from ref. 20. Copyright 2020 MDPI.

ordinate axis since the fragments designated by positive Δr values are repeated with the assigned negative Δr values. This is to model the PT reaction since the proton is attached to one of the O-centres and after the process to another O-centre. This symmetrisation reflects the equivalency of systems during the PT reaction of a homonuclear O-H···O hydrogen bond. The Δr parameter informs of the position of the proton during the PT reaction. For $\Delta r = 0$, the situation corresponds to the transition state of this process, while for long O···O distances the H-atoms are located far from their mid-points and close to one of the oxygen centres.

The broken line of Fig. VI.3 corresponds to the PT bond path calculated from the bond order (number) conservation rule.[30,31,39] This rule was discussed in Chapter IV (eq. IV.1) and it was presented as the sum of power-of-10 functions. Sometimes in various studies the sum of exponential terms occurs (eq. VI.2).

$$\exp(\Delta r_i / c) + \exp(\Delta r_j / c) = 1 \qquad \text{(VI.2)}$$

The explanation of this equation is in agreement with that presented in Chapter IV for eq. IV.1. The Δr_i and Δr_j in eq. VI.2 correspond to the $(r_0 - r_{OH})$ and $(r_0 - r_{H\cdots O})$ differences; r_0 is the typical single O-H bond length of the isolated Lewis acid unit that is not disturbed by any interaction, while r_{OH} and $r_{H\cdots O}$ are the O-H bond length and the H···O distance in the O-H···O bridge, respectively. The bond length of water in the gas phase equal to 0.957 Å was chosen in the study now discussed as the reference single O-H bond not disturbed by external interactions (the r_0 value).[20] The $\exp(\Delta r_i/c)$ term of eq. VI.2 may be treated as the definition of the bond order. The constant c for the O-H···O hydrogen bonds is determined by assuming that for the O-H···O linear system with the H-atom located in the O···O mid-point, and the O···O distance of 2.4 Å, two equivalent H···O distances possess a bond order equal to 0.5; in other words; $c = (0.957-1.2)/\ln(0.5)$. It was checked in several studies that the bond number conservation rule is often in excellent agreement with the experimental results.[20,34,39,40] It is also reflected by the results presented in Figure VI.3.

Let us consider the halonium cation transfer process that is similar to the PT reaction; this process was examined in the previous chapter. The cases of the $[NZN]^+$ systems analysed experimentally[41] where the Z^+ cation ($Z^+ = H^+$, Li^+, Na^+, F^+, Cl^+, Br^+, I^+, Ag^+ and Au^+) is inserted between nitrogen centres were discussed briefly in this chapter.

The halonium transfer may be modelled with the use of results of theoretical calculations performed recently on complexes of imidazol-2-ylidene ($C_3N_2H_4$) that acts as the Lewis base unit through the lone pair of the carbon centre with the XCCH, XCN and X_2 Lewis acid units (X = F, Cl, Br and I).[42] The latter species act as electrophiles through the σ-holes located at the X halogen centres in the extension of C-X or X-X bonds approximately. The calculations for these complexes were performed at the ωB97XD/aug-cc-pVDZ and ωB97XD/aug-cc-pVTZ levels. The optimised systems correspond to energetic minima since the frequency calculations were also performed and imaginary frequencies were not found. The singlet state of the imidazol-2-ylidene unit was chosen in calculations since it is lower in energy than the corresponding triplet state. In the singlet state the lone electron pair is located at the carbon that acts as the Lewis base centre.

For the complexes of imidazol-2-ylidene,[42] a broad spectrum of interactions is observed. For example, for the $C_3N_2H_4\cdots Cl_2$ and $C_3N_2H_4\cdots$ClCCH complexes the interaction energy is equal to –61.6 kcal/mol and –3.9 kcal/mol, respectively (ωB97XD/aug-cc-pVTZ level of calculations). The electron density at the C···Cl BCP (these BCPs are indicated in Fig. VI.4) for these complexes is equal to 0.216 au and 0.011 au, respectively. For the former complex the corresponding Laplacian of the electron density at BCP, $\nabla^2\rho_{BCP}$, is negative while for the latter complex the $\nabla^2\rho_{BCP}$ and H_{BCP} (the total electron energy density at BCP) values are positive. This clearly shows the covalent character of the interaction for the former complex and the weak Lewis acid — Lewis base interaction for the latter. Figure VI.4 presents molecular graphs of these two complexes. One can see that for the $C_3N_2H_4\cdots Cl_2$ complex, one of the atoms of the Cl_2 molecule is very close to the carbon centre of the imidazol-2-ylidene unit. The ωB97XD/aug-cc-pVTZ calculations show that the C···Cl distance in this complex is equal to 1.732 Å while the Cl-Cl bond length in the chlorine molecule is equal to 2.605 Å. Both distances are shorter than the corresponding sums of van der Waals radii (Table I.1 in Chapter I), 3.45 Å and 3.50 Å, respectively, but the geometrical parameters show that the Cl-atom is attached to the carbon centre. In other words, this is the $C_3N_2H_4Cl^+\cdots Cl^-$ complex. One may also state that the Cl^+ cation transfer is observed here, i.e., in the $C_3N_2H_4\cdots Cl_2 \rightarrow C_3N_2H_4Cl^+\cdots Cl^-$ reaction. There are the following characteristics of the Cl···Cl BCP: electron density at BCP is equal to 0.046 au, and the $\nabla^2\rho_{BCP}$ and H_{BCP} values are positive and negative, respectively. This shows partly covalent character of the Cl···Cl interaction. However, it is much weaker than the C···Cl interaction. It confirms that the chlorine atom is

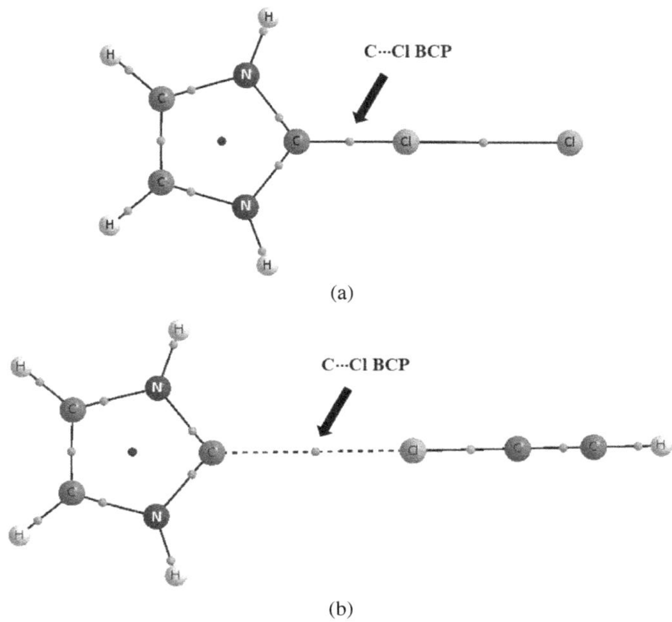

Figure VI.4 Molecular graphs of two complexes linked by halogen bonds: (a) $C_3N_2H_4 \cdots Cl_2$ complex, (b) $C_3N_2H_4 \cdots$ClCCH complex. The great circles correspond to atomic attractors, small circles to the bond and ring critical points. The continuous and broken lines correspond to bond paths.

attached to the carbon centre but not to the other chlorine atom. Thus, one can see that the results concerning the $C_3N_2H_4Cl^+ \cdots Cl^-$ complex discussed here indicate the evidence of chloronium cation transfer. In the case of the $C_3N_2H_4 \cdots$ClCCH complex (Fig. VI.4), the C···Cl distance is equal to 3.151 Å that is close to the corresponding sum of van der Waals radii but it is still a smaller distance. Such geometry corresponds to weak interactions.

The whole sample of the imidazol-2-ylidene complexes was applied to analyse the more general process of halonium transfer, i.e., the C···X-R ⇔ $C\text{-}X^+ \cdots R^-$ process (R is the -X, -CCH or -CN fragment of the Lewis acid unit). Figure VI.5 presents the relationship between the C···X contact and the R-X bond. The Wiberg[43,44] and DI[45,46] indices are considered here for this contact and for this bond. These indices were discussed in Chapter II. The delocalisation index, δ(A,B) (or marked as DI), defined by eq. II.35, corresponds to the number of electrons delocalised between

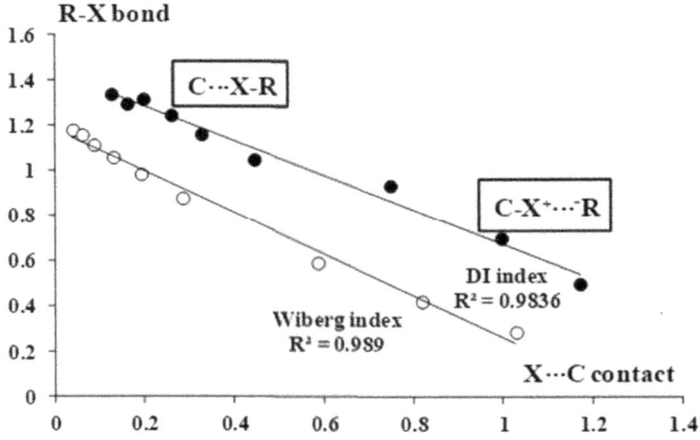

Figure VI.5 Relationship between the index of X···C distance and the index of R-X bond; full circles correspond to DI index and empty circles to the Wiberg index. Reprinted with permission from ref. 42. Copyright 2023 The Royal Society of Chemistry.

two atoms, say A and B. It may be understood as the number of shared electrons or as a measure of degree of covalency. The Wiberg index corresponds to the bond order, and it may also be treated as the number of shared electron pairs between two centres. Thus, both indices correspond to the strength of interaction and very often an excellent correlation is observed between them; such a correlation is presented in Figure II.10 (Chapter II), for example.

The relationships presented in Figure VI.5 for both Wiberg and DI indices express the same phenomenon of halonium cation transfer. For the greater value of the index (Wiberg or DI), a stronger local interaction between the atomic centres occurs. Thus, for the low index values of the C···X contact, greater distances between the imidazol-2-ylidene species and the Lewis acid unit are observed; simultaneously great values of the index for the R-X bond are observed and the bond lengths are rather short. This situation occurs for the C···X-R systems. If the R-X bond elongates, the C···X distance is shortened that corresponds to the decrease of the value of the R-X index and the increase of the value of the C···X index — up to the situation when the X^+ halonium cation is attached to the C-centre of the imidazol-2-ylidene species. The latter situation corresponds to the C-X^+···R^- system where the low value of the R-X bond ($R^- \cdots X^+$ contact) index and the high value of the index for the C···X contact (C-X^+ bond) are observed.

Let us consider other types of reactions, like the S_N2 reaction for which the tetrel bond may be treated as its preliminary stage. Figure VI.6 may be treated as the modelling of the S_N2 reaction. This figure concerns the tetrel-bonded and pnicogen-bonded complexes and it presents the relationships between the Lewis acid centre (tetrel or pnicogen) — Lewis base centre distance and the α-angle. The complexes considered are linked by linear F-Z···B and H-Z···B interactions (Z is the tetrel or pnicogen centre, B is the Lewis base centre), thus the α-angle may be defined as the angle between the line containing the F-Z···B or H-Z···B bridge and one of the other F-Z or H-Z Lewis acid unit bonds. The latter parameter is defined in Fig. VI.6 for two types of tetrahedral Lewis acid species, neutral tetrel moieties and the pnicogen cations. Two complexes, $SiFH_3 \cdots Cl^-$ and $NF_4^+ \cdots$ NCH, were used for this definition. It was mentioned in the previous chapter that the α angle for the ideal tetrahedron structure that is not disturbed by additional interactions

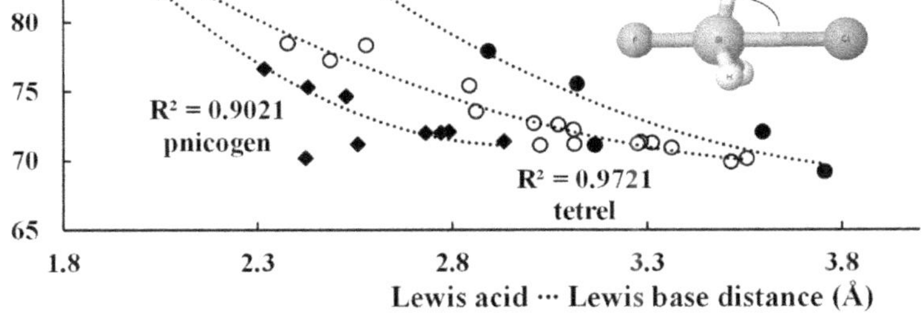

Figure VI.6 Relationships between the Lewis acid···Lewis base distance (Å), Z···B, and the α-parameter (defined at the right side of the figure) in A-Z···B tetrel and pnicogen bonds (Z-tetrel or pnicogen centre). Black and white circles correspond to the tetrel complexes linked by charge-assisted (CA) tetrel bonds and neutral tetrel bonds, respectively. Squares correspond to complexes linked by the pnicogen bond. Modified with permission from ref. 20. Copyright 2020 MDPI.

is equal to 70.5°. Very strong interactions change this tetrahedral structure into the trigonal bipyramid characterised by an α angle equal to 90°.

Two samples of complexes of tetrel units and one sample of complexes of pnicogen species are presented in Fig. VI.6. The geometrical results of these complexes were taken from other studies[47,48] where MP2/aug-cc-pVTZ calculations were performed. For one of the samples, the ZH_4, ZFH_3 and ZF_4 (Z = C, Si, Ge) Lewis acid units interact with the HCN and LiCN Lewis bases (nitrogen atom is the Lewis base centre here). There is a sample with complexes linked by the negatively charge-assisted tetrel bonds, and the same Lewis acid units interact with the chloride anion. For complexes linked by the pnicogen bond, the ZH_4^+, ZFH_3^+ and ZF_4^+ Lewis acid units (Z = N, P and As) interact with HCN and LiCN through the σ-hole at the Z-centre and through the lone electron pair at the nitrogen centre.

One can see (Figure VI.6) that for shorter Lewis acid — Lewis base distances that correspond to stronger interactions, the α-parameter is closer to 90°. Thus, the central part of the complex containing the Z-centre tends to planarity that is a typical geometrical characteristic of the transition state of the S_N2 reaction. Good correlations are observed here between the Lewis acid-base distance and the α-angle since the second-order polynomial dependences possess high values of the correlation coefficients. These relationships may be considered here as the reaction paths of S_N2 reactions. Thus, these reactions change the tetrahedral structures of tetrel neutral or pnicogen cation species into the trigonal bipyramid structures in complexes.

Another process (reaction) may be considered that changes the planar trigonal structure of the triel moiety into the tetrahedral structure. This may occur if the triel Lewis acid unit interacts with the electron-donating species and the corresponding triel bond that links both units is sufficiently strong. The following complexes were analysed theoretically in an early study:[49] $ZF_3\cdots N_2$ and $ZF_3\cdots NCH$ where Z is a triel centre (B, Al, Ga, In, Tl). The MP2/aug-cc-pVTZ level of calculations was applied to optimise these complexes; for indium and thallium atoms the aug-cc-pVTZ-PP basis set and the effective core potential were applied. Frequency calculations were also performed for these complexes to confirm that the optimised structures correspond to energetic minima.

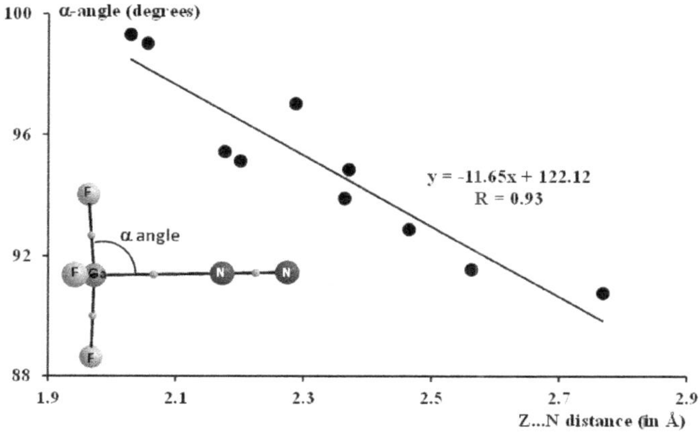

Figure VI.7 Relationship between the Z···N distance (Å) and the α-parameter (°) for the $ZF_3 \cdots N_2$ and $ZF_3 \cdots NCH$ complexes (Z is the triel centre: B, Al, Ga, In, Tl). Reprinted with permission from ref. 49. Copyright 2014 Wiley-VCH Verlag.

Figure VI.7 presents the relationship between the Lewis acid — Lewis base distance (this is the Z…N distance between the triel and nitrogen centres) and the α-angle. This angle is defined as that between the F-Z···N-N (or F-Z···NC-H) line and the Z-F bond which is not on this line. For the planar triel species that is not disturbed by the Lewis base unit, this angle is equal to 90° and it increases with the increase of the strength of interaction, i.e., with the decrease of the Z···N distance. In other words, a shorter distance corresponds to greater sharing of electrons in the Z···N interatomic region and to a geometry that is closer to the tetrahedron.

VI.3 Mechanisms Protecting the Octet (Doublet) Structure – Another Look at Interactions

"There is a close relationship between hydrogen bonding (HB), hypervalent bonding (HVB) and secondary bonding (SB): although often considered in isolation, they have key bonding aspects in common."[50] is the citation from the review article of Crabtree where different terms related to the Lewis acid — Lewis base interactions are discussed; these terms are also analysed in this book. The term "secondary bonding" is applied sometimes in various studies for interactions that are not classified as covalent bonds. In other words, this term, similar to "noncovalent interactions", has the same meaning as "Lewis acid — Lewis base interactions" used in this book (see Chapter I for more explanations of the application of the latter term).

In order to discuss the terms presented above, it is worth starting with the definition of hypervalency. Let us quote Crabtree's article again: "A hypervalent molecule is one that has more than four pairs of valence electrons around the central atom, E, in the Lewis diagram of a given molecule."[50] This definition refers to what is often named as the octet rule (that in the case of a hydrogen central atom is called the doublet rule). Centres which contain fewer than eight electrons in the valence shell are often named as hypovalent on the other hand.[51,52] The octet rule was introduced by Lewis who stated that "the atom tends to hold an even number of electrons in the shell, and especially to hold eight electrons."[53] Lewis also pointed out that "in the row of the periodic table comprising hydrogen and helium we have in place of the rule of eight the rule of two."[53] It is worth mentioning, however, that Abegg was the first who extended the concept of coordination into the valence, and who distinguished atoms into electron donors and electron acceptors, that led to the conclusion that the difference between maximum positive and maximum negative valences of an element is eight.[54] Lewis referred to these statements, naming them as Abegg's rule in the discussion of his concept of eight (two) electrons.[53] The name "octet rule" was introduced by Langmuir.[55]

It has been pointed out that N-E-L is a convenient label to classify hypervalent centres:[50] N is the number of valence electrons belonging formally to the central element according to the Lewis electron structure, E is the identity of the centre considered and L is the number of atoms bonded to E. The number of lone electron pairs is calculated as (N−2L)/2. In such a way, for the PF_5 species where the hypervalent P-centre is marked as 10–P–5, there is no lone electron pairs. For the oxygen centre in the water molecule, H_2O, the O-centre that obeys the octet rule has the following label, 8–O–2, thus the number of lone electron pairs is equal to $(8 - 2 \cdot 2)/2 = 2$.

The term "hypervalent bond" is related to "strong secondary bond". "Secondary bonding is best thought of as incipient hypervalency and occurs when a main group element with an octet structure, EX_x, interacts with the lone pairs of one or more neighbouring groups, Y, that bind weakly to E to give rise to E···Y distances of intermediate length, lying between full bonds and mere contacts."[50] This definition may suggest that secondary bonding occurs for centres obeying the octet (doublet) rule and that it is incipient (preliminary stage) of hypervalency. More precisely, if the interaction is strong enough it leads to hypervalency and the interaction is classified as a hypervalent bond. However, the triel trihalides and trihydrides that possess hypovalent electron structures may interact with electron-donating species. These are triel bonds and, if they are sufficiently strong, will lead to structures where triel centres obey the octet rule. If they are weak, the triel centres will still be hypovalent and may also be classified as secondary bonds. In the case of very strong triel bonds, they possess numerous characteristics of covalency and may be classified as covalent bonds, more or less polarised. This occurs for the $BF_3 \cdots F^-$ interaction, for example, that leads to the BF_4^- anion where all B-F bonds are equivalent in the gas phase.[56] Another example concerns species where multivalent halogen centres possessing hypervalent structures interact with electron donors forming complexes linked by halogen bonds.[12] The $BF_3 \cdots NCH$ and $BF_5 \cdots NCH$ complexes are examples of such a situation. Thus, one can see that not only centres obeying octet (doublet) rule may participate in additional secondary interactions.

There are numerous inaccuracies and inconsistencies regarding secondary interactions. For instance, it is pointed out incorrectly sometimes that they occur only in hypervalent species. It should be simply explained that this term is applied sometimes instead of terms such as non-bonding interactions or Lewis acid — Lewis base interactions. One can refer to the study of Silvi and coworkers where it is stated that "The concept of secondary bond covers a wide range of non-covalent interactions involving an acceptor (or electrophilic) molecule and an electron donor (or nucleophilic) one. It involves triel, tetrel, pnictogen, chalcogen, halogen, and aerogen bonds as well as hydrogen bonds. Such interactions yield complexes in which the internuclear distance of the electrophilic and nucleophilic centers is intermediate between the sums of the covalent and van der Waals radii of these atoms."[57]

Hence another picture of interactions with the use of the terms presented in this section may be described in the following way by considering some ideas of Crabtree.[50] Stable Lewis acid systems may interact with electron-donating species through their centres which often obey octet (doublet) rule. If such interactions are weak and they do not change the stable structure, they are named secondary interactions. If such interactions are strong enough to change the stable structure and if this is the octet (doublet) structure, a hypervalent centre is formed and the central atom ends up with more than eight electrons in the valence shell. Such interactions that change the octet (doublet) structure into hypervalency are called hypervalent bonds. The weak and medium secondary bonds are characterised by the electrostatic interaction that is the most important attractive interaction energy component, well described by the σ-hole and π-hole concept.[3,4] In the case of very strong interactions, among them the hypervalent bonds, the covalent contribution to the interaction is very important. This covalency is well described by orbital-orbital interactions; the n (lone electron pair) → σ* (antibonding orbital) overlap is the most important in the case of the hydrogen bond, halogen bond and numerous other interactions (see Chapter II for descriptions).

One can see from the above descriptions that very often, but not always, the phenomenon of hypervalency is related to strong interactions. There are two models of hypervalency at least.[52] It is assumed, in the first model, that additional d atomic orbitals are used to enlarge the valence shell. However, numerous calculations show that d-type orbitals are characterised by much higher energies than s- and p-type orbitals, thus d-orbitals do not contribute to the valence shells. The second, ionic resonance model assumes high ionic character of bonds and the so-called resonance delocalisation. The latter term for the XAX system (A is the central hypervalent atom here) may be described by the occurrence of resonance structures given in eq. VI.3.

$$\text{X-A}^{+}\ {}^{-}\text{:X} \Leftrightarrow \text{X:}^{-}\ {}^{+}\text{A-X} \qquad \text{(VI.3)}$$

A similar equation may be presented for the A-H···B hydrogen bridges (eq. VI.4) since for the strong hydrogen bonds their covalent character was discussed that is equivalent to the hypervalent character of the H-centre.

$$\text{A-H}^{+}\cdots{}^{-}\text{:B} \Leftrightarrow \text{A:}^{-}\cdots{}^{+}\text{H-B} \qquad \text{(VI.4)}$$

Particularly, in the case of the very strong hydrogen-bonded $(FHF)^-$ system, where both H···F contacts possess covalent character, the previous equation may be presented in the following form:

$$\text{F-H :F}^{-} \Leftrightarrow \text{F:}^{-}\ \text{H-F} \qquad \text{(VI.5)}$$

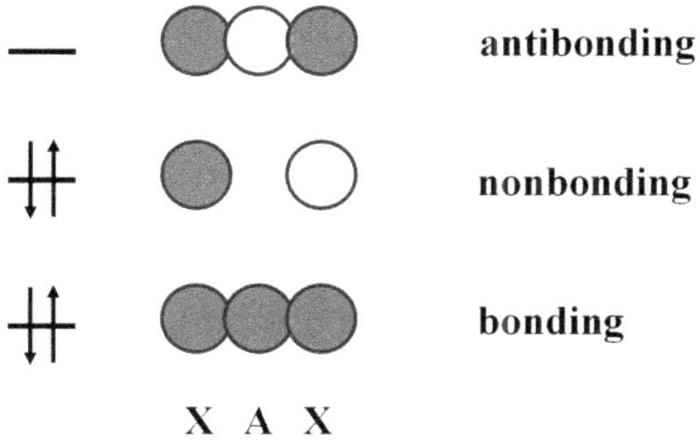

Scheme VI.3 Valence shell molecular orbitals for the $(XAX)^-$ system. Modified with permission from ref. 6. Copyright 2005 Cambridge University Press.

For the $(FHF)^-$ system, the H-centre possesses all characteristics typical of hypervalency. Coulson pointed out that the ionic resonance described in the latter equations is equivalent to the three-centre four-electron (3c-4e) model[58] where electrons occupy the bonding and nonbonding orbitals while the antibonding level is unoccupied (Scheme VI.3). Both ionic resonance and 3c-4e models assume that the links of the central atom possess ionic character; in such a way there is outflow of the electron charge from the central atom to ligands that protect this centre from an excess of electrons and consequently from hypervalency. For both models the role of d atomic orbitals to form molecular orbitals in the hypervalent species is not important, and is even negligible, according to theoretical calculations as mentioned above.

Another model related to the 3c-4e concept assumes the occurrence of so-called hyperbonds,[52] which are designated by ω in the monograph of Weinhold and Landis.[6] Hence, for example, the XAX or $(FHF)^-$ systems are classified as ω_{XAX} and ω_{FHF} hyperbonds, respectively. There are the following characteristics of hyperbonds (say ω_{ABC}):[52] linear A-B-C systems (or nearly so), and A-B and B-C bonds are longer than usual bonds in non-hypervalent cases. These bonds are long due to the high population of σ_{AB}^* and σ_{BC}^* antibonding orbitals. The resonance weighting of A···BC and AB···C forms are equal, or nearly so. The cationic character of the B-centre and the anionic character of A and C centres are observed.

One can see that the ionic resonance model presented here as well as the 3c-4e and hyperbond concepts explain partly the mechanisms that protect the octet (doublet) structures. These are, in the case of the A-centre of the AX_n species, electron charge shifts from the A-centre to the X atoms (ligands). In other words, polarisation of the A-X bonds occurs. Let us analyse various types of hypervalent species, but particularly various types of the secondary bonds. The examples presented below show that not only the systems that obey octet (doublet) rule participate in such interactions, but also hypervalent and hypovalent centres may be involved in additional interactions, i.e., secondary bonding. Figure VI.8 presents the selected characteristics of the BrF_3 and BrF_5 species where Br is a hypervalent centre.[1,12] The bromine centre in the BrF_3 species possesses formally ten electrons in the valence shell, three σ-orbitals corresponding to the Br-F bonds and two lone electron pairs. Figure VI.8a presents the molecular graph and reactive surface that is characterised by the Laplacian value equal to zero, $\nabla^2\rho = 0$, for the BrF_3 species. The reactive surface indicates the accumulation of electron charge at fluorine centres and there are areas of such accumulation above and below the plane of the molecule that possesses C_{2v} symmetry. The latter accumulations correspond to two lone electron pairs. The BCPs corresponding to the B-F bond paths are located outside of the area of negative values of $\nabla^2\rho$, hence the B-F bonds are not "purely" covalent in nature, but seem to be strongly polarised bonds. The distribution of atomic charges (Fig. VI.8b) confirms the latter assumption; the QTAIM charge for bromine

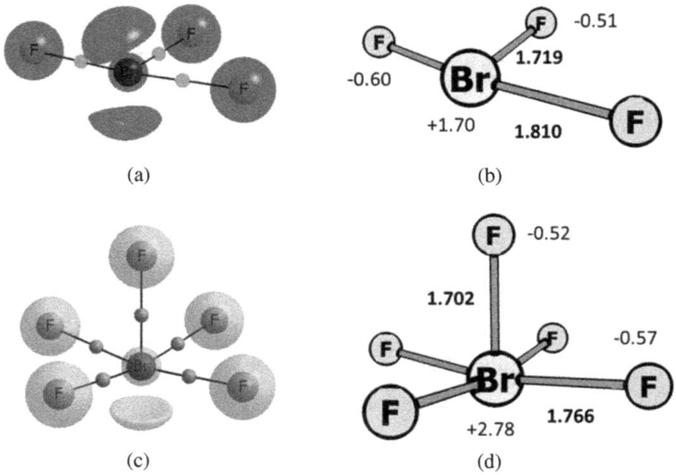

Figure VI.8 Selected characteristics of the BrF_3 (a and b) and BrF_5 (c and d) species. Reactive surfaces of the (a) BrF_3 and (c) BrF_5 molecules; selected parameters are presented, i.e., bond lengths (Å) and QTAIM atomic charges (au). Reprinted with permission from ref. 1. Copyright 2017 The Royal Society of Chemistry.

is equal to +1.70 au while for fluorine atoms, –0.60 au for those located in the F-Br-F closely linear fragment, and –0.51 au for the fluorine centre located at the C_2 symmetry axis. The length of the Br-F bond located at the C_2 axis is equal to 1.719 Å while the bond lengths of the F-Br-F system are equal to 1.820 Å. It clearly shows that the Br-F bonds are strongly polarised, and those of the F-Br-F system close to linearity are more polarised than the bond located at the 2-fold axis of symmetry. These polarisations are related to the outflow of the electron charge from bromine. This outflow tries to protect the octet electron structure of the valence shell of bromine. In other words, this is the 3c-4e system, or it may be classified as the hypervalent ω_{FBrF} bond.

The BrF_5 molecule is also characterised by a hypervalent bromine centre since it possesses 12 electrons in the valence shell, five σ-orbitals of the Br-F bonds and one lone electron pair. This molecule possesses C_{4v} symmetry. Figure VI.8c presents the molecular graph and reactive surface for the BrF_5 species. One can observe the areas of the accumulation of electron charge at F-centres as well as the area corresponding to the lone electron pair, similar to the BrF_3 molecule. This lone electron pair is located at the 4-fold axis of symmetry, in the extension of the axial F-Br bond. Figure VI.8d shows that the latter Br-F bond located at the 4-fold symmetry axis is equal to 1.702 Å, and it is shorter than the four remaining Br-F bonds, 1.766 Å. The fluorine atomic charge of this shorter bond is less negative, -0.52 au, than the remaining fluorine charges, -0.57 au. The positive charge of bromine, +2.78 au, indicates strong polarisation of all Br-F bonds, with smaller polarisation for the bond located at the C_4 axis of symmetry than for the remaining bonds. One can see here two F-Br-F systems that are close to linearity that may be classified as the ω_{FBrF} hypervalent bonds.

Figure VI.9 presents the complexes of the BrF_3 and BrF_5 moieties discussed above with hydrogen cyanide, HCN, and with chloride anion, Cl^-, acting as the Lewis base units (HCN plays the role of Lewis base through the nitrogen centre). Thus, these are the secondary bonds, i.e., halogen bonds, where the multivalent halogen centres of Lewis acid units are characterised by hypervalency. The halogen bond for the $F_3Br\cdots NCH$ complex is not so strong since the binding energy, E_{bin}, is equal to –6.9 kcal/mol. The $Br\cdots N$ distance is equal to 2.693 Å that is fewer than the corresponding sum of van der Waals radii, 3.38 Å (Table I.1). The complexation leads to electron charge transfer from the HCN unit to the hypervalent structure. The system tries to defend itself against the additional excess of the electron charge, hence there is further electron

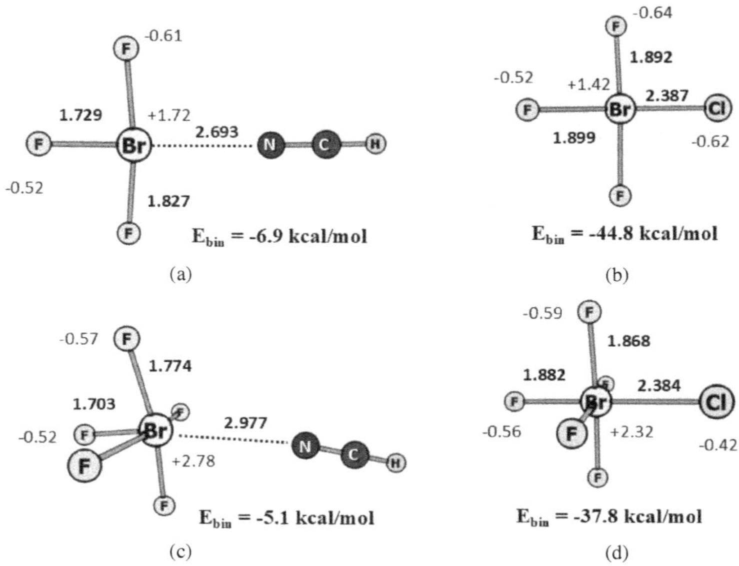

Figure VI.9 Complexes of BrF_3 and BrF_5 with hydrogen cyanide and chloride anion; binding energies (in kcal/mol), bond lengths (Å) and charges (au) are presented. Reprinted with permission from ref. 1. Copyright 2017 The Royal Society of Chemistry.

charge transfer from the bromine centre to the fluorine atoms. The bromine centre is more positive, and the fluorine centres are more negative in this complex than in the isolated BrF_3 molecule (Figs. VI.9a and VI.8b, respectively). The Br-F bonds in this complex are longer in comparison with the isolated hypervalent unit. However, these differences of charges and of bond lengths are rather small that shows the Br···N halogen bond does not change significantly the electron configuration of the hypervalent bromine. Another situation is observed for the strong interaction in the $BrF_3\cdots Cl^-$ complex (Fig. VI.9b). The binding energy is equal to –44.8 kcal/mol that indicates very strong interaction, and the Br···Cl distance equal to 2.387 Å is much shorter than the corresponding sum of van der Waals radii, 3.58 Å (Table I.1). The Laplacian of the electron density at BCP, $\nabla^2\rho_{BCP}$, is positive but the total electron energy density at BCP, H_{BCP}, is negative that confirms at least partly covalent character of this interaction. All atoms of the Lewis acid unit in the $BrF_3\cdots Cl^-$ complex (Fig. VI.9b) possess smaller atomic charges than for the isolated BrF_3 species (Fig. VI.8b) since fluorines are slightly more negative while the bromine centre is much less positive than in the isolated BrF_3 unit. Formally due to the additional $Br\cdots Cl^-$ contact, the bromine centre increases the number of electrons in the valence shell, from 10 electrons in the isolated Lewis acid unit to 12 electrons in the complex with the chloride anion. However, the elongation of Br-F bond lengths due to the complexation is much more significant than in the case of the complex with hydrogen cyanide. For the $BrF_3\cdots Cl^-$ complex two hyperbonds are observed, F-Br-F and F-Br-Cl, designated as ω_{FBrF} and ω_{FBrCl}, respectively.

The $BrF_5\cdots NCH$ and $BrF_5\cdots Cl^-$ complexes are also presented in Fig. VI.9. For the former complex (Fig. VI.9c), changes of charges of fluorine centres in comparison with the BrF_5 unit not involved in interactions are not observed. The E_{bin} is equal to –5.1 kcal/mol, thus it means that such interaction is not sufficient to make any changes in the Lewis acid unit. However, the $Br\cdots N$ distance is equal to 2.977 Å that is less than the corresponding sum of van der Waals radii, 3.38 Å (Table I.1). The $F\text{-}Br\cdots NCH$ arrangement is not linear due to the location of the lone electron pair (see Fig. VI.8c). This location results in the greatest positive EP at the molecular surface outside the straight extension of the F-Br bond.[12]

The greater changes of the BrF_5 unit occur in the $BrF_5\cdots Cl^-$ complex (Fig. VI.9d) where, due to the strong interaction, a linear F-Br-Cl arrangement is observed. A slight decrease of the atomic charges of four F-centres (they become "more negative") and a greater decrease of the positive charge of Br-centre in comparison with the BrF_5 not involved in interactions are observed. In the case of the F-centre located at the F-Br-Cl line, its charge is slightly less negative than in the isolated molecule. The great decrease of the positive charge of bromine results from the great electron charge transfer from the chloride anion to the BrF_5 unit as a result of complexation. The chlorine centre charge in the complex is equal only to –0.42 au, in comparison to –1 au in the isolated chloride anion. The E_{bin} value is equal to –37.8 kcal/mol here and partly covalent interaction is observed since the H_{BCP} at the BCP of $Br\cdots Cl$ bond path is negative ($\nabla^2\rho_{BCP}$ is positive). The elongation of the Br-F bonds is a result of complexation, much greater than for the $BrF_5\cdots NCH$ complex. The $Br\cdots Cl$ distance in the $BrF_5\cdots Cl^-$ complex equal to 2.384 Å is much shorter than the corresponding sum of van der Waals radii, 3.58 Å (Table I.1); this is a very similar situation to that occurring for the $BrF_3\cdots Cl^-$ complex. One can see three hypervalent bonds here: two F-Br-F bonds,

ω_{FBrF}, and one F-Br-Cl bond, ω_{FBrCl}. The deformed octahedral molecular structure of this complex is observed.

Figure VI.10 presents molecular graphs and reactive surfaces for two conformations of the $Cl_3B\cdots NCH$ complex for which MP2/aug-cc-pVTZ calculations were performed.[2,59] This is an example of a hypovalent species that acts as the Lewis acid unit through the π-hole located in the area parallel to the plane of the molecule interacting with the HCN Lewis base unit; the latter acts as the electron donor through the lone electron pair at the nitrogen centre. It was found that for some triel trihalides that interact with electron donors, there are two conformations related to the local energetic minima:[59] the conformation linked by weak interaction for the long distance between Lewis acid-base units, and the conformation linked by strong interaction where such a distance is very short. Such a situation occurs for the $Cl_3B\cdots NCH$ complex that is linked by the secondary interactions classified as π-hole and triel bonds. For one of the conformers the B···N distance is equal to 2.817 Å, for another 1.628 Å, while the corresponding sum of van der Waals radii is equal to 3.47 Å. For the conformer with the short B···N

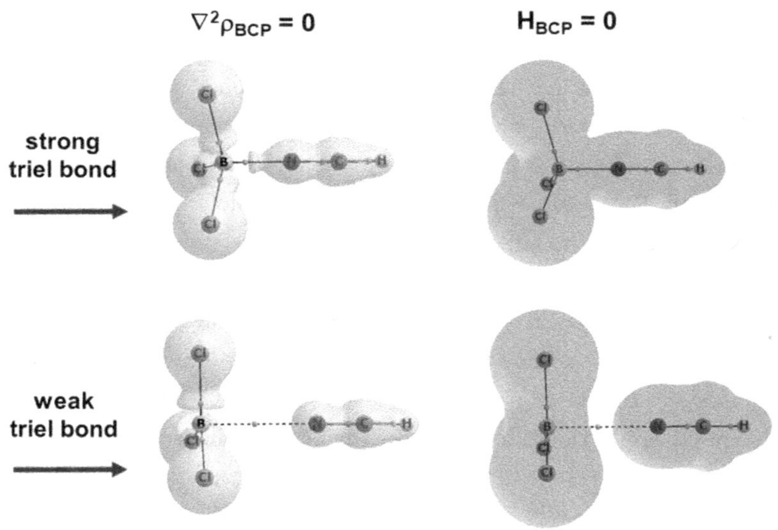

Figure VI.10 Reactive surfaces based on the Laplacian of the electron density and on the total electron energy density for two conformations of the $Cl_3B\cdots NCH$ complex (MP2/aug-ccpVTZ results). Reprinted with permission from ref. 2. Copyright 2021 MDPI.

distance, the interaction and binding energies are equal to –23.6 kcal/mol and –3.8 kcal/mol, respectively. This indicates a great deformation energy of +19.8 kcal/mol (see Chapter I, eq. I.2). And indeed, the trigonal planar BCl_3 system becomes tetrahedral in the complex with hydrogen cyanide (Fig. VI.10). For the complex with the long B···N distance, the interaction and binding energies are equal to –3.5 kcal/mol and –3.2 kcal/mol, respectively, thus negligible deformation energy is equal to 0.3 kcal/mol. The BCl_3 unit in this conformation of the complex is still planar.

The reactive surfaces based on the Laplacian of the electron density show that for both conformers the B···N BCP is located outside the area of the negative Laplacian, $\nabla^2\rho$. In the case of reactive surfaces based on the total electron energy density, *H*, for the conformer characterised by strong interactions the B···N BCP is located inside the area of the negative *H*. This conformer is linked by the partly covalent triel bond. In the case of the weaker triel bond, the B···N BCP is situated outside the *H* reactive surface; *H* is positive for this BCP.

Figure VI.11 presents the energetic diagram for the reaction of the splitting of hydrogen at the sulfur centre of the SFH species in the gas phase.[60] This is the SFH + H_2 reaction that, through a few stages, leads to the final products corresponding to the global minimum, i.e., to the H_2S···HF complex linked through the hydrogen bond. For reactants (stage E1) corresponding to the local energetic minimum, the S-centre obeys the octet rule since there are two σ-orbitals corresponding to the S-F and S-H bonds and two lone electron pairs. The sulfur centre is weakly connected with the molecular hydrogen through the chalcogen bond. The Lewis acid S-centre interacts through the σ-hole located in the extension of the F-S bond approximately, and the molecular hydrogen donates electron charge through its σ-bond. This is the S···σ chalcogen bond (secondary interaction). Next are the E2 (transition state) and E3 (local minimum) stages where the S-centre is hypervalent due to the secondary interaction mentioned above. There are ten electrons in the valence shell of the S-centre for these stages. The molecular structures correspond here to the trigonal bipyramid, approximately. The lone electron pair is directed to one of the vertices of this bipyramid. For the E2 and E3 stages an F-S-H system close to linearity is observed that corresponds to the ω_{FSH} hyperbond. In the E2 stage the trigonal bipyramid is disturbed more than for the E3 stage. However, the E2 stage is not energetically stable, and it corresponds to a transition state. For the E4 stage, F^-+H_3S^+ system, the S-centre obeys the octet rule (three S-H

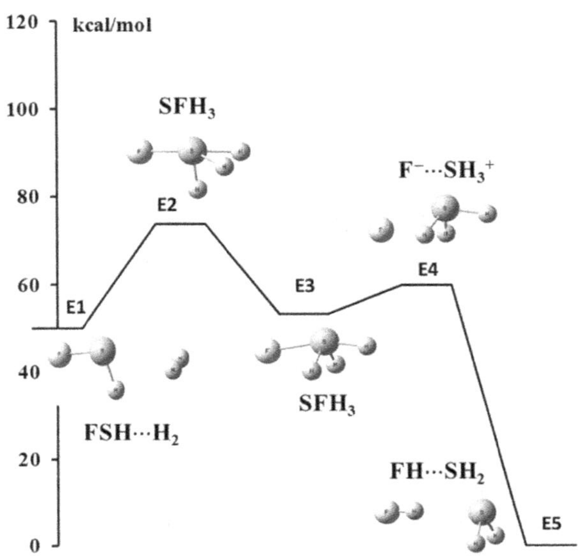

Figure VI.11 Reaction of the splitting of hydrogen for SFH + H_2 reactants. Energies of species on the vertical axis are related to the global minimum of the $H_2S\cdots HF$ system (right side of the figure); the case of the gas phase. Reprinted with permission from ref. 60. Copyright 2015 The Royal Society of Chemistry.

bonds and one lone electron pair). However, this system is not energetically stable, and it corresponds to a transition state. For the last stage, E5, the $H_2S\cdots HF$ complex is formed that is linked by the F-H···S hydrogen bond; here the sulfur centre also obeys the octet rule.

One can see from these examples (Figs. VI.8–11) that secondary reactions are observed not necessarily for systems obeying the octet rule that leads to hypervalent species. The hypervalent and hypovalent species may also be involved in secondary reactions; in the latter case the secondary interactions lead to systems that obey the octet rule.

VI.4 The VSEPR Model and the Classification of Interactions

It has been described in the previous chapter that numerous interactions lead to changes of molecular structure of interacting units. For example, the tetrahedral structures of tetrel, pnicogen or chalcogen centres may interact strongly as the Lewis acid sites with electron donors; in consequence these electrophilic centres change their structures into trigonal bipyramids. The triel trigonal centre may change its configuration into a tetrahedron if it interacts with a Lewis base unit. The only condition is that such interactions must be strong enough to change configurations. It is worth mentioning that these geometrical changes agree with the valence shell electron pair repulsion model (VSEPR).[61,62]

It is not the aim of this book to present a full description of the VSEPR model and all its assumptions. However, those that are important to explain the processes of the change of structures are presented here. Only the main group elements are considered in the discussion here and, for the clarity of explanations, only cases where the centres considered are linked with other atoms by single bonds. This means that only the s- and p-type atomic orbitals of the centre discussed are considered. Nevertheless, this seems to be justified since numerous theoretical calculations show that d orbitals participate negligibly in atomic hybrid orbitals that are responsible for the structures of centres considered.

According to the VSEPR model, the configuration of the centre considered depends on the number of valence electron pairs; these are electron pairs of bonds and lone electron pairs. The location of valence electron pairs around the centre is to maximise their distances apart. It is also assumed that non-valence electrons (core electrons) with the nucleus possess spherical symmetry that practically does not change during the changes of geometries and electron configurations.

The following structures specified below that consider valence electrons and that are in agreement with the VSEPR model occur most often; the number of electron pairs (bonds and lone electron pairs) increases in this specification from two to six. These are: linear (L), trigonal (Tr), tetrahedral (Td), trigonal bipyramid (TBP), and octahedron (Oc) structures. The abbreviations come from the study of Martin.[63]

The structures mentioned above were analysed for the $Be(NCH)_n^{2+}$ and $Mg(NCH)_n^{2+}$ clusters where wB97XD/aug-cc-pVTZ calculations were performed up to the number of NCH ligands, n, equal to six.[64] The n coordination number determines the type of configuration. The beryllium clusters were discussed briefly in Chapter I and presented in Figure I.17. This figure shows their molecular graphs with the maps of EP. One can see that for the $Be(NCH)_2^{2+}$, $Be(NCH)_3^{2+}$ and $Be(NCH)_4^{2+}$ systems the L, Tr and Td structures, respectively, are observed. Similar structures occur for the magnesium clusters characterised by the corresponding coordination number n, i.e., the number of NCH ligands. For the $Be(NCH)_5^{2+}$ cluster (Fig. I.17), the structure can be assigned as deformed TBP. However, in the case of the $Mg(NCH)_5^{2+}$ system (Fig. I.18 in Chapter I), two configurations corresponding to energetic minima are observed; one is characterised by the TBP structure, another is like a deformed Td structure containing four NCH ligands with the additional, fifth NCH ligand. For both $Be(NCH)_6^{2+}$ and $Mg(NCH)_6^{2+}$ clusters, the Oc structure occurs.

Figure VI.12 presents the structures discussed above as well as interactions that can lead to a change of their configurations. It is worth mentioning that these bonds and lone electron pairs are presented schematically only; the shape of lone electron pairs is different than in this figure. It is presented in Fig. VI.12 that the Td structure with four bond electron pairs that interacts, through the tetrel centre, with the electron donor may be changed into TBP; this interaction is known as the tetrel bond. However, there are other Td structures that contain one, two and three lone electron pairs for pnicogen, chalcogen and halogen centres, respectively, that interacting with electron donors change structures into TBPs. For all these TBP structures, the centre playing the role of the Lewis acid site is characterised by hypervalency, while the centres of Td structures obey the octet rule.

It is useful to introduce here the notation applied by Crabtree for clarity of further discussion.[50] This notation is also applied in Fig. VI.12 to mark the structures (configurations) presented. The AX_nE_m designates the species with the A-centre that usually acts as the Lewis acid site, n is the number of X-ligands (differences between ligands are not specified) and m is the number of lone electron pairs, Es. This notation may be useful to mark and name various types of interactions since their current markings and names seem to be inappropriate.[2,65] Present names concern whole groups of elements (halogen bond, chalcogen bond and others), others are related to single elements (hydrogen bond, carbon bond and others), and usually the element that gives the name to the interaction is the Lewis acid centre. However, these interactions occur between the electron acceptors and electron donors, thus the names of the latter should also be used to designate interactions. The main reservation

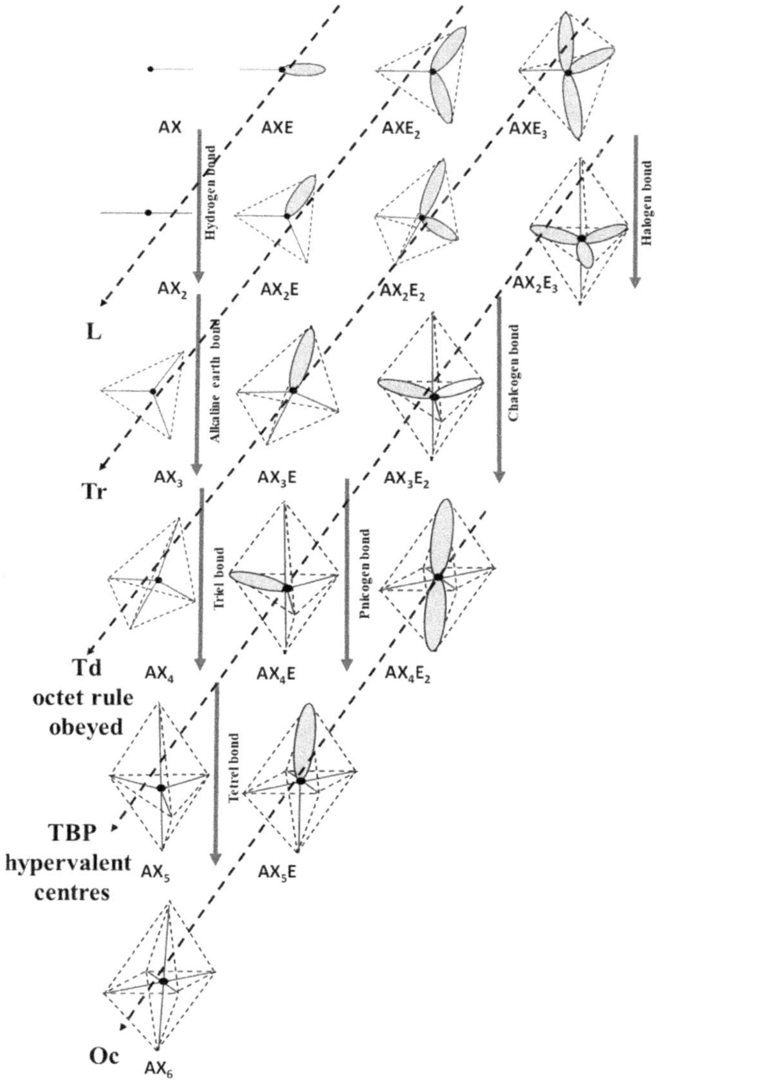

Figure VI.12 VSEPR structures and their interrelations with interactions. The lone electron pairs presented as ellipsoids do not correspond to their shapes; they are introduced to show schematically their localisations according to the VSEPR model. The bonds are designated by solid lines. The black circle corresponds to the centre considered. AX_nE_m marks the type of structure where n and m subscripts correspond to the number of ligands (X) and the number of electron pairs (E), respectively (n or m is omitted if it is equal to one). The arrows show interactions that change one of the structures into another. Modified with permission from ref. 2. Copyright 2021 MDPI.

to the commonly used names of interactions may concern those that are related to groups of elements. For example, the halogen bond concerns systems where the element of the 17th group acts as the Lewis acid site but various properties of elements of this group are detected; e.g., the iodine centre is usually a much better (stronger) Lewis acid site than chlorine.[3–5]

The tetrel bond that changes the Td configuration into the TBP may be designated as Td → TBD. However, pnicogen, chalcogen and halogen bonds that also change Td into TBD may be designated in the same way. Maybe it is better to apply here the notation of Crabtree[50] to mark the tetrel bond as $AX_4 \rightarrow AX_5$ and the other Td → TBD interactions as $AX_3E \rightarrow AX_4E$, $AX_2E_2 \rightarrow AX_3E_2$, and $AXE_3 \rightarrow AX_2E_3$ for the pnicogen, chalcogen and halogen bonds, respectively. It is worth mentioning here that for the TBP structures containing lone electron pairs, the following names are applied: see-saw structure for the AX_4E configuration and T-shaped for the AX_3E_2 one.

Another proposal for names of interactions[2] seems to be much better than other markings since it is based on the centres being in contact. For example, the S···O interaction is the chalcogen bond where the sulfur atom plays the role of the Lewis acid site while the oxygen centre belongs to the Lewis base unit. In other words, the S···O designation concerns the chalcogen bond. From this designation one knows the centres involved, whereas the name chalcogen bond is related to several electron donors and acceptors. The only matter for this proposal is to establish the convention that the Lewis acid site is located on the left of the marking. Besides, this new type of designation is in line with several definitions of interactions where only two centres being in contact are specified; for example, the two-sites hydrogen bond definition, 2sHB (see Chapter IV for more descriptions),[66,67] or the definition of Politzer and coworkers concerning the halogen bond.[3] In the case of the hydrogen bond with the oxygen proton acceptor the following designation may be applied, H···O. Such a designation concerns various proton donors; thus it may concern the O-H···O, N-H···O or C-H···O arrangements that are commonly accepted markings. However, the "two sites" designations may be slightly extended to indicate the centres linked to those being in contact, thus the (O)H···O(C) designation may be applied, which may concern the hydrogen bond between carboxylic groups (see an example in Chapter IV, Fig. IV.5).

Let us come back to Figure VI.12. Broken arrows show the type of structure (configuration), L, Tr, Td, TBP and Oc. The Td → TBP change was discussed. For the latter structure, hypervalent centres are observed. In one case all electron pairs concern the bond molecular orbitals, while for the remaining structures the lone electron pairs occur apart from the covalent bonds. It is important that the lone electron pairs occupy the equatorial positions of the trigonal bipyramid but not the axial ones, which are occupied by bonds. Why? This may be explained by the differences between electron pairs of bonds and lone electron pairs; the latter need more room to be located. And the equatorial positions of TBP possess more room than the axial ones.[51] There is another explanation related to the phenomenon of hypervalency. The lone electron pair orbitals are usually fully occupied; the occupancy is close to two. In the case of the σ-bond molecular orbitals their occupancy is often lower than two, sometimes even significantly. Besides, in the case when three atoms occupy two axial positions and the central position of TBP, say X1, X2 and X3 atoms, we have the X1-X2-X3 linear system where the ω_{X1X2X3} hypervalent bond may be formed. Such an arrangement protects the octet rule since the outflow of electron charge from the X2 centre occurs that leads to strong polarisation of the X1-X2 and X2-X3 bonds and to the decrease of negative charge of the X2 centre. The covalent bonds localised at equatorial positions of TBP cannot form hypervalent bonds since there is no tri-atomic linear arrangement here. In the case of the Oc structure, the occurrence of three ω_{X1X2X3} hyperbonds is possible.

Special attention should be paid to the hydrogen and halogen bonds that are indicated in Fig. VI.12 as those, if they are sufficiently strong, change the AX and AXE_3 structures into AX_2 and AX_2E_3, respectively. In the case of hydrogen bonds the triatomic A-H···B, or in another designation (A)H···B, systems are usually linear or nearly so. Thus, one can consider the occurrence of ω_{AHB} hypervalent bonds. However, it occurs only for very strong interactions, for example for the $(FHF)^-$ anion[29] where without any doubt the hypervalent H-centre, and consequently the hypervalent ω_{FHF} bond, is observed. There are numerous examples of other very strong hydrogen bonds where hypervalency occurs (see Chapter IV). The question arises as to whether there are changes of the linear (A)H···B hydrogen-bonded systems, i.e., AX_2 structures where A is the hydrogen centre, into AX_3 structures (A in all AX_nE_m marks is any centre considered and it does not have the same meaning as A in the A-H···B hydrogen bond designation). This may occur for the bifurcated proton donor hydrogen bond. The latter was mentioned by Jeffrey and Saenger in their monograph.[68] It was explained in Chapter IV that in this type of interaction the H-centre is involved in the covalent bond and additionally it is linked with two proton acceptors. However such interactions are rather weak, and named as partial interactions.[69] For these systems the H-centre coordination is three but usually they do not have regular trigonal geometry.

The change of AX_2 configuration into AX_3 is more probable for other centres than for hydrogen. One can mention the study on beryllium bonds and hydrogen bonds.[70] B3LYP/6-31+G(d,p) calculations were performed to optimise, among others, the complexes of BeH_2 and BeF_2 with one and two imidazole molecules, respectively. The same B3LYP functional and a more extended Pople-style basis set, 6-311++G(3df,2p), were applied to perform more accurate energetic calculations. It was found that the BeH_2 and BeF_2 species, due to the positive EP at the Be-centre, act as the Lewis acid units. If they interact with one imidazole molecule, the $AX_2 \rightarrow AX_3$ change occurs where the imidazole acts as the Lewis base through the nitrogen centre. This change results in trigonal structures. If each of the BeH_2 and BeF_2 species interacts with two imidazole molecules, tetrahedral structures occur for the beryllium centre that is connected with two H or F atoms and additionally with two N-centres. Thus, the following change of structures is observed for these beryllium species: $AX_2 \rightarrow AX_3 \rightarrow AX_4$. In both structural changes the beryllium bonds play an important role.

It is worth mentioning here that the structures presented in Fig. VI.12 correspond to ideal geometries. It occurs most often that the changes of structures are not so perfect; the L, Tr, Td, TBP and Oc configurations are often deformed, and the ligands and lone electron pairs are not located in a way corresponding to the ideal geometries of structures. This occurs also for the halogen-bonded systems.

Let us consider the monovalent halogen centre that interacts with the Lewis base unit. This halogen centre, if not involved in any interaction, is characterised by the AXE_3 configuration with sp^3 hybridisation that concerns one single bond to the halogen and three lone electron pairs. This is the idealised situation, similar to the water molecule where the O-centre is characterised by the same hybridisation with two O-H bonds and two lone electron pairs directed to the vertices of the tetrahedron. However, two lone electron pairs are not equivalent as it should occur for the idealised sp^3

hybridisation.[71] In the case of the monovalent halogen centre acting as the Lewis acid site that interacts with electron donors, the $AXE_3 \rightarrow AX_2E_3$ change of structure occurs. This happens for the (C)Cl···O interaction that is observed in the crystal structure of 1-methylpyrrol-2-yl trichloromethyl ketone (Fig. I.8 in Chapter I).[72] The other examples presented in Fig. VI.4, with the (Cl)Cl···C and (C)Cl···C halogen bonds in the $C_3N_2H_4 \cdots Cl_2$ and $C_3N_2H_4 \cdots ClCCH$ complexes, respectively, correspond to the $AXE_3 \rightarrow AX_2E_3$ change.

Different changes occur for the multivalent halogens that possess properties of Lewis acid centres. The interactions of bromine trifluoride, BrF_3, and bromine pentafluoride, BrF_5, species (Fig. VI.8) with hydrogen cyanide, HCN and chloride anion, Cl^-, acting as the Lewis base units (Fig. VI.9) were discussed earlier. For the BrF_3 and BrF_5 complexes, there are the following designations for changes of structures: $AX_3E_2 \rightarrow AX_4E_2$ and $AX_5E \rightarrow AX_6E$, respectively. In the former case the TBP structure is changed into Oc. In the latter case the Oc structure is changed, but the resulting structure is not shown in Fig. VI.12. The electron charge distribution for the AX_6E configuration is more complicated. However, it is worth to note that in the case of the strong halogen bond in the $BrF_5 \cdots Cl^-$ complex, the Br··· Cl^- interaction possesses characteristics of a covalent bond, thus one may say of the BrF_5Cl^- anion that the Br-F and Br-Cl bonds form the Oc structure and the additional lone electron pair is located at the bromine centre and it possesses mainly s-character.[12]

Let us consider the case of Tr structures that are involved in triel bonds that lead to Td structures, $AX_3 \rightarrow AX_4$. Figure VI.10 presents two conformations of the $BCl_3 \cdots NCH$ complex, where B···N interaction is observed and for the conformer characterised by the strong triel bond the structure of the complex is very close to tetrahedral. It is important that the tetrahedral triel centres may participate in further interactions with the electron-donating species that leads to the TBP structures (AX_5). Figure I.14 (Chapter I) presents two examples of complexes of aluminum trichloride-tetrahydrofuran[73] and aluminum trihydride-tetrahydrofuran[74] that possess TBP configurations in crystal structures.

Let us return to Figure VI.11 that presents the mechanism of the splitting of molecular hydrogen. For the E1 stage the FSH species possesses the tetrahedron structure, Td,

AX_2E_2 according to notation applied in Fig. VI.12. This structure is only slightly disturbed due to the interaction with the dihydrogen, i.e., the chalcogen bond. For E2 and E3 stages the species is characterised by the TBP structure, AX_4E configuration. For the E4 transition state the disturbed tetrahedron occurs, AX_3E; the deformation of the tetrahedron structure is observed due to the relatively strong interaction of the SH_3^+ unit with the fluoride anion. The final E5 stage corresponding to the global minimum concerns the H_2S molecule possessing the tetrahedral AX_2E_2 structure disturbed by the (F)H···S hydrogen bond. Thus, one can see that the notations seem to be useful to describe various processes.

It is worth to note that all classifications and changes of structures discussed in this chapter are mainly based on the Lewis acid centres. The Lewis base centres are not discussed in detail, and similarly studies concerning Lewis base centre configurations are rather rare. For example, Echeverría analysed crystal structures containing methyl groups acting as electron donors in interactions with triel centres.[75] Very recently, the triel bonds were analysed theoretically where the C-centre of the methyl group of $Be(CH_3)_2$ or $Mg(CH_3)_2$ is linked with the triel centre of triel trihydride or triel trihalide.[76] It was found that the carbon Lewis base centre possessing the tetrahedral structure, Td, changes into trigonal bipyramid, TBP, when the triel bond is formed. These interactions are very strong sometimes in this sample of complexes despite the weak Lewis base properties of the methyl groups. In the case of complexes with the $Be(CH_3)_2$ species, the absolute value of the energy of interaction exceeds 22 kcal/mol sometimes, while for some complexes of the $Mg(CH_3)_2$ species it exceeds 68 kcal/mol. Numerous interactions for this sample of complexes are covalent in nature since the Laplacian of the electron density at C···triel BCP, $\nabla^2\rho_{BCP}$, is negative or at least the total electron energy density at BCP, H_{BCP}, is negative.[76] There are other examples of studies where centres possessing Lewis base properties are analysed; however, this is not the scope of this book, so only a few examples were presented here.

Summary

The various interactions, among them the hydrogen bond, lead to numerous processes and reactions. It is well known that the absolute values of all interaction energy terms, resulting from decomposition schemes, increase with the increase of the strength of interaction. The Pauli repulsive interaction that certainly also increases with the strength of interaction, and thus with the decrease of distance between the Lewis acid and Lewis base units, initialises numerous other processes. For example, it leads to the outflow of electron charge from the Lewis acid centre, say Z, being in contact with the Lewis base unit to other parts of the electron acceptor unit. In general, in the case of the A-Z···B interaction, the outflow of electron charge from the Z-centre to the A-part of the unit leads to the increase of polarisation of the A-Z bond and to the increase of positive charge of the Z-centre. The latter is particularly evident in the case of hydrogen bond formation where it relates to the increase of positive charge of the H-centre.

These changes in the electron charge of the Z (or particularly H) centre relate to the protection of its octet (doublet) structure since the excess of electron charge leads to hypervalent structures. Another way to protect the octet (doublet) structure is the occurrence of triatomic linear systems that correspond to hyperbonds. For such triatomic systems, say A-B-C, the outflow of electron charge from B to A and C centres increases the polarisation of A-B and B-C bonds that are connected with the protection of the octet (doublet) rule.

It is important that interactions lead to change of configurations of the interacting centres if they are strong enough; these configurations follow VSEPR. Hence, the interactions may be classified in terms of changes of configurations. The configurations that occur most often are linear, L, trigonal, Tr, tetrahedral, Td, trigonal bipyramid, TBP and octahedral, Oc. In the case of TBP and Oc structures, the hyperbonds may occur to protect the octet (doublet) rule.

Appendix

The following tools were applied to construct figures presented in this chapter.

Figure VI.4

AIMAll program — Reference: Keith, T.A. *AIMAll (Version 11.08.23)*; TK Gristmill Software: Overland Park, KS, **2011**.

Schemes VI.2,3 and Figures VI.1,2,3,5,6,7,8,9,10,11,12
Reprinted with permission.

Scheme VI.1
PowerPoint.

References

1. Grabowski, S.J. Hydrogen bonds, and σ-hole and π-hole bonds — mechanisms protecting doublet and octet electron structures. *Phys. Chem. Chem. Phys.* **2017**, *19*, 29742–29759.
2. Grabowski, S.J. Classification of so-called non-covalent interactions based on VSEPR model. *Molecules* **2021**, *26*, 4939.
3. Politzer, P.; Murray, J.S.; Clark, T. Halogen bonding: an electrostatically-driven highly directional noncovalent interaction. *Phys. Chem. Chem. Phys.* **2010**, *12*, 7748–7757.
4. Politzer, P.; Murray, J.S.; Clark, T. Halogen bonding and other σ-hole interactions: a perspective. *Phys. Chem. Chem. Phys.* **2013**, *15*, 11178–11189.
5. Clark, T. σ-Holes. *WIREs Comput. Mol. Sci.* **2013**, *3*, 13–20.
6. Weinhold, F.; Landis, C. *Valency and Bonding: A Natural Bond Orbital Donor-Acceptor Perspective*, Cambridge University Press, Cambridge, **2005**.
7. Grabowski, S.J. Hydrogen and halogen bonds are ruled by the same mechanisms. *Phys. Chem. Chem. Phys.* **2013**, *15*, 7249–7259.
8. Landrum, G.A.; Goldberg, N.; Hoffmann, R.; Minyaev, R.M. Intermolecular interactions between hypervalent molecules: Ph_2IX and XF_3 (X=Cl, Br, I) dimers. *New J. Chem.* **1998**, *22*, 883–890.
9. O′ Hair, R.A.J.; Williams, C.M.; Clark, T. Neighboring group stabilization by σ-holes. *J. Mol. Mod.* **2010**, *16*, 559–565.
10. Wang, W. Halogen bond involving hypervalent halogen: CSD search and theoretical study. *J. Phys. Chem. A* **2011**, *115*, 9294–9299.
11. Cheng, N.; Bi, F.; Liu, Y.; Zhang, C.; Liu, C. The structures and properties of halogen bonds involving polyvalent halogen in complexes of FXO_n (X = Cl, Br; n = 0–3)–CH_3CN. *New J. Chem.* **2014**, *38*, 1256–1263.
12. Grabowski, S.J. Halogen bond with the multivalent halogen acting as the Lewis acid center. *Chem. Phys. Lett.* **2014**, *605–606*, 131–136.
13. Kitaura, K.; Morokuma, K. A new energy decomposition scheme for molecular interactions within the Hartree-Fock approximation. *Int. J. Quantum Chem.* **1976**, *10*, 325–340.

14. Ziegler, T.; Rauk, A. CO, CS, N_2, PF_3, and $CNCH_3$ as σ donors and π acceptors. A theoretical study by the Hartree-Fock-Slater transition-state method. *Inorg. Chem.* **1979**, *18*, 1755–1759.
15. von Hopffgarten, M.; Frenking, G. Energy decomposition analysis. *WIREs Comput. Mol. Sci.* **2012**, *2*, 43–62.
16. Sokalski, W.A.; Roszak, S. Efficient techniques for the decomposition of intermolecular interaction energy at SCF level and beyond. *J. Mol. Struct. (Theochem)* **1991**, *234*, 387–400.
17. Sokalski, W.A.; Roszak, S.; Pecul, K. An efficient procedure for decomposition of the SCF interaction energy into components with reduced basis set dependence. *Chem. Phys. Lett.* **1988**, *153*, 153–159.
18. Jeziorski, B.; Moszyński, R.; Szalewicz, K. Perturbation theory approach to intermolecular potential energy surfaces of van der Waals complexes. *Chem. Rev.* **1994**, *94*, 1887–1930.
19. Grabowski, S.J.; Sokalski, W.A. Are various σ-hole bonds steered by the same mechanisms? *ChemPhysChem* **2017**, *18*, 1569–1577.
20. Grabowski, S.J. Hydrogen bond and other Lewis acid–Lewis base interactions as preliminary stages of chemical reactions. *Molecules* **2020**, *25*, 4668.
21. Sobczyk, L.; Grabowski, S.J.; Krygowski, T.M. Interrelation between H-bond and pi-electron delocalization. *Chem. Rev.* **2005**, *105*, 3513–3560.
22. Warshel, A.; Papazyan, A. Energy considerations show that low-barrier hydrogen bonds do not offer a catalytic advantage over ordinary hydrogen bonds. *Proc. Natl. Acad. Sci. USA* **1996**, *93*, 13665–13670.
23. Cleland, W.W.; Frey, P.A.; Gerlt, J.A. The low barrier hydrogen bond in enzymatic catalysis. *J. Biol. Chem.* **1998**, *273*, 25529–25532.
24. Leiserowitz, L. Molecular packing modes. Carboxylic acids. *Acta Cryst.* **1976**, *B32*, 775–802.
25. Nygren, C.L.; Wilson, C.C.; Turner, J.F.C. On the solid state structure of 4-iodobenzoic acid. *J. Phys. Chem. A* **2005**, *109*, 2586–2593.
26. Karothu, D.P.; Weston, J.; Desta, I.T.; Naumov, P. Shape-memory and self-healing effects in mechanosalient molecular crystals. *J. Am. Chem. Soc.* **2016**, *138*, 13298–13306.
27. Zundel, G.; Metzger, H. Energy bands tunneling excess protons in liquid acids. IR spectroscopic study of the nature of $H_5O_2^+$ groups. *Z. Phys. Chem.* **1968**, *58*, 225–245.
28. Golubev, N.S.; Melikova, Shchepkin, D.N.; Shenderovich, I.G.; Tolstoy, P.M.; Denisov, G.S. Interpretation of hydrogen/deuterium isotope effects on NMR chemical shifts of $[FHF]^-$ ion base on calculations of nuclear magnetic shilding tensor surface. *Z. Phys. Chem.* **2003**, *217*, 1549–1563.
29. Grabowski, S.J. $[FHF]^-$—the strongest hydrogen bond under the influence of external interactions. *Crystals* **2016**, *6*, 3.

30. Dunitz, J.D. *Analysis of the Structure of Organic Molecules*, Cornell University Press, Ithaca, **1979**.
31. Bürgi, H.B. Stereochemistry and reaction paths as determined from crystal structure data — a relationship between structure and energy. *Angew. Chem. Int. Ed.* **1975**, *14*, 460–473.
32. Bürgi, H.B.; Dunitz, J.D. From crystal statics to chemical dynamics. *Acc. Chem. Res.* **1983**, *16*, 153–161.
33. Bürgi, H.B.; Dunitz, J.D.; Shefter, E. Geometrical reaction coordinates. II. Nucleophilic addition to a carbonyl group. *J. Am. Chem. Soc.* **1973**, *95*, 5065–5067.
34. Grabowski, S.J. What is the covalency of hydrogen bonding? *Chem. Rev.* **2011**, *11*, 2597–2625.
35. Grabowski, S.J.; Krygowski, T.M. The PT path for C=O...H-O systems modelled from crystal structure data. *Chem. Phys. Lett.* **1999**, *305*, 247–250.
36. Groom, C.R.; Bruno, I.J.; Lightfoot, M.P. Ward, S.C. The Cambridge structural database. *Acta Cryst.* **2016**, *B72*, 171–179.
37. Wong, R.; Allen, F.H.; Willett, P. The scientific impact of the Cambridge Structural Database: A citation-based study. *J. Appl. Cryst.* **2010**, *43*, 811–824.
38. Wilson, C.C. *Single Crystal Neutron Diffraction From Molecular Materials*, World Scientific, Singapore, **2000**.
39. Brown, I.D. Bond valences — a simple structural model for inorganic chemistry. *Chem. Soc. Rev.* **1978**, *7*, 359–376.
40. Gilli, P.; Bertolasi, V.; Ferretti, V.; Gilli, G. Covalent nature of the strong homonuclear hydrogen-bond — study of the OH---O system by crystal-structure correlation methods. *J. Am. Chem. Soc.* **1994**, *116*, 909–915.
41. Reiersølmoen, A.C.; Battaglia, S.; Øien-Ødegaard, S.; Gupta, A.K.; Fiksdahl, A.; Lindh, R.; Erdélyi, M. Symmetry of three-center, four-electron bonds. *Chem. Sci.* **2020**, *11*, 7979–7990.
42. Grabowski, S.J. Halogen bonds with carbenes acting as Lewis base units: complexes of imidazol-2-ylidene: theoretical analysis and experimental evidence. *Phys. Chem. Chem. Phys.* **2023**, *25*, 9636–9647.
43. Wiberg, K. Application of the Pople-Santry-Segal CNDO method to the cyclopropylcarbinyl and cyclobutyl cation and to bicyclobutane. *Tetrahedron* **1968**, *24*, 1083–1096.
44. Mayer, I. Bond order and valence indices: a personal account. *J. Comput. Chem.* **2007**, *28*, 204–221.
45. Fradera, X.; Austen, M.A.; Bader, R.F.W. The Lewis model and beyond. *J. Phys. Chem. A* **1999**, *103*, 304–314.

46. Fradera, X.; Poater, J.; Simon, S.; Duran, M.; Solà, M. The calculation of electron localization and delocalization indices at the Hartree–Fock, density functional and post-Hartree–Fock levels of theory. *Theor. Chem. Acc.* **2002**, *107*, 362–371.
47. Grabowski, S.J. Tetrel bond — σ-hole bond as a preliminary stage of the S_N2 reaction. *Phys. Chem. Chem. Phys.* **2014**, *16*, 1824–1834.
48. Grabowski, S.J. σ-Hole bond versus hydrogen bond: from tetravalent to pentavalent N, P and As atoms. *Chem. Eur. J.* **2013**, *19*, 14600–14611.
49. Grabowski, S.J. Boron and other triel Lewis acid centers: From hypovalency to hypervalency. *Chem. Phys. Chem.* **2014**, *15*, 2985–2993.
50. Crabtree, R.H. Hypervalency, secondary bonding and hydrogen bonding: siblings under the skin. *Chem. Soc. Rev.* **2017**, *46*, 1720–1729.
51. Gillespie, R.J.; Silvi, B. The octet rule and hypervalence: two misunderstood concepts. *Coord. Chem. Rev.* **2002**, *233–234*, 53–62.
52. Ref. 6, pp. 275–351.
53. Lewis, G.N. The atom and the molecule. *J. Am. Chem. Soc.* **1916**, *38*, 762–785.
54. Abegg, R. Die Valenz und das periodische System. Versuch einer Theorie der Molekularverbindungen. *Zeitsch. für Anorg. Chem.* **1904**, *39*, 330–380.
55. Langumir, I. The arrangement of electrons in atoms and molecules. *J. Am. Chem. Soc.* **1919**, *41*, 868–934.
56. Grabowski, S.J. Hydrogen bonds with BF_4^- anion as a proton acceptor. *Crystals* **2020**, *10*, 460.
57. Silvi, B.; Alikhani, E.; Ratajczak, H. Towards a unified chemical model of secondary bonding. *J. Mol. Model.* **2020**, *26*, 62.
58. Coulson, C.A. The nature of the bonding in xenon fluorides and related molecules. *J. Chem. Soc.* **1964**, 1442–1454.
59. Grabowski, S.J. Two faces of triel bonds in boron trihalide complexes. *J. Comput. Chem.* **2018**, *39*, 472–480.
60. Grabowski, S.J. Cleavage of hydrogen by activation at a single non-metal centre — towards new hydrogen storage materials. *Phys. Chem. Chem. Phys.* **2015**, *17*, 13539–13546.
61. Gillespie, R.J.; Hargittai, I. *The VSEPR Model of Molecular Geometry*. Allyn & Bacon, Boston, **1991**. Reprinted: Gillespie, R.J.; Hargittai, I. *The VSEPR Model of Molecular Geometry*. Dover Publications, New York, **2012**.
62. Gillespie, R.J. The electron-pair repulsion model for molecular geometry. *J. Chem. Educ.* **1970**, *47*, 18–23.
63. Martin, J.C. "Frozen" transition states: pentavalent carbon et al. *Science* **1983**, 221, 509–514.

64. Grabowski, S.J. Coordination of Be and Mg centres by HCN ligands — Be…N and Mg…N interactions. *ChemPhysChem* **2018**, *19*, 1830–1840.
65. Taylor, R. Aerogen bond, halogen bond, chalcogen bond, pnictogen bond, tetrel bond, triel bond ... why so many names? *Cryst. Growth Des.* **2024**, *24*, 4003–4012.
66. Grabowski, S.J. *Understanding Hydrogen Bonds: Theoretical and Experimental Views*, The Royal Society of Chemistry, Cambridge, **2021.**
67. Grabowski, S.J. Hydrogen bond types which do not fit accepted definitions. *Chem. Comm.* **2024**, 60, 6239–6255.
68. Jeffrey, G.A.; Saenger, W. *Hydrogen Bonding in Biological Structures*, Springer, Berlin, **1991**.
69. Chęcińska, L.; Grabowski, S.J. Partial hydrogen bonds — structural studies on thioureidoalkylphosphonates. *J. Phys. Chem. A* **2005**, *109*, 2942–2947.
70. Mó, O.; Yáñez, M.; Alkorta, I.; Elguero, J. Modulating the strength of hydrogen bonds through beryllium bonds. *J. Chem. Theory Comput.* **2012**, *8*, 2293–2300.
71. Clauss, A.D.; Nelsen, S.F.; Ayoub, M.; Moore, J.W.; Landis, C.R.; Weinhold, F. Rabbit-ears hybrids, VSEPR sterics, and other orbital anachronisms. *Chem. Educ. Res. Pract.* **2014**, 15, 417–434.
72. Bilewicz, E.; Rybarczyk-Pirek, A.J.; Dubis, A.T.; Grabowski, S.J. Halogen bonding in crystal structure of 1-methylpyrrol-2-yl trichloromethyl ketone. *J. Mol. Struct.* **2007**, *829*, 208–211.
73. Cowley, A.H.; Cushner, M.C.; Davis, R.E.; Riley, P.E. Crystal and molecular structure of the 1:2 aluminum trichloride-tetrahydrofuran complex $AlCl_3 \cdot 2THF$. *Inorg. Chem.* **1981**, *20*, 1179–1181.
74. Gorrell, I.B.; Hitchcock, P.B.; Smith, J.D. Preparation and crystal structures of aluminium hydride adducts with tetrahydrofuran $[AlH_3 \cdot OC_4H_8]_2$ and $AlH_3 \cdot 2OC_4H_8$. *J. Chem. Soc. Chem. Commun.* **1993**, 189–190.
75. Echeverría, J. Alkyl groups as electron density donors in π-hole bonding. *CrystEngComm* **2017**, *19*, 6289–6296.
76. Wang, X.; Cheng, Y.; Li, Q.; Scheiner, S. Triel bonds with methyl groups as electron donors. A pentacoordinate C atom. *ChemPhysChem* **2025**, *26*, e202400931.

CHAPTER VII

Intramolecular Interactions

Abstract: Intramolecular interactions are analysed and compared with intramolecular hydrogen bonds. Particularly, the five and six-membered rings closed by such interactions are described. Special attention is paid to the intramolecular halogen bond. The so-called resonance-assisted hydrogen bonds (RAHBs), particularly intramolecular hydrogen bonds, were a subject of numerous studies, hence the σ-hole and π-hole bonds that are analogous to intramolecular RAHBs are described. The same concern intramolecular interactions that are analogous to the proton sponges. The distances between the Lewis acid and Lewis base centres are discussed as rough measures of the strength of intramolecular interactions.

Keywords: intramolecular hydrogen bond, intramolecular halogen bond, resonance-assisted interaction, halide and hydride sponges, intramolecular tetrel bond

VII.1 Resonance-Assisted and Other Intramolecular Halogen Bonds

It was mentioned in Chapter IV that intramolecular hydrogen bonds, especially those that contain the C-H proton-donating bond, C-H···B (B is the proton acceptor centre), are categorised into two groups:[1] those where the closure by the hydrogen bond occurs for a longer chain of connected atoms and those where this chain is rather short because, enclosed in a ring, it contains five, six atoms or even less. Hydrogen bonds from the first group have the same properties as intermolecular hydrogen bonds, because the proton donor and acceptor can position themselves "freely" relative to each other. In the case of the second group, geometric constraints caused by short chain closure impose interaction properties that may differ from properties of intermolecular hydrogen bonds.

This chapter describes intramolecular interactions which are often counterparts of intramolecular hydrogen bonds. Hence, it is worth mentioning so-called resonance-assisted hydrogen bonds (RAHBs)[2,3] and proton sponges.[4,5] Let us start with the halogen bonds that are similar to RAHBs. The simple RAHB system, the malonaldehyde molecule, is presented in Scheme IV.6 (Chapter IV). One of the main assumptions of the RAHB concept is that for some hydrogen bonds (and it does not concern only intramolecular systems), the real structure may be treated as a mixture of resonance structures.[2,3] However, several studies contest this statement.[6–8] The systems classified as RAHBs are characterised by π-electron delocalisation, not by the occurrence of resonance forms.[8] For example, for malonaldehyde and its simple derivatives (Scheme IV.6), this delocalisation leads to the equalisation of the C=O and C-O bonds on one hand and the C=C and C-C bonds on the other hand. Hence such systems should be classified as π-electron delocalisation-assisted hydrogen bonds rather. However, due to the RAHB name that is used in numerous studies to describe inter- and intramolecular interactions characterised by specific properties,

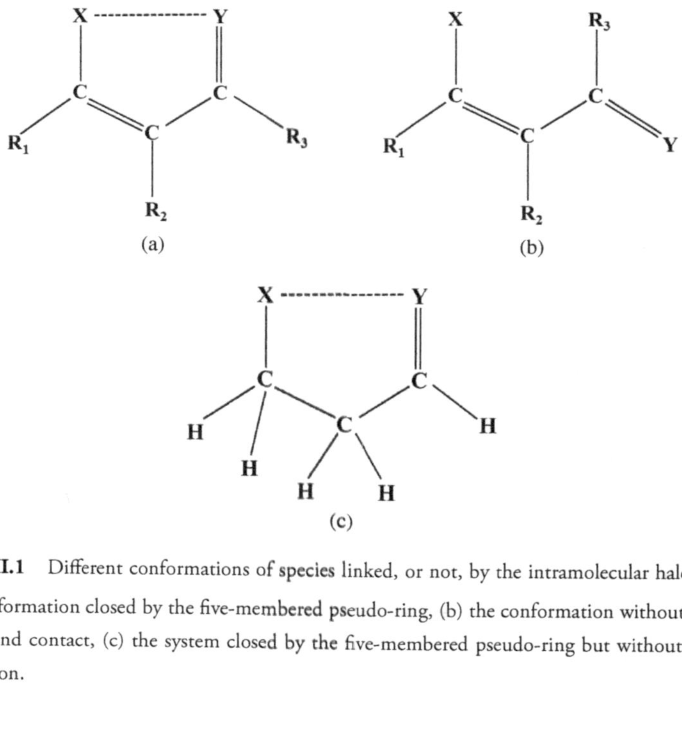

Scheme VII.1 Different conformations of species linked, or not, by the intramolecular halogen bond: (a) the conformation closed by the five-membered pseudo-ring, (b) the conformation without the X···Y halogen bond contact, (c) the system closed by the five-membered pseudo-ring but without π-electron delocalisation.

it is also applied here, despite this term not being correct. The same concerns other types of so-called resonance-assisted interactions, for example the resonance-assisted halogen bonds (RAXBs).

Scheme VII.1a presents a five-membered ring closed by an X···Y contact where X marks the halogen centre while Y is the electron donor.[9,10] MP2/6-311++G(d,p) calculations were performed on 3-chloropropenal and its fluoro derivatives.[9] All combinations of one, two and three F-substituents in R_1, R_2 and R_3 positions (Scheme VII.1a) were taken into account, and the H-atoms occupy remaining R-positions that are not occupied by F-substituents. These are so-called closed conformations characterised by short Cl···O distances. The other conformations were obtained by rotation of the CYR_3 group by 180° around the C–C single bond. After such rotations the conformations, named as "open", were optimised (see Scheme VII.1b). In such a way all closed and corresponding open conformations were fully optimised without any geometrical constraints. And all of them correspond to the local energetic minima since imaginary frequencies were not observed.

If a negative value of the energy difference between the closed conformation (Scheme VII.1a) and the corresponding open conformation (Scheme VII.1b) is observed, it may prove the stabilising influence of the Cl···O interaction that may be assigned to the intramolecular halogen bond. However, the results are inconclusive for a few reasons. The differences are sometimes negative and sometimes positive and their absolute values are fewer than 3 kcal/mol. If the fluorine substituent occupies the R_3 position, this difference is negative. It means that, in this case, the close conformations with the Cl···O contacts are more energetically stable than the open conformations with the Cl···F contacts. However, if the hydrogen centre occupies the R_3 position, this difference is positive. It means that, in this case, the open conformations with the Cl···H contacts are more energetically stable than the closed conformations with the Cl···O contacts. It seems that the change of conformation results in numerous geometrical changes and this is more important, as it results in the change of electron charge distribution in the whole structure. Therefore, the mentioned difference, resulting from the energies of conformers, does not concern only the impact of Cl···O interaction or other contacts, but also other effects.

Quantum Theory of Atoms in Molecules (QTAIM)[11] calculations were also performed for the sample of species discussed here.[9] Molecular graphs of the closed conformations reveal Cl···O bond paths with corresponding bond critical points (BCPs). Additionally, for all closed conformations, a linear relationship was found between the Cl···O distance and the electron density at the corresponding BCP. This may indicate the stabilising nature of the Cl···O interaction that increases with the shortening of the Cl···O distance. However, it was pointed out in several studies that the bond paths do not always correspond to stabilising interactions.[12–14] At least it was earlier a subject of polemics and discussions.[15,16]

It is worth mentioning that the closed and open conformations discussed here correspond to rigid and planar "skeletons"; however, these are neither RAXB systems nor systems characterised by π-electron delocalisation. The equalisation of C-C and C=C bonds is not observed, and the elongations of C=O bonds do not occur. These are effects that are observed for so-called RAHB systems where π-electron delocalisation occurs (see Chapter IV).

To check the role of intramolecular Cl···O contacts in crystals the Cambridge Structural Database (CSD)[17,18] search was performed[9] to find structures corresponding to Scheme VII.1a. The following search criteria were applied: error-free crystal structures of organic and metalloorganic compounds, polymeric and powder structures excluded, and R-factor set at no more than 10%. The Cl···O distance was limited to values fewer than 3.2 Å, which corresponds approximately to the sum of van der Waals radii of these centres, 3.27 Å (Table I.1 in Chapter I). 26 systems that meet the search criteria described above were found. For these systems the following mean parameters were observed: C=O bond length of 1.217 Å, and C-C and C=C bonds not equalised since the mean values are equal to 1.482 Å and 1.345 Å, respectively. The mean Cl···O distance is equal to 2.979 Å but the shortest distance amounts to 2.904 Å for the crystal structure of a-chloro-3,4-methylenedioxy-benzylidene-tetrahydrofuran-2,4,5-trione.[19]

Summarising, all theoretical results from MP2/6-311++G(d,p) calculations as well as experimental data resulting from the CSD search do not indicate clearly the occurrence of intramolecular Cl···O halogen bonds. The QTAIM approach may indicate the existence of such interactions but, as pointed out above, the occurrence of bond paths with corresponding BCPs is not strong evidence. In a later study, a more extended sample of species was considered.[10] MP2/aug-cc-pVTZ calculations were performed for the five-membered ring systems closed by X···Y intramolecular interactions (Scheme VII.1a) where X = F, Cl, Br and Y = O, S. The following substituents are located at the R-positions: H, NO_2 and CH_3. Since these systems are characterised by double C=C and C=O (C=S) bonds, similar species containing intramolecular X···Y contacts but only single bonds (Scheme VII.1c) were considered, i.e., systems where π-electron delocalisation does not occur without any doubt. Numerous theoretical approaches were applied in this study such as the QTAIM approach,[11] the natural bond orbital method (NBO),[20] the non-covalent interactions method (NCI),[21,22] and others such as electrostatic potentials (EPs) at molecular surfaces were also calculated. Such a broad spectrum of approaches applied allowed more decisive conclusions to be reached. Let us describe the results concerning the EP for molecular surfaces characterised by the electron density of 0.001 au considered in this study. It was found that for the bromine and chlorine centres, σ-holes in extension of C-Br and C-Cl bonds that possess the positive EPs occur. In the case of the fluorine centre, σ-holes are not observed but the positive EP occurs at the "cusp point", i.e., at the elongations of C-F bonds. These are the maximum positive EP values observed for the halogen centres. However, the "belt points", according to names applied by the authors, are in contact with the negative EP sites of oxygen and sulfur centres. These "belt points" of halogens for all RAXB structures are characterised by positive EP values. Hence the electrostatic X···Y interactions in systems are attractive according to the EP values of sites being in contact. This concerns both types of systems, those which contain double bonds in the skeleton and those which contain only single bonds. It is worth mentioning that according to these results the Cl···O interaction in the system where H-atoms occupy the R-positions is attractive,[10] thus the same system analysed earlier at the MP2/6-311++G(d,p) level is "closed" by the attractive intramolecular halogen bond.[9]

The authors apply the RAXB term for systems with double C=C and C=O(S) bonds,[10] but as pointed out earlier, such resonance structures are not detected. In results presented in this study, the C=C and C-C bonds as well as the C=O bonds were not discussed. The classification of these systems as RAXB is based on findings that X···Y distances are shorter for species containing double bonds than those with single bonds only.

However, this situation, i.e., X···Y distances, is forced by structures of skeletons. For systems with double bonds the planar systems with shorter X···Y distances are forced.

The QTAIM approach for these RAXB systems shows the occurrence of the bond paths and corresponding BCPs for 75 species (the whole sample contains 78 systems).[10] And there is a linear relationship between the electron density at the X···Y BCP and the X···Y distance; for shorter distances, greater values of electron density are observed.

The NCI results, similar to the EP values at molecular surfaces, indicate the occurrence of X···Y attractive interactions.[10] It is worth mentioning that the Laplacian of the electron density at the BCP, $\nabla^2\rho_{BCP}$, may be decomposed into three eigenvalues of the Hessian matrix (the NCI approach was described in Chapter II). The second eigenvalue, λ_2, is characterised by a negative value for attractive interactions like hydrogen bonds, other σ-hole bonds as well as π-hole bonds, and for van der Waals interactions. It is positive for repulsive interactions. Hence it is convenient to apply the $(\text{sign}\lambda_2)\rho$ parameter to separate the attractive and repulsive interactions. The electron density in BCP itself is not a parameter that allows such a distinction to be made.

Figure VII.1 presents 3D NCI plots of the X···Y interaction regions for some example structures.[10] The plots corresponding to intramolecular hydrogen bonds are also included in this figure for comparison. It is worth mentioning that the interpretations of plots in this figure differ slightly from interpretations presented in the original study.[10]

Figure VII.1A presents the 3D NCI plot for the malonaldehyde species where an intramolecular O-H···O hydrogen bond occurs. The dark blue surface in the H···O region appears that confirms the occurrence of a strong hydrogen bond. The green surface with reddish traces occurs within the closed ring of the malonaldehyde species that is also typical for other ring systems, aromatic ones for example.[23] Figure VII.1B presents the plot of 3-hydroxypropanal; the conformation containing the intramolecular O-H···O hydrogen bond is shown. This is the analogue of malonaldehyde, but in this case the double bonds do not occur. The green in the H···O region indicates a medium-strength hydrogen bond. This is a weaker interaction than in the case of malonaldehyde since in the latter the H···O distance is shorter than in the 3-hydroxypropanal species. The H···O distances in both systems (Figs. VII.1A and VII.1B) result mainly from the geometrical constraints of molecular skeletons. The separate surface within the

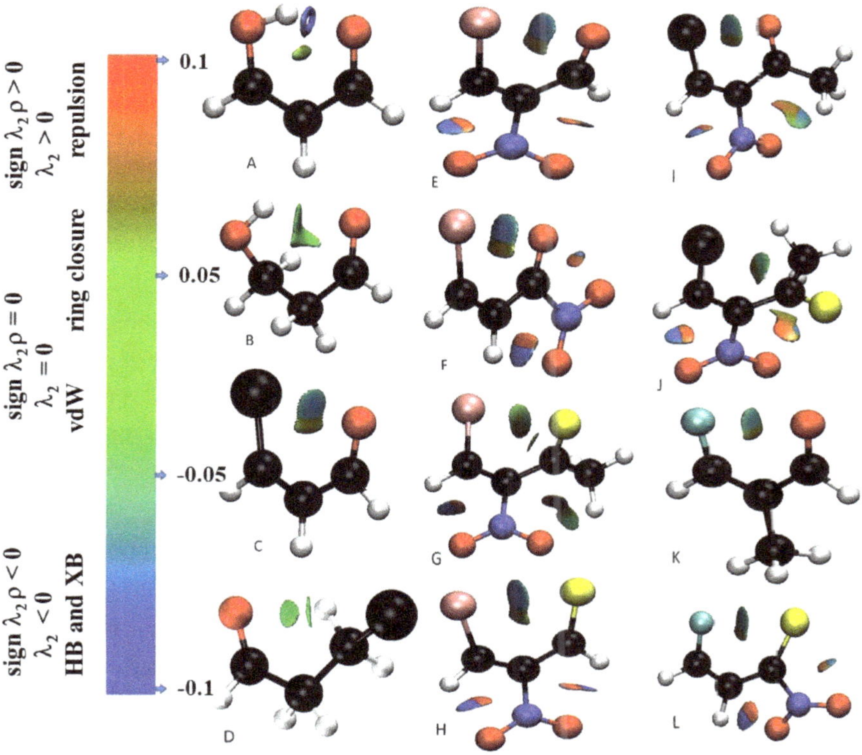

Figure VII.1 3D NCI plot of the X···Y interaction regions for some example structures. The following colours are applied: white for hydrogen, small black for carbon, large black for chlorine, red for oxygen, pink for bromine, yellow for sulfur and blue for nitrogen. Modified with permission from ref. 10. Copyright 2015 The Royal Society of Chemistry.

closed ring of this species does not occur (Fig. VII.1B). In contrast, the curved shape of the green surface is located partly in the H···O region and partly close to the centre of the ring.

It is worth mentioning that for numerous weak intramolecular hydrogen bonds, the ring critical point resulting from the closure of the molecular ring is located close to the bond path and its BCP, corresponding to the proton-acceptor contact. Hence the 3D NCI surfaces for weak intramolecular interactions (hydrogen and halogen bonds) discussed here (Fig. VII.1) are located sometimes in the H···O or X···Y region and they are extended to the centre of the ring.

Figure VII.1C presents the system with intramolecular Cl···O interaction. The green part of the surface located between the chlorine and oxygen centres indicates attractive forces while the green-reddish part located closer to the centre of the ring may indicate partly repulsive forces between C=O and C-Cl bonds. These results of the NCI approach confirm the existence of the intramolecular Cl···O halogen bond on one hand and indicate unique properties of such type of interactions on the other hand. The intermolecular halogen bonds concern links through σ-holes at halogen centres while intramolecular analogues (in small five- or six-membered rings) do not; this is through regions of positive EP but not through σ-holes. The analogue system to that of Fig. VII.1C is presented in Fig. VII.1D, but only single bonds occur for this ring. One can see that the optimisation led to the conformation with the C-H···O hydrogen bond but not with the Cl···O contact. The green surface at the H···O region indicates weak attractive interaction.

It is worth to note here that all remaining 3D NCI plots, Figs. VII.1E-L, indicate attractive intramolecular halogen bond interaction (except for Fig. VII.1J) since the green surfaces appear in the X···Y regions and for these structures, most often, the repulsive/attractive forces between the C-X and C=Y bonds occur. Besides, for a few structures, intramolecular C-H···O hydrogen bonds may exist since the green/red surfaces corresponding to attractive/repulsive forces are detected in the H···O regions. These are interactions between the C-H bond and the O-centre of the nitro group; such interactions occur for the structures presented in Figs. VII.1E, F, G, H, I, J, L.

In the structures presented in Figs.VII.1F and VII.1L, repulsive interactions with small additional attractive interactions are observed; these are O···O and S···O interactions, respectively. For the species presented in Fig. VII.1J, C-H···O weak hydrogen bond occurs, and there is the repulsive S···O interaction and the interaction between the Cl-centre and the C-C bond; the expected Cl···S contact does not occur.

It is worth mentioning here the study on intramolecular halogen-halogen interactions that may be named as dihalogen bonds.[24] MP2/aug-cc-pVTZ calculations were performed on the perhalogenated ethanes, $X_3C–CY_3$ (X, Y = F, Cl), in this study and analysis of the electron density on these species was carried out. The interactions between the substituents, i.e., the halogen centres, were analysed.[24] BCPs between halogen atoms are not observed, and also the concept of σ-holes is not useful here to explain the nature of halogen-halogen contacts since these holes are not directed in agreement to the halogen-halogen lines. The authors applied the NCI approach to detect the stabilising nature of these contacts and they classified them as strong van der Waals-type interactions.

It is worth to note that the intramolecular dihalogen bonds occurring in the abovementioned systems concern four-membered rings. However, summarising the results discussed in this section, one can see that the intramolecular halogen bonds that probably occur in small four and five-membered rings are not formed through the σ-holes of the halogen centres. Hence these interactions are characterised by unique properties that differ from those of intermolecular halogen bonds.

VII.2 Intramolecular Dihydrogen Bonds

It is worth to note that the concept of resonance-assisted interactions was probably first analysed for dihydrogen-bonded (DHB) systems after being introduced for the hydrogen-bonded systems.[25,26] DHB interactions may be treated as a sub-class of hydrogen bonds.[27] They were briefly mentioned in Chapter IV, but intramolecular DHBs were not analysed and, more importantly, not analysed in the context of the RAHB model.

The short and simple definition of the dihydrogen bond states that it is an interaction between H-centres of opposite charges.[28] To be more correct, this should concern opposite EPs of hydrogen centres. However, it is difficult to find (and imagine) the small surface of the hydrogen centre that is characterised by areas of both positive and negative EPs. The term "dihydrogen bond" was introduced in the middle of the nineties of the last century,[29] and it seems that two groups of researchers performed pioneering studies on crystal structures where inter- and intramolecular attractive H···H interactions are observed: the group of Morris[30,31] and the group of Crabtree.[29,32,33]

It seems that the study of Zachariasen and Mooney on the crystal structure of ammonium hypophosphite from 1934 was the first on DHB interactions.[34] It is stated in this study that "the hydrogen atoms of the hypophosphite group behave toward ammonium as if they were H^- ions". Other studies on dihydrogen bonds, although

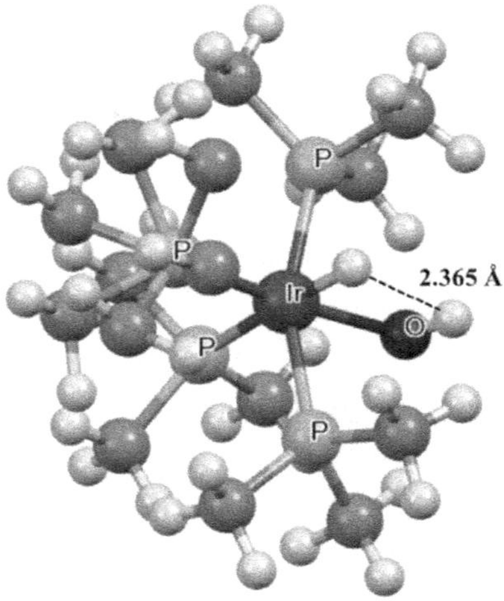

Figure VII.2 A fragment of the crystal structure of *cis*-hydrido-hydroxy-tetrakis(trimethylphosphine)-iridium(iii) hexafluorophosphate. Lighter (smaller) and darker (greater) gray circles correspond to hydrogen and carbon atoms, respectively.

this name was not used, appeared from this time occasionally. It is not the aim of this book to refer to early studies on DHBs, but one can mention the study of Stevens and coworkers on the crystal structure of hydridohydroxy *cis*-$[IrH(OH)(PMe_3)_4]PF_6$ measured at 20 K with the use of single-crystal neutron diffraction.[35] The authors postulated the occurrence of an attractive $H^{-\delta}\cdots{}^{+\delta}H$ interaction between the hydridic Ir-H bond and the electrophilic hydrogen of the OH group. Figure VII.2 presents a fragment of this crystal structure; one can see the intramolecular DHB that occurs in the four-membered ring. However, it seems to be a rather weak interaction since the $H^{-\delta}\cdots{}^{+\delta}H$ distance is equal to 2.37 Å that is slightly more than the corresponding sum of van der Waals radii, 2.2 Å (Table I.1). In a later study in 1994, Ir-$H^{-\delta}\cdots{}^{+\delta}H$-N intramolecular links were analysed in crystal structures and it was stated that this is a new kind of interaction.[30] It can therefore be noticed that the first DHBs described in crystal structures were most often intramolecular interactions.

HF and MP2 calculations were performed on two conformations of the (1Z)-2-borylethen-1-ol species, on so-called "closed" and "open" conformations that are presented in Schemes VII.2a and VII.2b, respectively.[25] Several basis sets were applied in these calculations, 6-311++G(d,p), 6-311++G(2d,2p) and 6-311++G(3d,3p);

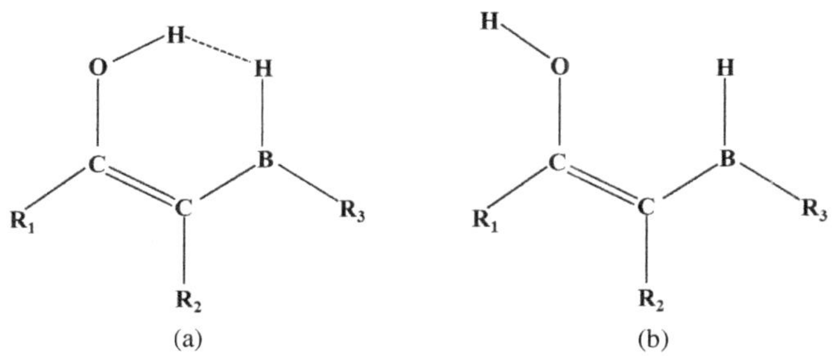

Scheme VII.2 Two conformations of (1Z)-2-borylethen-1-ol: (a) closed and (b) open.

in all levels of calculations these conformations, after optimisations, correspond to energetic minima since imaginary frequencies were not observed. The hydrogen centres occupy the R-positions (see Scheme VII.2). The same calculations were performed for a derivative where the R_1 position is occupied by the chlorine substituent, R_2 by hydrogen and R_3 by sodium. The open conformation was obtained from the closed after 180° rotation of the O–H bond around the C–O bond followed by full optimisation of this species. The closed conformation for both species with all H-substituents, called henceforth the (H,H,H) system, and with Cl, H and Na substituents, called henceforth the (Cl,H,Na) system, is characterised by a lower total energy than the open conformation.

This energetic difference between closed and open conformations may be treated as a rough estimation of the energy of DHB. This is only a rough estimation since the change of conformation from closed to open conformation is accompanied by numerous other geometrical changes that affect the energetic difference. This rough DHB energy calculated at the MP2/6-311++G(3d,3p) level for the (H,H,H) system is equal to –4.6 kcal/mol while for the (Cl,H,Na) system it is equal to –6.5 kcal/mol. The reason for the stronger DHB for the latter system is connected with the greater π-electron delocalisation than for the (H,H,H) system. The electron-withdrawing Cl-substituent localised at the R_1 position leads to the loss of electron charge at the C=C bond and consequently to its elongation. On the other hand, the Na-substituent that donates electron charge causes the addition of electron charge to the C-B bond and its shortening. It is worth to note that the π-electron delocalisation should be greater for the closed conformations. This was detected for numerous intramolecular RAHBs (see Chapter IV).

Let us look at the C=C and C-B bond lengths. All geometrical results are at the same MP2/6-311++G(3d,3p) level. For the open conformation of the (H,H,H) system the C=C and C-B bond lengths are equal to 1.355 Å and 1.528 Å, respectively. In the case of the closed conformation these bond lengths are equal to 1.364 Å and 1.516 Å, respectively. It means that the closure of the ring that leads to the formation of the intramolecular DHB results also in the elongation of the C=C bond and the shortening of the C-B bond that is exactly due to the increase of π-electron delocalisation. The H···H distance and the O-H···H angle of the DHB triatomic arrangement is equal to 1.933 Å and 140.0°, respectively.

In the open conformation of the (Cl,H,Na) system, the C=C and C-B bond lengths are equal to 1.356 Å and 1.547 Å, respectively. For the closed conformation these bond lengths are equal to 1.367 Å and 1.534 Å, respectively. The H···H distance and the O-H···H angle is equal here to 1.674 Å and 146.7°, respectively. These geometrical parameters suggest stronger intramolecular interaction for the (Cl,H,Na) system than for (H,H,H) and indeed it is in agreement with the rough estimation of the intramolecular DHB strengths described above. For the closed conformation of both systems, the bond paths occur for the H···H contacts with the corresponding BCPs. The electron density at the BCP for these systems is equal to 0.016 au and 0.027 au, respectively. This also shows that the DHB is stronger for (Cl,H,Na) than for (H,H,H).

Similar studies were performed for species where, instead of the -BHR_3 group, the -$BHR_3R_4^-$ group was introduced.[26] Hence anions where H···H intramolecular interactions occur for the closed conformations were considered. The R_1, R_2, R_3 and R_4 positions are occupied by hydrogen and fluorine centres. The closed and open conformations of these anionic systems, similar to the neutral ones presented in Scheme VII.2, were calculated at the MP2/6-311++G(d,p) level.

One may conclude that negatively charge-assisted DHBs occur for these systems. And one may expect that these are stronger interactions than for the neutral closed conformation of the (1Z)-2-borylethen-1-ol species and its derivative. For the anionic species with all R_1, R_2, R_3 and R_4 positions occupied by the hydrogen atoms, i.e., for the (H,H,H,H) system, the H···H distance is equal to 1.704 Å. For the open conformation of this system the C=C and C-B bond lengths are equal to 1.343 Å and 1.626 Å, respectively; these bond lengths for the closed conformation are equal to 1.345 Å and 1.625 Å, respectively. Thus, meaningless C=C bond elongation and C-B bond shortening are observed because of π-electron delocalisation. The latter effect related to the closure of the ring is very small here, smaller than for the neutral analogue.

One can refer to other interesting systems where intramolecular DHBs occur, such as the (1Z,4Z)-1,4-dipentene-3-bora-1,5-diol (DBD) species and its derivatives where H-atoms are replaced by fluorine or methyl groups.[36] Various combinations of the occupation of the H-atom positions, i.e., the R-positions, were considered (see Scheme VII.3). Three conformations of each species characterised by the occupation of R-positions were calculated. The closed conformation is characterised by the occurrence of two rings that means two intramolecular DHBs. Since there is one electron donor for such systems, the hydridic H-centre is bonded to the boron, so these interactions may be classified as the proton acceptor bifurcated intramolecular DHBs. In the case of closed-open configuration only one ring is observed with the corresponding intramolecular DHB, and for the open conformation rings are not formed and DHBs do not occur. All systems are planar, and only the hydrogen atoms of methyl groups are located outside the planes of molecules. In the case of closed conformations, if the H-atoms of methyl groups are not considered, the systems have C_{2v} symmetry. Hence the electron density and geometrical properties within both rings of the closed system are the same, according to this symmetry. This is the situation like that where two rings closed by the O-H···O hydrogen bonds coexist (Scheme VII.4).[37] The properties of intramolecular DHBs in closed conformation (Scheme VII.3) are very similar to the properties of hydrogen bonds in the double-ring system (Scheme VII.4).

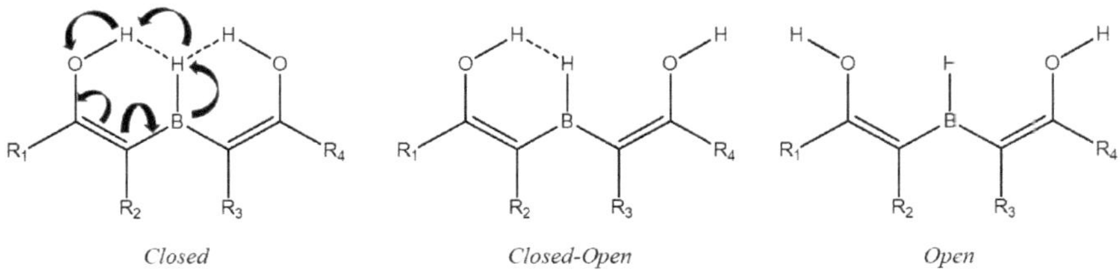

Scheme VII.3 Three conformers of two-sequence systems; for the closed conformer the possible electron charge shifts in one of the rings are indicated. Reprinted with permission from ref. 36. Copyright 2024 Wiley-VCH Verlag GmbH & Co. KGaA, Weinheim.

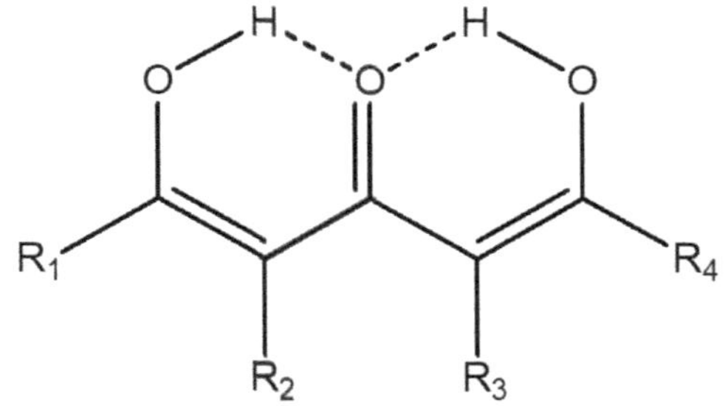

Scheme VII.4 Two-ring RAHB system. Reprinted with permission from ref. 36. Copyright 2024 Wiley-VCH Verlag GmbH & Co. KGaA, Weinheim.

For the systems mentioned above, MP2/6-311++G(d,p) calculations were performed and all these systems correspond to energetic minima.[36] For those systems where H···H contacts are observed with corresponding DHB interactions, the bond paths with BCPs also occur. In the case of rings of closed-open and closed conformations, changes of geometrical and other parameters are observed in comparison with the open conformations. This is the effect of π-electron delocalisation. The changes observed for the ring in the closed-open conformation seem to be greater than those in rings of the closed conformation.

Let us present selected parameters for the system with H-atoms as the R-substituents. For the open conformation, the C=C and C-B bond lengths are equal to 1.356 Å and 1.542 Å, respectively. In the closed-open conformation, in the ring formed by the DHB interaction these bond lengths are equal to 1.363 Å and 1.531 Å, respectively. One can see that the elongation of the C=C bond and the shortening of the C-B bond resulting from the closure of the ring and consequently the π-electron delocalisation occur here. The H···H distance in the ring is equal to 1.928 Å, and the OHH angle at 141.6° is rather far from linearity due to the geometrical constraints. The electron density at the BCP of the corresponding bond path, ρ_{BCP}, is equal to 0.0147 au.

The parameters for the closed configuration are as follows. The C=C and C-B bond lengths amount to 1.363 Å and 1.529 Å, respectively, which only slightly differ from those of closed-open conformation. However, the intramolecular DHB seems to be weaker here than in the closed-open conformation since the H···H distance in the ring is equal to 1.962 Å, and the OHH angle is equal to 140.2°. The electron density at the BCP of the corresponding bond path, ρ_{BCP}, is equal to 0.0134 au. The H···H distance and the electron density, ρ_{BCP}, may be treated as measures of the strength of interaction, where shorter distances and greater electron densities at BCPs are observed for stronger interactions.

Only those related to π-electron delocalization of intramolecular DHBs are discussed here, but there are others. The early theoretical study of Kulkarni, where DFT calculations were performed on simple four-membered rings closed by H···H contacts, are the preliminary stages of dehydrogenation reactions.[38] In another study, B3LYP/6-311G(d,p) calculations were performed on the conformations of amino acids where closed rings are observed due to intramolecular DHBs.[39] One can also mention the study on the role of intramolecular DHBs in the intermediates of processes of the BH_3- and BF_3-catalysed epoxide ring opening.[40]

VII.3 Analogues of Proton Sponges

Proton sponges were mentioned in Chapter IV as systems which are characterised by high basicity due to the centres having lone electron pairs located close to each other.[41] The 1,8-bis(dimethylamino)naphthalene species is an example of a proton sponge that has a basicity ten million times higher than that of N,N-dimethylaniline where a single nitrogen centre with a lone electron pair occurs.[41] The proton sponges are not limited only to compounds which, like 1,8-bis(dimethylamino)naphthalene, are classified as peridisubstituted naphthalenes but they are also observed for other systems containing aromatic moieties.[41]

It is interesting that a proton inserted between the lone electron pairs of two centres located close to each other leads to the formation of a very stable system. This is also connected with the formation of the intramolecular hydrogen bond, N-H···N, in the case of the protonated 1,8-bis(dimethylamino)naphthalene. The question arises as to whether there are other systems similar to proton sponges where other cations or anions are located between two centres that are close to each other. One can mention the crystal structure of potassium (m_2-hydrido-B,B')-naphthalene-1,8-diyl-bis (dimethylborane) dioxane solvate where the 1,8-naphthalenediylbis(dimethylborane) system occurs with the hydride anion, H^-, inserted between two boron centres (Fig. VII.3a).[42] This is the X-ray diffraction measurement of the crystal structure; the estimated standard deviations for lengths of bonds containing H-atoms are rather large.

Hence the position of the hydride anion is not exact. The B···H distances are equal to 1.49(5) Å and 1.20(5) Å while the BHB angle is equal to 142(4)°,[43] thus estimated standard deviations for these distances and for the angle are equal to 0.05 Å and 4° that corresponds to very low accuracy. However, this is evidence that this anion is located between two boron centres.

Each of two trivalent B-centres is characterised by six valence electrons, thus the octet rule is not obeyed. The hypovalency for the boron centre results in its Lewis acid properties. Hence the B···H···B arrangement may be classified as a hydride bond on one hand and the C-B···H as the triel bond on the other hand (Fig. VII.3a). The name of the former interaction is based on the Lewis base centre, while the latter name is based on the Lewis acid site that is more common in names of interactions. Referring to the names concerning different types of hydrogen bonds, the C-B···H···B-C system may be named as the bifurcated triel bond interaction since both B-centres are linked with the same nucleophilic hydride anion. However, if the name hydride bond is applied, it is the "inverse situation" to that occurring for proton sponges where the cation (proton) is inserted between lone electron pairs. In the hydride sponges, the anion (hydride ion) is inserted between two vacant lone electron pair orbitals.

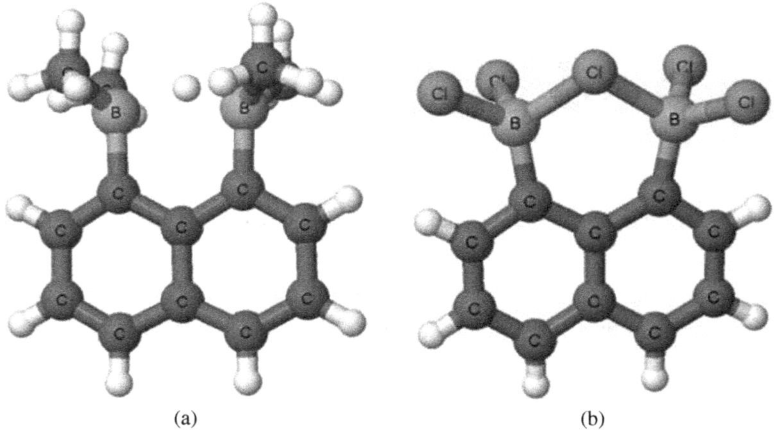

(a) (b)

Figure VII.3 (a) 1,8-Naphthalenediylbis(dimethylborane) from the crystal structure of potassium (μ_2-hydrido-B,B')-naphthalene-1,8-diyl-bis(dimethylborane) dioxane solvate (ref. 42). (b) 1,8-naphthalenediyl-bis(dichloroborane) chloride from the crystal structure of bis(triphenylphosphine)-iminium 1,8-naphthalenediyl-bis(dichloroborane) chloride (ref. 44). The white unlabelled circles correspond to the H-atoms.

The crystal structure of bis(triphenylphosphine)-iminium 1,8-naphthalenediyl-bis(dichloroborane) chloride is another example of a system similar to the proton sponge. This is the 1,8-naphthalenediyl-bis(dichloroborane) chloride species (Figure VII.3b) where the chloride anion is inserted between two boron centres.[44] Thus it is also an example of the bifurcated triel bond where one Cl^- anion is the nucleophilic centre that interacts with the Lewis acid sites of boron centres. On the other hand, this interaction may be named as the halide (or chloride) bond if the Lewis base centre is the basis for its name. The B···Cl distances are equal to 1.92(1) Å and 2.01(1) Å and the B-Cl-B angle is equal to 102.0(4)°.[44] It is interesting to consider if systems analysed in this section, including typical proton sponges, are classified as intramolecular interactions or "only" bifurcated interactions. This is the hydride bond interaction of the 1,8-naphthalenediylbis(dimethylborane) unit with the hydride anion or the triel bond for the structure presented in Fig. VII.3a, and the halide bond interaction of the 1,8-naphthalenediyl-bis(dichloroborane) unit with the chloride anion or the triel bond for the structure presented in Fig. VII.3b.

CSD searches were performed to find structures similar to those of Fig. VII.3.[45] Two types of compounds were prototypes for these searches: 1,2-bis(dichloroboryl)benzene hydride (and halide) and 1,8-bis(dichloroboryl)naphthalene hydride (and halide). The species presented in Fig. VII.3 are similar to the latter prototype. In other words, the aim of these searches was to find a special kind of systems that contain $B \cdots H^- \cdots B$ arrangements and their halide $B \cdots X^- \cdots B$ analogues. These are only prototypes since, in the CSD searches, not only boron atoms were taken into account as the Lewis acid centres but also other triel atoms. There are additional requirements concerning these searches. The centre located between the triel centres is a halogen atom in its halide form, or a hydrogen atom (its deuterium isotope is also taken into account) in its anionic form. Two boron centres with chlorine substituents were mentioned above in the prototypes but the triel centres attached to benzene or naphthalene are connected with a variety of substituents, not only with chlorines. Also, H-atoms of the benzene and the naphtalene could be replaced by more or less complex substituents.

The results of searches are shown in Table VII.1 (2019 CSD April release being in accordance with the CSD resources used in the publication described here).[45] The following requirements related to the accuracy of results were applied: no structures with unresolved errors, no powder structures, $R \leq 10\%$, and no restrictions for

Table VII.1 The T(H/X⁻)T systems of the halide and hydride sponges from CSD. Geometrical parameters are presented: the T···H/X distances (in Å) (subscripts 1 and 2 correspond to two distances in the same system), and the T-(H/X)-T angle (in °). Refcodes of crystal structures and corresponding references are included. Modified with permission from supplementary material of ref. 45. Copyright 2019 MDPI.

Refcode	System	$(T\cdots H/X)_1$	$(T\cdots H/X)_2$	T-(H/X)-T	Ref.
		The naphthalene skeleton			
AGIPAI	BFB	1.599	1.606	125.7	46
CUTFUS	BHB	1.487	1.201	142	42
CUTFUS10	BHB	1.487	1.201	142	43
EZAZOU	BFB	1.632	1.586	126	47
FIGRAO*	BClB	1.920	2.012	102	44
LUFCEV**	BHB	1.309	1.370	142.4	48
LUFCEV**	BHB	1.443	1.415	121.3	48
PERZES**	BFB	1.635	1.636	126.6	49
YIRHIQ	GaClGa	2.370	2.366	77.3	50
YIRHIQ	GaClGa	2.396	2.379	76.5	50
		The benzene ring skeleton			
ECUGUF**	BFB	1.487	1.487	113.2	51
ETIMAX	BFB	1.596	1.62	111.2	52
	BFB	1.611	1.617	111.3	52
	BFB	1.589	1.624	111.9	52
GAYRUV	BClB	1.999	2.039	94.1	53
GOTNEI	AlClAl	2.281	2.278	98.3	54
	AlClAl	2.289	2.272	99.1	54
IQAJOZ	BClB	2.040	2.033	94.3	55
IQAJOZ01	BClB	2.036	2.041	94.3	52

*Disorder concerns chlorine atoms bound to triel centre.

**Disorder occurs but it neither concerns triel atoms nor their neighbours.

estimated standard deviations of bond lengths. Such not-very-stringent accuracy requirements were applied since the aim of the searches was not to analyse accurate geometrical parameters, but to present structures with unique types of triel bonds.

Nevertheless, only 19 systems fulfilling described conditions were found, nine systems containing a benzene skeleton and ten containing a naphthalene skeleton. Among systems containing the naphthalene skeleton, there are eight species with the B···H/X···B bifurcated triel bonds and only two species with the Ga···Cl···Ga arrangements that may be classified as triel bonds and also as halide bonds (chloride bonds). There are no similar systems with the naphthalene skeleton that contain other triel centres.

If one refers to the names of interactions that take into account the Lewis base centres, in the case of the B···H/X···B bifurcated triel bonds that occur in species containing the naphthalene skeleton, there are four systems with the B···H···B hydride bonds, as well as four systems with the halide bonds (three with B···F···B arrangements and one with B···Cl···B). The range of B···H/X···B angle is between 102° and 142°. The smallest angle is observed for the B···Cl···B system and the greatest angles occur for the B···H···B hydride bonds. In the case of two Ga···Cl···Ga links, the angle is equal to 77° which is caused by steric effects (greater sizes of the chloride anion and Ga-centres). For these ten systems containing the naphthalene skeleton, the hydride or halide anion is not always equidistant, or nearly equidistant, from the triel centers. Particularly, the H···B distances are not equal in the case of hydride bonds.

In the case of species containing the benzene skeleton, the triel centres are attached to the neighbouring benzene carbon centres, thus the T···X···T (T is the triel centre and X is the halogen) angle is rather small, far from linearity (the range of this angle is between 94° and 113°). There are only nine species containing the benzene skeleton and among them there are only B···F···B (four cases), B···Cl···B (three cases) and Al···Cl···Al (two cases). In three cases the halide anion is equidistant, or nearly equidistant, from the triel centers, i.e., the difference between the T···X distances is less or equal to 0.005 Å. There are no systems with the benzene skeleton that contain the hydride ion, i.e., T···H···T arrangements do not occur.

While there are only 19 such systems with arrangements like those occurring in proton sponges, only triel centres linked with naphthalene and benzene skeletons were taken into account. No extended and systematic studies have yet been conducted on proton sponge-like systems that contain other aromatic fragments.

It is worth noting that theoretical analyses on species similar to those containing the bifurcated triel bonds and found in CSD were also performed.[45] MP2/aug-cc-pVTZ calculations were carried out for chlorides, fluorides and hydrides of 1,2-bis(dichloroboryl) benzene and 1,8-bis(dichloroboryl)naphthalene. Frequency calculations have been performed at the same level and the lack of imaginary frequencies confirms that the optimised structures correspond to energetic minima.

Six systems were analysed theoretically, and the interaction energies, E_{int} (see eq. I.1, Chapter I), the binding energies, E_{bin} (eq. I.2), and consequently the deformation energies, E_{def} (eq. I.2), were calculated for these species. The Basis Set Superposition Error, BSSE (eq. II.11, Chapter II), for each of these systems was calculated, thus all energies mentioned above are corrected for BSSE. The monomers to calculate E_{int} and E_{bin} values are defined for each of the six species. The hydride or halide anion is one of the monomers and the remaining part of the species is the other monomer. In this way, for each complex, the values of energies referred to such monomers "cover" two C-B···H/X triel bonds (both are named as bifurcated interactions).

Table VII.2 presents the values of energies as well as other geometrical and topological parameters. The absolute value of the interaction energy, $|E_{int}|$, is between

Table VII.2 The energetic (kcal/mol) and geometrical (Å,°) parameters from MP2/aug-cc-pVTZ calculations. The interaction and binding energies corrected for BSSE are presented, E_{int} and E_{bin}, as well as the deformation energy, E_{def}. The $(B\cdots X/H)_{1/2}$ distances in the B···H/X···B system (subscripts 1 and 2 correspond to two distances within one system considered, X marks the halogen atom) and the B-H/X-B angle are included. Topological parameters (au) corresponding to the MP2/aug-cc-pVTZ optimisations: the electron density at BCP, ρ_{BCP}, and the total electron energy density at BCP, H_{BCP}, are included. These parameters for both H/X...B distances in the complex considered are equal or almost equal to each other. In the latter case, mean values are presented. Modified with permission from ref. 45. Copyright 2019 MDPI.

System	E_{int}	E_{bin}	E_{def}	$(B\cdots H/X)_1$	$(B\cdots H/X)_2$	B-X/H-B	ρ_{BCP}	H_{BCP}
				napthalene skeleton				
BClB	-104.1	-49.1	55.1	1.968	1.952	102.3	0.093	-0.077
BFB	-148.5	-96.7	51.8	1.540	1.545	125.3	0.102	-0.052
BHB	-161.0	-113.3	47.6	1.315	1.315	139.1	0.111	-0.099
				benzene skeleton				
BClB	-105.7	-53.7	51.9	1.991	1.991	96.5	0.087	-0.071
BFB	-149.0	-96.7	52.4	1.569	1.568	113.3	0.098	-0.052
BHB	-159.1	-106.0	53.1	1.360	1.360	121.3	0.106	-0.092

104 kcal/mol and 161 kcal/mol, approximately. It means that there are 52 kcal/mol and more per single C-B···H/X interaction and that this interaction should possess covalent or partly covalent character, at least. This expectation is supported by the topological parameters; the electron density at the BCP, located at the B···H/X bond path, is between 0.09 au and 0.11 au. These are values that may occur for covalent bonds. Table VII.2 also shows that the total electron energy density at BCPs, H_{BCP}, is negative for all six systems (the corresponding Laplacian values, not presented in the table, $\nabla^2\rho_{BCP}$, are positive). Thus, the topological and energetic parameters indicate that the B···H/X triel bonds are partly covalent interactions.

Table VII.2 shows that the absolute value of the binding energy, $|E_{bin}|$, is between 49 kcal/mol and 113 kcal/mol, much less than the interaction energy. It indicates great values of deformation energy, E_{def}; for each complex this value is close to 50 kcal/mol, approximately. This means that the inclusion of the hydride or halide anion between two boron centres causes enormous geometrical deformations. The B-H/X-B angle is between 97° and 139°, far from linearity. However, in the case of intramolecular interactions, systems that are far from linear are typical. The B-H/X-B angle is greater for species containing the naphthalene skeleton than for the corresponding complex with the benzene skeleton. However, other parameters are very similar for both systems. The differences between the B···H/X distances do not exceed 0.04 Å. For example, for the "naphthalene" B···Cl···B arrangement the B···Cl distances are equal to 1.97 Å and 1.95 Å, while for the "benzene" arrangement these distances are equal to 1.99 Å and 1.99 Å. For all B···Cl contacts in these two complexes, the ρ_{BCP} value is equal to 0.09 au approximately. The interaction energies for these complexes are equal to −104 kcal/mol and −106 kcal/mol, respectively.

For these six complexes, energy decomposition analysis was carried out according to the scheme applied in the ADF program.[56] The BP86-D3/TZ2P calculations applying the scheme were performed for these complexes,[45] with geometries resulting from optimisations carried out at the same level; at the same level the frequency calculations were performed and imaginary frequencies were not detected indicating that the complexes correspond to energetic minima. It was found that for all complexes the electrostatic and orbital energies are the most important attractive terms; both contribute almost equally to the energetic stabilisation of these systems. The dispersion attractive terms are rather negligible in comparison. MP2/aug-cc-pVTZ and BP86-D3/TZ2P calculations performed for the six complexes led to similar geometric results, and the differences between bonds and angles for both levels of calculations are negligible.[45]

Summarising, it is worth to note that the topic in this section was not extensively analysed in former studies. It is true that experimental and theoretical results on systems similar to proton sponges are reported from time to time. However, there is much room for further research. Several interesting works on this topic can be mentioned. For example, MP2/aug′-cc-pVDZ (aug′ means that the diffuse functions are removed from the H-atoms) calculations were performed on systems marked as 1,8-ZXH and 1,8-ZX_2 species.[57] It means that for the 1 and 8 positions of naphthalene, two ZXH or ZX_2 substituents are attached (Z is the P or As pnicogen centre while X marks the halogen centre). Five-membered rings are formed for such species closed by pnicogen-pnicogen links since the Z···Z bond paths occur with the corresponding BCPs. The electron densities at these BCPs are between 0.02 au and 0.05 au, indicating that these dipnicogen bonds are medium-strength interactions.

In another study,[58] the 1,8-bis(dimethylboranyl)naphthalene (hydride sponge) was prepared and combined with 1,8-bis(dimethylamino)naphthalene (proton sponge). Hence the [hydride sponge - EH_n] [proton sponge - H] adducts are formed with the protic compounds EH_n = H_2O, NH_3, H_2S, PH_3, H_2Se, HCN. These adducts were analysed with the use of NMR spectroscopy.[58]

VII.4 Intramolecular Interactions – Theoretical and Experimental Studies

Several examples of various types of intramolecular interactions are presented in this section. It is worth repeating that the concept of resonance assistance is contested in several studies, especially in the case of hydrogen bonds,[6–8] and it is better to refer to π-electron delocalisation that occurs, particularly, if conjugated double bonds are present in the rings.

It is worth to note that such π-electron shifts were discussed for other intramolecular interactions, such as halogen bonds,[10] beryllium bonds,[59] magnesium bonds,[60] and triel bonds.[61] For triel interactions, MP2/aug-cc-pVTZ calculations were performed (aug-cc-pVTZ-PP basis set for the In heavy atom was applied to include relativistic effects partially) on the five-membered rings with C-C single and C=C double bonds and the S···T (T designates the Al, Ga or In triel atom) intramolecular triel bond that closes these rings. These are systems like that presented in Scheme VII.1a, but here the Y centre is a sulfur atom while X is replaced by the TH_2 group. NO_2, CH_3 and H were chosen as R-substituents in this study.

Figure VII.4b presents an example of the system with the sulfur Lewis base centre and the AlH_2 group acting as the electron acceptor.[61] Summarising, 42 systems closed by S···T triel bonds were analysed theoretically, where the S-centre acts as the electron donor through the lone electron pairs while T is an electron acceptor through the π-hole. Thus, these are π-hole bonds. The absolute values of interaction energies are detected for the range between 12 kcal/mol and 21 kcal/mol that indicates strong and very strong interactions. The S···T distances in the range between 2.4 Å and 2.7 Å are observed. The sum of van der Waals radii for the S···Al, S···Ga and S···In contacts is equal to 3.64 Å, 3.67 Å and 3.73 Å, respectively, thus one can see that they do not differ significantly. The distances are slightly shorter for the Ga-species than for the corresponding Al-species, by ~0.02 Å, while the In-systems are characterised by the longest distances, longer than the Ga-species by about ~0.2 Å.[61] This may suggest weakest interactions for the indium systems. However, the electron density at BCP, ρ_{BCP}, is a good indicator of the strength of interaction as it was justified in numerous studies that a greater ρ_{BCP} value is connected with stronger interactions.[62,63] In the case of S···T interactions, the greatest values of ρ_{BCP} are observed for the S···Ga bond paths, followed by S···In and S···Al. The same order was obtained from the values of energies of interactions.[61] For the intramolecular triel bonds, a $E_{int} = 0.48\,V_{BCP}$ relationship was found[61] as an approximation similar to that proposed for the O-H···O hydrogen bonds ($E_{int} = 0.5\,V_{BCP}$).[64] The V_{BCP} parameter is the potential electron energy density at the BCP. For all the systems, the Laplacian of the electron density at BCP, $\nabla^2\rho_{BCP}$, is positive that means the S···T interactions do not possess characteristics typical for covalent bonds. However, the total electron energy density at BCP, H_{BCP}, for the S···T interactions in all species is negative that means these systems have a significant degree of covalency. The range of H_{BCP} values for S···Al, S···Ga and S···In interactions are (–0.006; –0.004), (–0.019; –0.012) and (–0.015; –0.006), respectively. Only for the S···Ga contacts in two cases the H_{BCP} value is equal to –0.003 au and –0.004 au. These results show that the H_{BCP} values agree with the other parameters regarding the strength of interactions. The more negative the value, the stronger the interaction, so the following order of the strength of interaction is confirmed: Al < In < Ga.

It is stated in the study[61] that the strong triel interactions result from phenomena that are typical for RAHB systems. This is mainly π-electron delocalisation that is enhanced by the closure of the ring that leads to the formation of the triel bond interaction, as well as being enhanced due to the occurrence of the conjugated C=C double bond. Figure VII.4 illustrates the enhancement of π-electron delocalisation caused by various factors. The S=CH-CH_2-CH_2-AlH_2 system containing a five-membered ring where there is no conjugated C=C double bond is presented in Fig. VII.4a. The subsequent S=C, C-C, C-C, C-Al bond lengths in the ring are equal to 1.629 Å, 1.490 Å, 1.528 Å, and 2.003 Å, respectively. And the S···Al distance corresponding to the triel bond contact is equal to 2.502 Å. The situation changes if one of the single C-C bonds is replaced by a C=C double bond that leads to the S=CH-CH=CH-AlH_2 species. The subsequent S=C, C-C, C=C, C-Al bond lengths in the ring in this case are equal to 1.645 Å, 1.444 Å, 1.359 Å, and 2.005 Å, respectively; the S···Al distance is equal to

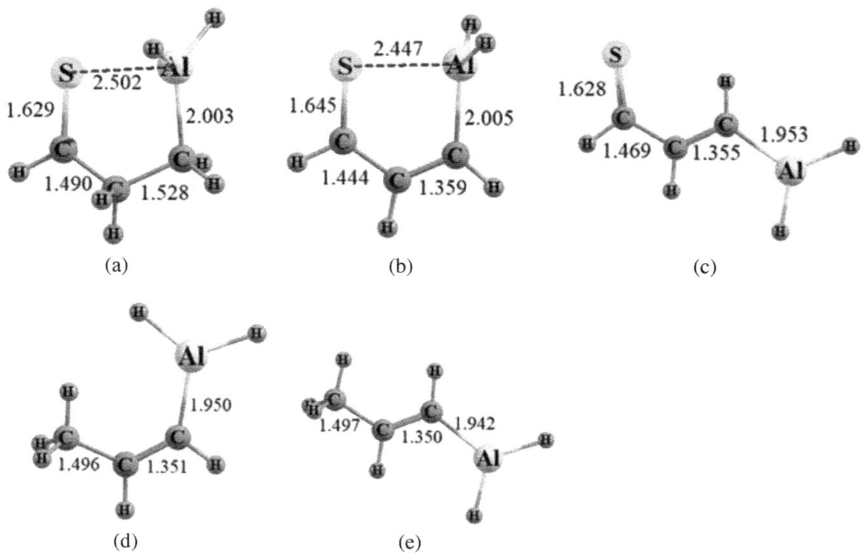

Figure VII.4 Bond lengths (Å) are presented for the following species: (a) alkyl $SCHCH_2CH_2AlH_2$ system, (b) *cis* and (c) *trans* conformation of $SCHCHCHAlH_2$ with conjugated double C=C bonds, (d) *cis* and (e) *trans* conformation of $CH_3CHCHAlH_2$. Reprinted with permission from ref. 61. Copyright 2022 The Royal Society of Chemistry.

2.447 Å. So one can see that the π-electron delocalisation due to the alternating occurrence of the double and single bonds enhances the strength of the intramolecular triel bond that is reflected by the shortening of the S···Al distance. The elongation of the S=C bond here in comparison with "non-conjugated" systems (Fig. VII.4a) as well as the small difference between the C-C and C=C bond lengths of ~0.08 Å illustrate the effects of π-electron delocalisation. These processes also occur for the "open" conformation of the S=CH-CH=CH-AlH_2 species (Fig. VII.4c); however, they are not as evident as for the closed conformation (Fig. VII.4b). In the conformation without the triel bond (Fig. VII.4c), the difference between the C-C and C=C bond lengths is 0.11 Å, the S=C bond length of 1.628 Å is almost the same as for the "non-conjugated" conformation, and the Al-C bond length is shorter than for the conformations where the triel bond is observed.

Figure VII.4c shows that the alternation of double and single bonds is significant in π-electron delocalisation. Additionally, a very weak intramolecular S···H(C) hydrogen bond may have a little influence on this phenomenon. In the structure presented in Fig. VII.4e, the CSH group is replaced by the CH_3 group, which leads to less π-electron delocalisation that is reflected by the C-C and C=C bond length difference of 0.15 Å. In the case of the "closed" conformation (Fig. VII.4d) that is the counterpart of the latter structure, the difference between CC bonds is practically the same, but an elongation of the C-Al bond is observed from 1.942 Å to 1.950 Å. This may result from the formation of the intramolecular C-H···H-Al DHB and the corresponding closure of the six-membered ring. The results lead to the conclusion that both the closure of the ring and the alternating bonds enhance the π-electron delocalisation and influence strongly the intramolecular interaction.

There are numerous other examples on studies concerning various intramolecular interactions. The intramolecular O···S chalcogen bond is observed in the dimer of the *trans* conformer of thiosalicylic acid.[65–67] Figure II.7 (Chapter II) presents the molecular graph of this dimer with electron density isolines and gradient paths. The ELF isosurfaces for this dimer are also presented in Fig. II.15. This is a centrosymmetric dimer, thus the monomers are linked by two equivalent O-H···O hydrogen bonds. In each monomer an additional intramolecular chalcogen bond occurs. For this dimer, ωB97XD/6-311++G(d,p) calculations were performed[67] and the calculated O···O intermolecular distance of the hydrogen bond is equal to 2.657 Å while the S···O intramolecular distance of the chalcogen bond is equal to 2.715 Å. These distances in dimers in the crystal structure are equal to 2.642 Å and 2.757 Å, respectively.[66] The intramolecular S···O chalcogen bond is between the oxygen Lewis base centre and the σ-hole located at the sulfur centre in the extension of the H-S bond (see Fig. II.7).

One can mention several other studies on intramolecular chalcogen bonds. 2-substituted and 2,6-disubstituted arylchalcogen compounds were analysed and special attention was paid to their conformations characterised by the occurrence of Se···O and Te···O intramolecular interactions.[68] The oxygen acts here as the Lewis base centre and the selenium or tellurium as the Lewis acid centre through the σ-hole. In this study the controlling role of the chalcogen bonds on the conformational states was discussed. The experimental results, among them X-ray diffraction, were supported by theoretical calculations.

In another study,[69] MP2/aug-cc-pVDZ calculations were performed on the 1-OX-8-YZ naphthalene derivatives where the oxygen centre at the 1-position of naphthalene is an electron donor connected with X that is H or CH_3. The Y-centre at the 8-position is the O, S or Se electron acceptor linked with Z that is H, F, or CN. Hence the Y···O links are classified as the intramolecular chalcogen bonds that close the five-membered rings. The interaction energies were estimated for these systems from isodesmic reactions and it was found that they are negative (stabilising) for the sulfur centre as the electron acceptor if it is linked with fluorine. This is because the electron-withdrawing F-substituents enhances the σ-hole at the S-centre; a greater positive EP is observed here. In the case of the selenium centre as the electron acceptor, stabilising interactions are observed if it is connected with F and CN electron-withdrawing substituents. The absolute values of interaction energies for the strongest interactions are equal to ~6-7 kcal/mol, approximately.[69]

The species where the indazolium backbone with the Se and Te centres are attached to a benzene ring and to a heterocycle ring, respectively, were analysed in terms of the occurrence of the intramolecular Te···Se triel bond.[70] The Te-centre with a phenyl substituent attached acts as the electron acceptor through the σ-hole while the selenium is the Lewis base centre. Two crystal structures were presented where such intramolecular Te···Se interactions occur. In one case the Se-centre is linked by a double bond with

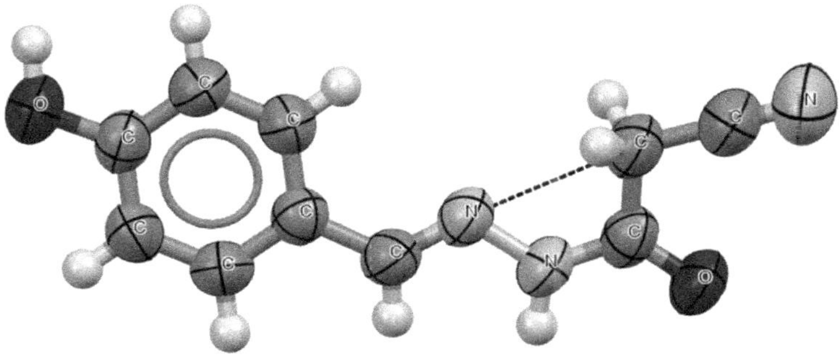

Fig. VII.5 2-cyano-N′-[(1E)-(4-hydroxyphenyl)methylidene]-acetohydrazide (ref. 70). The ellipsoids are shown for non-hydrogen atoms (probability level 75%). The N···C tetrel bond is designated by a broken line.

the heterocycle, and in the second case it is linked by a single bond since the methyl group is attached to the Se-centre. The Te···Se distance is equal to 3.3157(7) Å and 3.488(2) Å, respectively. The corresponding sum of van der Waals radii is equal to 3.96 Å. In the second structure, with the methyl group attached to the selenium centre, disorder is observed thus the Se and Te positions characterised by higher occupancy (75%) were taken into account. For the other positions (25% of occupancy), this distance amounts to 3.429(5) Å.[70]

It is worth mentioning here the case of the intramolecular tetrel bond. The tetrel centre that acts as the electron acceptor is often characterised by sp^3 hybridisation and a tetrahedral (or nearly so) structure. This is why links to the Lewis acid tetrel centres are often difficult due to the tetrel substituents, particularly if the tetrel centre is carbon, the smallest in the 14th group of elements. The substituents obscure it. Figure VII.5 presents the 2-cyano-N'-((E)-4-hydroxybenzylidene)acetohydrazide species taken from the crystal structure.[71] The N···C distance corresponding to the intramolecular tetrel bond is equal to 2.761(2) Å. This is much less than the corresponding sum of van der Waals radii amounting to 3.25 Å (Table I.1). It is also very interesting that the intramolecular tetrel bond discussed here closes the four-membered ring. Usually five and six-membered rings are considered. The calculations performed at the ωB97-XD/6-311+G(d,p) level and with the application of the NBO approach[20] indicate an interaction energy corresponding to the n(N) → σ*(C-CN) overlap of ~1 kcal/mol only.

Thus, it means that the C≡N electron-withdrawing substituent does not enlarge the σ-hole significantly. These are only some examples briefly described in this section; the number of studies on intramolecular Lewis acid — Lewis base interactions is large.

Summary

There are numerous analogies between intramolecular hydrogen bonds and other intramolecular interactions. For example, in the case of RAHBs, π-electron delocalisation leads to enhancement of the strength of such interactions as well as to the equalisation of the single and double bonds. This is not a "full equalisation" but the elongation of the double bonds and the shortening of the single bonds occur. The same is observed for the σ-hole and π-hole bonds that are classified as so-called interactions with resonance assistance. Particularly, these processes were discussed for the intramolecular halogen and triel bonds characterised by five-membered rings closed by such interactions.

Special attention was paid in this chapter to the intramolecular DHBs. Not only single rings closed by these interactions were considered here but also intramolecular DHB interactions where there is one proton acceptor (hydridic H-centre) and two proton donors (O-H bonds); such intramolecular bifurcated DHBs possess numerous interesting characteristics.

The analogues to the proton sponges were also discussed. In these species the proton is inserted between two lone electron pairs, most often electron pairs of nitrogen centres. Such systems are classified as intramolecular hydrogen bonds. However, an anion may also be inserted between two Lewis acid centres, say triel centres. Such systems containing triel centres and hydride or halide anion inserted between them were analysed here. These systems may be considered as those where bifurcated triel bonds occur that can also be named as intramolecular hydride and halide bonds. One can say that they are inverse to the proton sponges since the anion (hydride or halide) is inserted here between centres characterised by a vacancy of electron pairs. These systems were analysed theoretically, and the crystal structures where they occur were presented as well. Other examples of intramolecular interactions were briefly discussed, such as chalcogen bonds or the tetrel bond characterised by the four-membered ring.

Appendix

The following tools were applied to construct figures presented in this chapter.

Figure VII.3

Jmol program — Reference: Hanson, R.M. Jmol SMILES and Jmol SMARTS: specifications and applications. *J. Cheminform.* **2016**, *8*, 50.

Figures VII.2,5

Graphical tools of the Cambridge Structural Database — References: (1) Groom, C.R.; Bruno, I.J.; Lightfoot, M.P.; Ward, S.C. The Cambridge structural database. *Acta Cryst.* **2016**, *B72*, 171–179; (2) Wong, R.; Allen, F.H.; Willett, P. The scientific impact of the Cambridge Structural Database: A citation-based study. *J. Appl. Cryst.* **2010**, *43*, 811–824.

Schemes VII.3,4 and Figures VII.1,4

Reprinted with permission.

Schemes VII.1,2

PowerPoint.

References

1. Desiraju, G.R.; Steiner, T. *The Weak Hydrogen Bond in Structural Chemistry and Biology*, Oxford University Press, New York, **1999**.
2. Gilli, G.; Bellucci, F.; Ferretti, V.; Bertolasi, V. Evidence for resonance-assisted hydrogen bonding from crystal-structure correlations on the enol form of the β-diketone fragment. *J. Am. Chem. Soc.* **1989**, *111*, 1023–1028.
3. Bertolasi, V.; Gilli, P.; Ferretti, V.; Gilli, G. Evidence for resonance-assisted hydrogen bonding. 2. Intercorrelation between crystal structure and spectroscopic parameters in eight intramolecularly hydrogen bonded 1,3-diaryl-1,3-propanedione enols. *J. Am. Chem. Soc.* **1991**, *113*, 4917–4925.
4. Alder, R.W. Strain effects on amine basicities. *Chem. Rev.* **1989**, *89*, 1215–1223.
5. Staab, H.A.; Saupe, T. "Proton sponges" and the geometry of hydrogen bonds: aromatic nitrogen bases with exceptional basicities. *Angew. Chem. Int. Ed. Engl.* **1988**, *27*, 865–879.

6. Alkorta, I.; Elguero, J.; Mó, O.; Yañez, M.; Del Bene, J.E. Do coupling constants and chemical shifts provide evidence for the existence of resonance-assisted hydrogen bonds? *Mol. Phys.* **2004**, *102*, 2563–2574.
7. Alkorta, I.; Elguero, J.; Mó, O.; Yañez, M.; Del Bene, J.E. Are resonance-assisted hydrogen bonds 'resonance assisted'? A theoretical NMR study. *Chem. Phys. Lett.* **2005**, *411*, 411–415.
8. Góra, R.W.; Maj, M.; Grabowski, S.J. Resonance-assisted hydrogen bonds revisited. Resonance stabilization vs. charge delocalization. *Phys. Chem. Chem. Phys.* **2013**, *15*, 2514–2522.
9. Palusiak, M.; Grabowski, S.J. Do intramolecular halogen bonds exist? Ab initio calculations and crystal structures' evidences. *Struct. Chem.* **2007**, *18*, 859–865.
10. Pandiyan, B.V.; Deepa, P.; Kalondaivel, P. Do resonance-assisted intramolecular halogen bonds exist without a charge transfer and a σ-hole? *Phys. Chem. Chem. Phys.* **2015**, *17*, 27496–27508.
11. Bader, R.F.W. *Atoms in Molecules: A Quantum Theory*, Oxford University Press, Oxford, **1990**.
12. Cioslowski, J.; Mixon, S.T. Universality among topological properties of electron density associated with the hydrogen–hydrogen nonbonding interactions. *Can. J. Chem.* **1992**, *70*, 443–449.
13. Jabłoński, M. Bond paths between distant atoms do not necessarily indicate dominant interactions. *J. Comput. Chem.* **2018**, *39*, 2183–2195.
14. Jabłoński, M. On the uselessness of bond paths linking distant atoms and on the violation of the concept of privileged exchange channels. *ChemistryOpen* **2019**, *8*, 497–507.
15. Poater, J.; Solá, M.; Bickelhaupt, F.M. Hydrogen–hydrogen bonding in planar biphenyl, predicted by atoms-in-molecules theory, does not exist. *Chem. Eur. J.* **2006**, *12*, 2889–2896.
16. Bader, R.F.W. Bond paths are not chemical bonds. *J. Phys. Chem. A* **2009**, *113*, 10391–10396.
17. Groom, C.R.; Bruno, I.J.; Lightfoot, M.P. Ward, S.C. The Cambridge structural database. *Acta Cryst.* **2016**, *B72*, 171–179.
18. Wong, R.; Allen, F.H.; Willett, P. The scientific impact of the Cambridge Structural Database: A citation-based study. *J. Appl. Cryst.* **2010**, *43*, 811–824.
19. Begley, M.J.; Crombie, L.; Havard, R.G.; and Reynolds, D.P. Cyclisation of arylpropiolyl chloro-oxalyl anhydrides: the chemistry of aryl(chloro) methylenetetrahydrofuran-2,4,5-triones and the *X*-ray crystal structure of a 3,4-methylenedioxybenzylidene representative. *J. Chem. Soc., Perkin Trans. 1* **1977**, 138–145.

20. Weinhold, F.; Landis, C. *Valency and Bonding: A Natural Bond Orbital Donor — Acceptor Perspective*, Cambridge University Press, Cambridge, **2005**.
21. Johnson, E. R.; Keinan, S.; Mori-Sanchez, P.; Contreras-Garcia, J.; Cohen, A. J.; Yang, W. Revealing noncovalent interactions. *J. Am. Chem. Soc.* **2010**, *132*, 6498–6506.
22. Contreras-Garcia, J.; Johnson, E. R.; Keinan, S.; Chaudret, R.; Piquemal, J.-P.; Beratan, D. N.; Yang, W. NCIPLOT: a program for plotting noncovalent interaction regions. *J. Chem. Theory Comput.* **2011**, *7*, 625–632.
23. Narth, C.; Maroun, Z.; Boto, R.A.; Chaudret, R.; Bonnet, M-L.; Piquemal, J-P.; Contreras-Garcia, J. A Complete NCI Perspective: From New Bonds to Reactivity, in Chauvin, R.; Lepetit, C.; Silvi, B.; Alikhani, E. (Eds.), *Applications of Topological Methods in Molecular Chemistry*, Springer, **2016**, pp. 491–527.
24. Johansson, M.P.; Swart, M. Intramolecular halogen–halogen bonds? *Phys. Chem. Chem. Phys.* **2013**, *15*, 11543–11553.
25. Grabowski, S.J. Theoretical evidence for a new kind of intramolecular DHB. *Chem. Phys. Lett.* **2000**, *327*, 203–208.
26. Wojtulewski, S.; Grabowski, S.J. Ab initio and AIM studies on intramolecular DHBs. *J. Mol. Struct.* **2003**, *645*, 287–294.
27. Cybulski, H.; Pecul, M.; Sadlej, J. Characterization of dihydrogen-bonded D–H…H–A complexes on the basis of infrared and magnetic resonance spectroscopic parameters. *J. Chem. Phys.* **2003**, *119*, 5094–5104.
28. Grabowski, S.J. What is Common for DHB and H…σ Interaction — Theoretical Analysis and Experimental Evidences, in Scheiner, S. (Ed.), *Noncovalent Forces*, Springer, **2015**.
29. Richardson, T.B.; de Gala, S.; Crabtree, R.H. Unconventional hydrogen bonds: intermolecular B-H…H-N interactions. *J. Am. Chem. Soc.* **1995**, *117*, 12875–12876.
30. Lough, A.J.; Park, S.; Ramachandran, R.; Morris, R.H. Switching on and off a new intramolecular hydrogen-hydrogen interaction and the heterolytic splitting of dihydrogen. Crystal and molecular structure of [Ir({H(η^1-SC_5H_4NH)}2(PCy_3)$_2$]BF_4.2.7CH_2C1_2. *J. Am. Chem. Soc.* **1994**, *116*, 8356–8357.
31. Ramachandran, R.; Morris, R.H. A new type of intramolecular H…H…H interaction involving N-H…H(Ir)…H-N atoms. Crystal and molecular structure of [IrH(η^1-SC_5H_4NH) $_2$(η^2-SC_5H_4N) (PCy_3)]BF_4.0.72CH_2C1_2. *J. Chem. Soc. Chem. Commun.* **1994**, 2201–2202.
32. Wessel, J.; Lee Jr, J.C.; Peris, E.; Yap, G.P.A.; Fortin, J.B.; Ricci, J.S.; Sini, G.; Albinati, A.; Koetzle, T.F.; Eisenstein, O.; Rheingold, A.L.; Crabtree, R.H. An unconventional intermolecular three-center N-H...H_2Re hydrogen bond in crystalline [$ReH_5(PPh_3)_3$]-indole-C_6H_6. *Angew. Chem. Int. Ed. Engl.* **1995**, *34*, 2507–2509.

33. Crabtree, R.H.; Siegbahn, P.E.M.; Eisenstein, O.; Rheingold, A.L.; Koetzle, T.F.A. A new intermolecular interaction: unconventional hydrogen bonds with element-hydride bonds as proton acceptor. *Acc. Chem. Res.* **1996**, *29*, 348–354.
34. Zachariasen, W.H.; Mooney, R.C.L. The structure of hypophosphite group as determined from the crystal lattice of ammonium hypophosphite. *J. Chem. Phys.* **1934**, *2*, 34–37.
35. Stevens, R.C.; Bau, R.; Milstein, D.; Blum, O.; Koetzle, T.F. Concept of the H(δ+)··· H(δ–) interaction. A low-temperature neutron diffraction study of *cis*-[IrH(OH)$(PMe_3)_4$] PF_6. *J. Chem. Soc. Dalton Trans.* **1990**, 1429–1432.
36. Gholami, S.; Aarabi, M.; Grabowski, S.J. Theoretical insights into bifurcated intramolecular DHBs. *ChemPhysChem* **2024**, *25*, e202300849.
37. Fazli, M.; Raissi, H.; Chahkandi,B.; Aarabi, M. The effect of formation of second hydrogen bond in adjacent two-ring resonance-assisted hydrogen bonds — Ab initio and QTAIM studies. *J. Mol. Struct.* **2010**, *942*, 115–120.
38. Kulkarni, S.A. Intramolecular DHBing in main group elements. connection with dehydrogenation reactions. *J. Phys. Chem. A* **1999**, 103, 9330–9335.
39. Pakiari, A.H.; Jamshidi, Z. Intra-molecular DHB in the amino acid. *J. Mol. Struct.* **2004**, *685*, 155–161.
40. Dixit, V.A.; Rathi, P.C.; Bharatam, P.V. Intramolecular DHB: A new perspective in Lewis acid catalyzed nucleophilic epoxide ring opening reaction. *J. Mol. Struct.* **2010**, *962*, 97–100.
41. Staab, H.A.; Saupe, T. "Proton sponges" and the geometry of hydrogen bonds: aromatic nitrogen bases with exceptional basicities. *Angew. Chem. Int. Ed. Engl.* **1988**, *27*, 865–879.
42. Katz, H.E. Hydride sponge: 1,8-Naphthalenediylbis(dimethylborane). *J. Am. Chem. Soc.* **1985**, *107*, 1421–1423.
43. Katz, H.E. Hydride sponge: complexation of 1,8-naphthalenediylbis (dimethylborane) with hydride, fluoride, and hydroxide. *J. Org. Chem.* **1985**, *50*, 5027–5032.
44. Katz, H.E. 1,8-Naphthalenediylbis(dichloroborane) chloride: the first bis boron chloride chelate. *Organometallics* **1987**, *6*, 1134–1136.
45. Grabowski, S.J. Bifurcated triel bonds—hydrides and halides of 1,2-bis(dichloroboryl) benzene and 1,8-bis(dichloroboryl)naphthalene. *Crystals* **2019**, *9*, 503.
46. Dorsey, C.L.; Jewula, P.; Hudnall, T.W.; Hoefelmeyer, J.D.; Taylor, T.J.; Honesty, N.R.; Chiu, C-W.; Schulte, M.; Gabbaï, F.P. Fluoride ion complexation by a B_2/Hg heteronuclear tridentate Lewis acid. *Dalton Trans.* **2008**, 4442–4450.
47. Sole, S.; Gabbaï, F.P. A bidentate borane as colorimetric fluoride ion sensor. *Chem. Comm.* **2004**, 1284–1285.

48. Berke, H.; Jiang, C.; Blacque, O. Metal-free hydrogen activation and hydrogenation of imines with 1,8-bis(dipentafluorophenylboryl)naphthalene. *Chem. Comm.* **2009**, 5518–5520.
49. Melaimi, M.; Solé, S.; Chiu, C-W.; Wang, H.; Gabbai, F.P. Structural and electrochemical investigations of the high fluoride affinity of sterically hindered 1,8-bis(boryl)naphthalenes. *Inorg. Chem.* **2006**, *45*, 8136–8143.
50. Hoefelmeyer, J.D.; Brode, D.L.; Gabbaï, F.P. Synthesis and characterization of a trigallacycle. *Organometallics* **2001**, *20*, 5653–5657.
51. Chase, P.A.; Henderson, L.D.; Piers, W.E.; Parvez, M.; Clegg, W.; Elsegood, M.R.J. Bifunctional perfluoroaryl boranes: synthesis and coordination chemistry with neutral lewis base donors. *Organometallics* **2006**, *25*, 349–357.
52. Clegg, W. **2016**, *CSD Communication*, private communication.
53. Sgro, M.J.; Domer, J.; Stephan, D.W. Stoichiometric CO_2 reductions using a bis-borane-based frustrated Lewis pair. *Chem. Comm.* **2012**, *48*, 7253–7255.
54. Tschinkl, M.; Bachman, R.E.; Gabbaï, F.P. 1,2-Bis(chloromethylalumino) tetrafluorobenzene, an aluminium-based bifunctional Lewis acid with a perfluorinated backbone. *Chem. Comm.* **1999**, 1367–1368.
55. Lewis, S.P.; Taylor, N.J.; Piers, W.E.; Collins, S. Isobutene polymerization using a chelating diborane co-initiator. *J. Am. Chem. Soc.* **2003**, *125*, 14686–14687.
56. te Velde, G.; Bickelhaupt, F.M.; Baerends, E.J.; Fonseca Guerra, C.; van Gisbergen, S.J.A.; Snijders J.G.; Ziegler, T. Chemistry with ADF. *J. Comput. Chem.* **2001**, *22*, 931–967.
57. Sánchez-Sanz, G.; Trujillo, C.; Alkorta, I.; Elguero, J. Intramolecular pnicogen interactions in phosphorus and arsenic analogues of proton sponges. *Phys. Chem. Chem. Phys.* **2014**, *16*, 15900—15909.
58. Becker, C.; Schwabedissen, J.; Neumann, B.; Stammler, H-G.; Mitzel, N.W. Frustrated Lewis pair chemistry of hydride sponges. *Dalton Trans.* **2022**, *51*, 6547–6564.
59. Montero-Campillo, M.M.; Alkorta, I.; Elguero, J.; Yáñez, M. Large stabilization effects by intramolecular beryllium bonds in ortho-benzene derivatives. *Molecules* **2021**, *26*, 3401.
60. Sanz, P.; Montero-Campillo, M.M.; Mó, O.; Yáñez, M.; Alkorta, I.; Elguero, J. Intramolecular magnesium bonds in malonaldehyde-like systems: A critical view of the resonance-assisted phenomena. *Theor. Chem. Acc.* **2018**, *137*, 97.
61. Liu, N.; Li, Q., Scheiner, S.; Xie, X. Resonance-assisted intramolecular triel bonds. *Phys. Chem. Chem. Phys.* **2022**, *24*, 15015–15024.
62. Knop, O.; Rankin, K.N.; Boyd, R.J. Coming to grips with N–H···N bonds. 1. Distance relationships and electron density at the BCP. *J. Phys. Chem. A* **2001**, *105*, 6552–6566.

63. Knop, O.; Rankin, K.N.; Boyd, R.J. Coming to grips with N–H···N bonds. 2. Homocorrelations between parameters deriving from the electron density at the BCP. *J. Phys. Chem. A* **2003**, *107*, 272–284.
64. Espinosa, E.; Molins, E.; Lecomte, C. Hydrogen bond strengths revealed by topological analyses of experimentally observed electron densities. *Chem. Phys. Lett.* **1998**, *285*, 170–173.
65. Steiner, T. S-H···S hydrogen-bond chain in thiosalicylic acid. *Acta Crystallogr. Sect. C Cryst. Struct. Commun.* **2000**, *C56*, 876–877.
66. Pavan, M. S.; Sarkar, S.; Row, T. N. G. Exploring the rare S-H···S hydrogen bond using charge density analysis in isomers of mercaptobenzoic acid. *Acta Crystallogr. Sect. B Struct. Sci. Cryst. Eng. Mater.* **2017**, *B73*, 626–633.
67. Gholami, S.; Aarabi, M.; Grabowski, S.J. Coexistence of intra- and intermolecular hydrogen bonds: salicylic acid and salicylamide and their thiol counterparts. *J. Phys. Chem. A* **2021**, *125*, 1526–1539.
68. Selvakumar, K.; Singh, H.B. Adaptive responses of sterically confined intramolecular chalcogen bonds. *Chem. Sci.* **2018**, *9*, 7027–7042.
69. Sánchez-Sanz, G.; Trujillo, C.; Alkorta, I.; Elguero, J. Enhancing intramolecular chalcogen interactions in 1-hydroxy-8-YH-naphthalene derivatives. *J. Phys. Chem. A* **2017**, *121*, 8995–9003.
70. Maltz, L.T.; Gabbaï, F.P. Exploring the effects of Se basicity on a Te···Se interaction supported by a rigid indazolium backbone. *Organometallics* **2024**, *43*, 1246–1255.
71. Zhang, Z.; Wang, L.; Xuan, X. The C=N···C–X σ-hole interaction acts as a conformational lock. *New J. Chem.* **2017**, *41*, 42–46.

Concluding Remarks

Different inter- and intramolecular links that are classified as Lewis acid — Lewis base interactions have been analysed in this book. The electron charge transfer from the Lewis base unit to the Lewis acid is one of the main characteristics. The hydrogen bond and the halogen bond are examples of such interactions, as are σ-hole bonds and π-hole bonds that are the subject of this book. Weak interactions that are characterised by the domination of dispersion forces and for which the electron charge transfer does not occur, or is negligible at most, were not described in this book. They are often called van der Waals interactions and are not classified as Lewis acid — Lewis base interactions.

However, several types of Lewis acid — Lewis base interactions were not discussed here. One can mention interactions between ions, particularly monoatomic ions, links in sandwich compounds,[1] including their metallocene subclass, stacking interactions,[2] etc. These are great groups of interactions; in each case a great diversity of links is observed, thus these interactions deserve separate monographs.

Different names were introduced in studies on various kinds of interactions, the most common being names corresponding to the centre possessing Lewis acid properties. Such names may be related to elements, for example, the hydrogen bond, the carbon bond or the lithium bond, or to the group of elements, for example the halogen bond, the chalcogen bond, or the triel bond. The situations where the name of interaction is related to the Lewis base centre are rather rare; one can mention the hydride bond or the halide bond. New names appear from time to time; one can mention spodium bond,[3] matere bond[4] or other interactions. The question arises as to whether the multiplication of such names has any sense since they are not sufficiently informative. For example, the halogen bond covers halogen elements that in interactions play the role of the Lewis acid centres. The Lewis acidity increases with the increase of the atomic number of halogens, and in numerous studies it is pointed out that fluorine does not play the role of Lewis acid since its charge is negative and even the whole electrostatic potential surface is negative. However, the Lewis acid or Lewis base properties depend on numerous factors, thus there are studies describing Lewis acid properties of fluorine centres located in an appropriate environment. In any case, the properties of halogen bonds for different halogen elements vary significantly, for example, halogen bonds with F-centres differ significantly from interactions with I-centres.

Taking into account the above reservations, other ways of naming interactions can be proposed. The easiest way is to take into account the centres that are in contact, e.g., the designation Cl···O is much more informative than the term halogen bond. One just should remember the convention that the Lewis acid centre appears on the left side of the designation and the Lewis basic centre on the right.

This designation where only two centres are presented may be slightly modified; for example, the neighbouring atoms of the centres being in contact may be included. This may be important in the case of hydrogen bonds, because it is customary to designate three atoms of a hydrogen bridge. For example, the O-H···O interactions between carboxylic groups may be designated as (O)H···O(C). Another important topic related to interactions is that if they are strong enough, they can lead to a change of conformation. Such changes are in agreement with the valence-shell electron-pair repulsion model.[5] For example, the boron centre of the trigonal BF_3 species may interact with a fluoride anion that leads to the BF_4^- tetrahedral species. Hence maybe it would be useful to include the possible changes of conformations of interacting centres. In such a way the latter interaction could be designated as B···F(Tr-Td) since the boron centre changes its Tr (trigonal) conformation in the BF_3 species to the Td (tetrahedral) conformation in the BF_4^- unit. It seems to be more reasonable to apply such designations instead of the "old names" corresponding to a group of elements or to single elements. One more example is given below to justify the advisability of using new designations of interactions. The tetrahedral tetrel, pnicogen and chalcogen centres, if interacting strongly with the Lewis base ligands, may change their conformations into the trigonal bipyramid. These interactions are very similar to each other in spite of the Lewis acid centres belonging to different groups of the periodic table. On the other hand, the pnicogen bond, say As···F, differs significantly from the π-hole pnicogen bond between an NO_2 group and the electron donor despite the Lewis acid centres belonging to the same group of elements.

Only a restricted number of experimental approaches, those related to crystal structure determination, are presented in this book. The spectroscopic and NMR methods as well as others are not discussed, as well as new techniques that arose recently, for example atomic force microscopy[6] or small-molecule serial femtosecond X-ray crystallography.[7] The reason for the concentration on crystallographic methods is that the determination of the crystal structure usually provides complete information about the arrangement of molecules and consequently about the interactions occurring. And the number of crystal structures in the Cambridge Structural Database is over 1.3 million. The CSD is the largest source of information about interactions between various types of chemical units. It is very useful in various types of studies. It seems that not one experimental technique will provide such large and complete information in the near future as crystallographic techniques.

Bearing in mind such a large number of structures collected in the CSD, it is clear that a huge number of different interactions can be analysed with the use of this tool. Besides, there is a great number of well- or even very well-determined crystal structures in the CSD where very accurate geometrical parameters have been determined. Thus one can see that the crystallographic methods seem to be superior to other experimental techniques.

The theoretical methods that are applied most often in the analysis of interactions are presented in this book. In the discussion on intramolecular interactions and on intermolecular interactions in complexes or in larger aggregates such as clusters, in addition to the results of *ab initio* or DFT calculations, results obtained from other theoretical approaches are described. Approaches such as the Quantum Theory of Atoms in Molecules, Non-Covalent Interaction method, Natural Bond Orbital method, and sometimes the Electron Localisation Function and Car-Parrinello molecular dynamics are commonly used in various studies. All of them are described in this book. Theoretical studies on interactions that apply these methods are also often supported by the presentation of crystal structures. Hence one can see that this book describes both theoretical and experimental methods that are used in modern research on structures and interactions.

References

1. Goodwin, C.A.P. What is a sandwich complex? *Inorg. Chem.* **2024**, *63*, 9363–9365.
2. Martinez, C.R.; Iverson, B.L. Rethinking the term "pi-stacking". *Chem. Sci.* **2012**, *3*, 2191–2201.
3. Bauzá, A.; Alkorta, I.; Elguero, J.; Mooibroek, T.J.; Frontera, A. Spodium bonds: noncovalent interactions involving group 12 elements. *Angew. Chem. Int. Ed.* **2020**, *59*, 17482–17487.
4. Gomila, R.M.; Frontera, A. Matere bonds vs. multivalent halogen and chalcogen bonds: three case studies. *Molecules* **2022**, *27*, 6597.
5. Grabowski, S.J. Classification of so-called non-covalent interactions based on VSEPR model. *Molecules* **2021**, *26*, 4939.
6. Giessibl, F. Atomic resolution of the silicon (111)-(7x7) surface by atomic force microscopy. *Science* **1995**, *267*, 68–71.
7. Schriber, E.A. *et. al.* Chemical crystallography by serial femtosecond X-ray diffraction. *Nature* **2022**, *601*, 360–365.

Index

www.ingramcontent.com/pod-product-compliance
Lightning Source LLC
LaVergne TN
LVHW010329220926
848484LV00004BB/867

* 9 7 8 9 8 1 9 8 0 1 4 3 5 *